# Light Water Reactor Nuclear Fuel Cycle

Editors

**Raymond G. Wymer, Ph.D.**
Associate Division Director
Chemical Technology Division
Oak Ridge National Laboratory
Oak Ridge, Tennessee

**Benedict L. Vondra, Jr., M. S.**
(deceased)
Program Manager
Process Research and Development
Consolidated Fuel Reprocessing Program
Chemical Technology Division
Oak Ridge National Laboratory
Oak Ridge, Tennessee



CRC Press, Inc.
Boca Raton, Florida

**Library of Congress Cataloging in Publication Data**
Main entry under title:

Light water reactor nuclear fuel cycle.

Bibliography: p.
Includes index.
1. Nuclear fuels. 2. Light water reactors.
I. Wymer, Raymond G. II. Vondra, Benedict L., Jr.
TK9360.L53 621.48'335 80-12432
ISBN 0-8493-5687-3



Direct all inquiries to CRC Press, Inc., 2000 N.W. 24th Street, Boca Raton, Florida, 33431.



International Standard Book Number 0-8493-5687-3

Library of Congress Card Number 80-12432
Printed in the United States

# PREFACE

In writing this book an attempt was made to present the various elements of the light water reactor (LWR) nuclear fuel cycle, with an emphasis on the chemistry of the fuel cycle operations. To provide perspective, some historical information has been included. In addition, an overview of the entire fuel cycle has been included to make it easier for the reader to understand where the various elements fit into the overall picture. For the most part, those elements which have been commercialized in the U.S. have not been addressed in detail.

R. G. Wymer
Oak Ridge, Tenn.
1979

## THE EDITORS

**Raymond G. Wymer, Ph.D.**, is presently Associate Division Director, Chemical Technology Division, Oak Ridge National Laboratory, Oak Ridge, Tennessee.* He received his B.A. in 1950 from Memphis State College, and his M.A. and Ph.D. in 1953 from Vanderbilt University, Nashville.

His experience includes teaching nuclear separations technology as Associate Professor of Chemical Engineering at Georgia Institute of Technology, Atlanta, research in radiation effects on solvent extraction diluents at the Georgia Tech Engineering Experiment Station, Head of Nuclear Chemistry at Industrial Reactor Laboratories, Inc., Plainsboro, New Jersey, and several research and development and administrative positions in the Chemical Technology Division of Oak Ridge National Laboratory. His recent work has been in the development of high temperature gas-cooled reactor (HTGR) fuel cycle, especially, in HTGR fuel material preparation, head-end processes, off-gas clean up, and in light water reactor (LWR) fuel cycle studies.

Dr. Wymer's distinctions include participation as the U.S. representative to three International Atomic Energy Agency (IAEA) Panels held in Vienna, Austria, two of which were held on Sol-Gel processes for preparing nuclear fuels and the third, as a member of an advisory group at a meeting on fuel cycle demand, supply, and cost trends. He is presently serving a 3-year term on the Subcommittee on Nuclear and Radiochemistry, National Research Council, National Academy of Sciences.

Dr. Wymer is a member of the American Chemical Society, Tennessee Academy of Sciences, American Institute of Chemical Engineers, American Institute of Chemists (Chairman of the Tennessee Chapter), American Nuclear Society, and the honorary society Sigma Xi. He was elected an ANS Fellow in 1977.

Dr. Wymer's publications, which have been mainly in the area of nuclear fuel recycle, include a book *Chemistry in Nuclear Technology,* co-authored with Sigfred Peterson. He is one of five editors of the *Radiochimica Acta Journal.* He was one of four recipients of the Special Award given for "The Fuel Cycle" in 1977 by the American Nuclear Society. He is a member of the Editorial Advisory Board of *Nuclear Technology.*

**Benedict L. Vondra, Jr.** (deceased) was Program Manager, Process Research and Development, Consolidated Fuel Reprocessing Program, Oak Ridge National Laboratory, Oak Ridge, Tennessee. Mr. Vondra graduated from the University of Colorado, Boulder, in 1943 with a B.A. degree in chemistry and received an M.S. degree in chemistry from Purdue, Lafayette in 1948. He served as a chemistry instructor at Syracuse University, New York in 1948 to 1949 and on the Chemistry Faculty at Clarkson College, Potsdam, New York in 1949 to 1951. He spent 2 years at Pennsylvania State University, University Park working toward a doctorate degree before transferring to industry in 1949. Since then he has held the following positions: Supervisor Process Research and Development, Materials and Processes Lab, General Electric Company, 1952 to 1959; Technical Director, Research and Development, Nuclear Materials and Equipment Corporation (NUMEC), 1959 to 1967; Manager, Inorganic Chemistry, NUMEC (Atlantic Richfield), 1967 to 1971; Manager, Research and Development NUMEC (B & W), 1971 to 1973; Project Manager, Senior Staff NUMEC (B & W), 1973 to 1975; and Program Manager, Nuclear Fuel Cycle Process Research and Development, Oak Ridge National Laboratory, 1975 to 1980.

* Chemical Technology Division, Oak Ridge National Laboratory, Oak Ridge, Tennessee, operated by Union Carbide Corporation under Contract W-7405-eng-26 with the U.S. Department of Energy.

Mr. Vondra was a member of the American Nuclear Society, American Chemical Society, American Institute of Chemists, Electrochemical Society, and the honorary societies Alpha Chi Sigma and Sigma Xi. He was named a Fellow of the American Institute of Chemists in 1968 and received a letter of commendation from the U.S. Atomic Energy Commission in 1966, and ONR Fellowship in 1950, and an AEC Fellowship in 1951.

He has 10 patents and was the author of a number of journal articles on various aspects of the nuclear fuel cycle. His current research interests are mainly in reactor fuel reprocessing and in converting purified fuel solutions to oxide products for recycle to power reactors.

# CONTRIBUTORS

**Walter D. Bond, Ph.D.**
Group Leader
Experimental Engineering Section
Chemical Technology Division
Oak Ridge National Laboratory
Oak Ridge, Tennessee

**David O. Campbell, Ph.D.**
Group Leader
Chemical Technology Division
Oak Ridge National Laboratory
Oak Ridge, Tennessee

**Allen G. Croff, M.S.**
Nuclear Engineer
Chemical Technology Division
Oak Ridge National Laboratory
Oak Ridge, Tennessee

**Rex E. Leuze, M.S.**
Section Head
Experimental Engineering Section
Chemical Technology Division
Oak Ridge National Laboratory
Oak Ridge, Tennessee

**James C. Mailen, Ph.D.**
Group Leader
Chemical Development of Reactor Fuel Reprocessing
Chemical Technology Division
Oak Ridge National Laboratory
Oak Ridge, Tennessee

**Louis M. Toth, Ph.D.**
Research Staff Member
Chemical Technology Division
Oak Ridge National Laboratory
Oak Ridge, Tennessee

# TABLE OF CONTENTS

Chapter 1

# LIGHT WATER REACTOR FUEL CYCLE — PAST, PRESENT, FUTURE

**B. L. Vondra***

Nuclear power reactors require a large supporting fuel cycle industry to provide the fissile material needed for heat generation in the reactor. The principal elements in the complete light water reactor (LWR) fuel cycle are (1) mining and milling to obtain a uranium oxide concentrate, (2) refining and conversion of the ore concentrate to $UF_6$, (3) uranium isotopic enrichment, (4) fabrication of $UO_2$ fuel elements, (5) reprocessing of irradiated fuel from the reactor to recover fissionable materials ($^{235}U$ and Pu) for recycle, and (6) long-term waste management. At present this fuel cycle is incomplete in the U.S. There is neither active reprocessing of irradiated power reactor fuel nor production capability for managing high-level radioactive wastes for long-term, safe disposal.

Prior to April 1977, development and implementation of most elements of the fuel cycle were proceeding in concert with an expanding industry for producing LWRs. The large, privately operated uranium mining and milling industry in the western U.S. and the large, government-operated gaseous diffusion uranium isotopic enrichment plants, both of which had been started to support nuclear weapons production, had shifted primarily to support of these nuclear power reactors. Conversion of ore concentrates to $UF_6$ and fabrication of low-enriched ($\sim$3% $^{235}U)UO_2$ fuel elements were established industries. Although irradiated fuel reprocessing capability had lagged somewhat, by 1975 a large reprocessing plant was essentially completed by Allied-General Nuclear Services (AGNS) in Barnwell, S.C., and was awaiting licensing by the Nuclear Regulatory Commission. Also, Exxon Nuclear was actively pursuing plans for a still larger plant in Oak Ridge, Tenn. At the same time, the federal government was in the midst of an intensive study to identify geologically suitable sites for long-term storage of radioactive wastes.

One commercial fuel reprocessing plant, the Nuclear Fuel Services (NFS) plant in West Valley, N.Y., has operated in the U.S. The NFS plant, which processed about 650 tons of spent fuel, including about 300 tons of LWR fuel, was shut down in 1972 for modifications and enlargement; since then the NFS management has withdrawn its application for a license to reopen. In addition, General Electric built a fuel reprocessing plant at Morris, Ill., but this plant never operated because of unsolved problems encountered in cold testing. Early operation of the AGNS plant and construction of the proposed Exxon plant have been precluded by actions of the present federal administration.

In April 1977, U.S. President Jimmy Carter announced a new government policy concerning the development of nuclear power. The principal thrust of his policy was to indefinitely defer the use of plutonium as recycle fuel for LWRs and to delay a decision on its use in the liquid metal fast breeder reactors (LMFBR) that are under development. This policy was proposed as a means of minimizing opportunities for

* Chemical Technology Division, Oak Ridge National Laboratory, operated by Union Carbide Corporation under Contract W-7405-eng-26 with the U.S. Department of Energy.

using plutonium in the proliferation of nuclear weapons. The President announced the following decisions taken to implement his new nuclear policy:

1. Defer indefinitely the commercial reprocessing of nuclear fuels and recycle of plutonium for use in LWRs
2. Restructure the U.S. breeder reactor development program to give greater priority to alternative breeder designs and to delay commercial introduction of the breeder
3. Accelerate research into alternative nuclear fuel cycles that do not involve direct access to fissile materials usable in nuclear weapons
4. Explore the establishment of an International Nuclear Fuel Cycle Evaluation (INFCE) program aimed at developing alternative fuel cycles

The impact of the new policy on the nuclear fuel cycle was assumed to be primarily on reprocessing since the rate of LWR growth and use was assumed to be unaffected and the large supporting industries will still be required for the other elements of the fuel cycle, regardless of the fuel cycle chosen. At the time of this writing, the National Alternative Systems Assessment Program (NASAP) and INFCE study groups are evaluating established and alternative fuel cycles; it is expected that INFCE study groups will complete and provide information in 1980 which may strongly influence future U.S. and world-wide policy. (It should be recognized that fuel reprocessing is a key element in all of the proposed fuel cycles except the one which entails storage of the spent fuel itself as waste — the so-called once-through cycle for low-enriched uranium [LWR] or natural uranium [CANDU] reactor fuel cycles.)

Although fuel reprocessing has been deferred indefinitely in the U.S., France is presently operating LWR fuel reprocessing plants. In addition, new plants are planned in France, Great Britain, West Germany, Italy, India, and Canada. Small plants in Belgium and Japan are being updated and improved. France and Great Britain are also planning commercial capability for processing uranium-plutonium oxide fuels from LMFBRs.

The problem of processing irradiated reactor fuels to recover and purify fissionable material (plutonium) was first faced in the early 1940s, when a process was developed for recovering the plutonium for weapons use. This process operated by dissolving the uranium metal fuel in nitric acid; by carrying out repeated batch carrier precipitations the plutonium was purified. This precipitation process (the Bismuth Phosphate Process) was soon replaced by efficient, continuous liquid-liquid solvent extraction processes which recovered not only the plutonium but also the uranium (containing unspent $^{235}U$) in pure form from the fission products which accompanied them in the irradiated fuel. The most successful of these processes is the Purex Process which uses a kerosene or other hydrocarbon solution of tributylphosphate (TBP) as extractant. The Purex Process has been used successfully for more than 25 years at the plutonium production plants at Hanford and Savannah River in reprocessing aluminum-clad uranium-metal fuel; it was also used at NFS in reprocessing Zircaloy-clad-$UO_2$ LWR fuel. In fact, it is the process of choice for use in all present and planned reactor fuel reprocessing plants throughout the world.

Although the reprocessing of spent LWR fuels at NFS was conducted with reasonable success, future reprocessing plants must meet many regulatory requirements which did not exist when the NFS plant was licensed. These new constraints require addition of new operations to the fuel reprocessing plant and indeed dictate a whole new philosophy of plant design and operation. Some of the important changes from the requirements of the early NFS operation that must be considered today are

1. Of the volatile radionuclides, only retention of $^{131}I$ was required at NFS. Future

plants will be required to provide for retention of $^{129}I$, $^{85}Kr$, and perhaps also of $^{3}H$ and $^{14}C$.

2. Liquid wastes will probably not be discharged to the environment. Rather, recycle within the plant of all aqueous wastes will be practiced to the maximum extent possible; water will be evaporated up tall stacks to maintain the plant water balance.
3. Off-site shipments of plutonium must be with plutonium in solid form, rather than in solution; this necessitates addition of a conversion system to convert plutonium in solution to plutonium oxide.
4. High-level liquid wastes (HLW) must be solidified within 5 years and shipped from the reprocessing plant site within 10 years of their production.
5. New but as yet incompletely defined safeguards requirements for protecting plutonium and enriched uranium must be met.
6. There is a strong trend toward reduction in the allowable radiation exposure limits for plant personnel; the effect of this change will be to emphasize incorporation in the plant of increased remote operation and maintenance capability.

Obviously the above changes would greatly increase the complexity and cost of reprocessing plants.

To meet the above requirements and constraints, new technology is being developed as a hedge against the possibility that fuel reprocessing will one day be allowed. New methods (Chapter 5) are being developed for controlling tritium release and for removing $^{129,131}I$, $^{85}Kr$, and $^{14}C$ (as $CO_2$) from process off-gases. Although these processes have not been used commercially, it appears that those for iodine and krypton (and xenon) can provide adequate retention of these volatile radionuclides. Development of suitable methods for controlling tritium release, however, lags behind those for removing the other volatile contaminants. Both of the candidate tritium retention methods are difficult, and much development work should be done before the best choice could be made for inclusion in a commercial reprocessing plant.

Solidification of high-level liquid wastes by calcination and fixation in a glass (vitrification) has been studied extensively, and several calcination and glass melting systems have been demonstrated successfully (Chapter 6). However, the complete fixation process has not been adequately demonstrated. Preparations are presently being made at the Pacific Northwest Laboratories for an engineering-scale demonstration of the calcination-vitrification process with high-burnup LWR fuel wastes. Such a demonstration has already been carried out in France, and it appears that this technology is close to being ready for commercial application.

Preparation of mixed uranium-plutonium oxide may be accomplished by any of several possible coconversion methods. Although none of these has been adequately demonstrated, it appears certain that this technology will be available when needed.

Future plants which are to handle fissile material must accommodate as yet undefined safeguards requirements. Development of more accurate instrumentation and control techniques is needed for measuring and monitoring plant input and output, as well as measuring the plant inventory of fissionable materials in a current or "real time" way. It is entirely possible that safeguards requirements could have a large impact on the optimum design, cost, and operating procedures of such plants.

The April 1977 policy statements by President Carter concerning deferral of plutonium recycle and fuel reprocessing, and consideration of alternative fuel cycles, have necessitated a reevaluation of the situation with respect to the fuel cycles that may prevail and, consequently, of the technology that may be required for reprocessing.

The most practical approach to future reprocessing would be to upgrade and continue to use the present uranium-plutonium fuel cycle, but with tightened safeguards, improved effluent control, and due attention to minimizing effects of natural disasters such as earthquakes and tornadoes. In an extension of the present approach, coprocessing, that is, only partially separating the uranium from the plutonium, might be practiced. Following a coconversion step, this would lead to a mixed uranium-plutonium oxide product rather than to a product stream of pure plutonium oxide. It has also been suggested that fissile materials might be adulterated with radioisotopes, e.g., $^{60}Co$ or fission products to enhance their resistance to theft or to use in weapons production. None of these modified approaches appears to pose insuperable problems to the technologist, though some may question their real value in deterring unauthorized and undesirable use of fissile material.

With respect to the future, large expansions of the uranium mining and milling industry will be required, assuming that nuclear power grows at even a moderate pace. This is true regardless of the fuel cycle chosen, since inclusion of thorium in the fuel cycle will increase the near- and intermediate-term demand for $^{235}U$. Also, delayed introduction of breeder reactors will result in increased uranium demand. With a long delay in introduction of breeders, it is not clear that the mining and milling industry will be able to meet uranium demands. The trend in uranium milling will be toward the processing of even lower grade ores, with attendant increases in $U_3O_8$ costs. It seems unlikely that the processing methods will change significantly, except possibly for larger utilization of *in situ* leaching methods. The principal problem that must be solved is acceptable disposition of the huge amounts of slightly radioactive ore tailings. Large-scale adoption of the thorium fuel cycle would require evolution of a thorium mining and milling industry, which at present is virtually nonexistent.

The needed expansion of refinery capacity to furnish $UF_6$ for enrichment plants should not be a problem, and little change in the highly developed refinery processes would be expected. Enrichment capacity, however, must grow. Because of the much lower energy consumption of the gas centrifuge plants, and their greater flexibility with respect to incremental additions to the plant, planned capacity additions are to be met with gas centrifuge rather than gaseous diffusion plants. Laser enrichment is also being studied and conceivably could emerge as a competitive uranium isotope enrichment method.

Adequate expansion of plants for production of $UO_2$ and fabrication of fuel elements does not appear to pose any significant obstacles. Acceptance of the thorium fuel cycle, however, would dictate development and implementation of suitable commercial-scale methods for preparation of uranium-thorium oxide fuels for recycle. The fuel conversion and refabrication steps in this case must be accomplished in remotely operated and maintained facilities because of the high radioactivity that derives from a $^{232}U$ impurity in the recycled $^{233}U$.

If breeder reactors are deployed in the U.S., fuel reprocessing will eventually be needed, even though a decision were made to accept the once-through fuel cycle for LWRs. If complete plutonium recovery from spent fuel is to occur, then construction of a number of large fuel reprocessing plants would be required to reduce the stored fuel backlog while also keeping pace with the concurrent generation of spent fuel. Meanwhile, extensive research and development is needed to broaden the base of understanding of the process chemistry and equipment performance for reprocessing so that as yet undefined regulatory requirements and new fuel cycles may be accommodated if needed.

Because there is no high-level waste storage capability, a large expansion of spent fuel storage capacity will be required for an interim period with or without fuel reprocessing.

Chapter 2

# AN OVERVIEW OF THE LIGHT WATER REACTOR FUEL CYCLE IN THE U.S.

**R. E. Leuze***

## TABLE OF CONTENTS



* Chemical Technology Division, Oak Ridge National Laboratory, operated by Union Carbide Corporation under Contract W-7405-eng-26 with the U.S. Department of Energy.

## I. ELECTRICAL POWER GENERATION FROM NUCLEAR ENERGY IN THE U.S.

A substantial nuclear power industry has developed in the U.S. over the 20 years since the first commercial generation of electrical power at the Shippingport Atomic Station in December of 1957. The Shippingport nuclear unit is a 90-MWe pressurized water reactor designed by Westinghouse Electric and constructed by Stone and Webster. This power station was a joint project by Duquesne Light Company and the U.S. Atomic Energy Commission. It is still in operation today, and joint ownership remains with Duquesne Light Company and the U.S. Department of Energy.

The growth in electrical power generation and the steadily increasing role of steam nuclear units is shown in data from the *Statistical Abstract of the U.S. for 1977*[1] which is presented in Table 1. Electrical production from steam nuclear power stations had increased to 191 billion kWh by 1976, representing 9.4% of the total production by U.S. electric utilities. During 1977, electrical output from nuclear plants increased by 31% over the 1976 output to nearly 250 billion kWh, or about 12% of the total electrical power generated in the U.S.[2] Electrical power generation by nuclear stations has grown to the point where it is about the same as that generated by hydroplants. For the first time, electrical output from nuclear plants in 1977 exceeded that from hydroplants, which generated 284 billion kWh in 1976, but only 220 billion kWh in 1977. Forecasts of electrical power growth and especially that from nuclear power stations have varied over a wide range. More recent forecasts predict a smaller growth than those made several years ago. A study[3] made by the Energy Information Administration of the Department of Energy estimates that total electrical generation in 1985 will be about 3000 billion kWh and that nuclear power stations will supply about 560 billion kWh, or 19% of the total. By 1990, the forecast is a total electrical power generation of between 3385 and 3615 billion kWh, with nuclear power stations supplying about 26% (920 to 940 billion kWh).

In addition to those reactors already licensed and operating, more than 70 reactors are under construction and more than 100 others have been ordered or announced.[4] These power stations are concentrated primarily in the midwest, northeast, and south, with a few located in the west and northwest. Start-up of the first nuclear plant in the southwest is planned for 1980. A summary of existing and planned nuclear electrical capacity by geographical region is presented in Figure 1.

All but two of the nuclear reactors used for generation of electrical power in the U.S. are light water reactors, which are usually referred to simply as LWRs. Light water reactors will continue to dominate the field of nuclear power for the next 25 years. These reactors derive their name from the fact that ordinary or "light" water (not $D_2O$) is used both as a heat transfer medium and as the moderator to slow down fission neutrons. Fuel for the reactors consists of uranium dioxide pellets fabricated from uranium which has been enriched to contain from 2 to 4% of the fissile isotope $^{235}U$, compared to only 0.711% found in natural uranium. The pellets are clad in metal tubes of Zircaloy, which are assembled into bundles to form fuel elements. Two basic types of LWRs are offered commercially — boiling water reactors (BWRs) by General Electric and pressurized water reactors (PWRs) by Westinghouse Electric, Babcock and Wilcox, and Combustion Engineering. Approximately two thirds[4] of the current nuclear capacity is furnished by PWRs and the other third by BWRs.

## II. FUEL CYCLE DESCRIPTION

A large industrial complex is required to support nuclear power stations, both to

## Table 1
## THE ROLE OF STEAM NUCLEAR PLANTS IN U.S. ELECTRIC UTILITIES INSTALLED GENERATING CAPACITY AND ELECTRIC ENERGY PRODUCTION[1]

| | Installed electric capacity | | | | Production | | |
|---|---|---|---|---|---|---|---|
| | | Nuclear | | | | Nuclear | |
| Calendar year | Total GW(e) | GW(e) | % of total | Number of steam nuclear units at end of year | Total (billion kWh) | billion kWh | % of total |
| 1955 | 115 | 0 | 0 | 0 | 547 | 0 | 0 |
| 1960 | 168 | <0.5 | <0.3 | 5 | 753 | 1 | 0.1 |
| 1965 | 236 | 1 | 0.4 | 9 | 1055 | 4 | 0.4 |
| 1970 | 340 | 6 | 1.8 | 16 | 1492 | 22 | 1.4 |
| 1972 | 398 | 15 | 3.8 | 26 | 1750 | 54 | 3.1 |
| 1973 | 442 | 21 | 4.8 | 30 | 1861 | 83 | 4.5 |
| 1974 | 479 | 32 | 6.7 | 38 | 1867 | 114 | 6.1 |
| 1975 | 508 | 40 | 7.9 | 45 | 1918 | 173 | 9.0 |
| 1976 | 532 | 43 | 8.1 | 48 | 2037 | 191 | 9.4 |

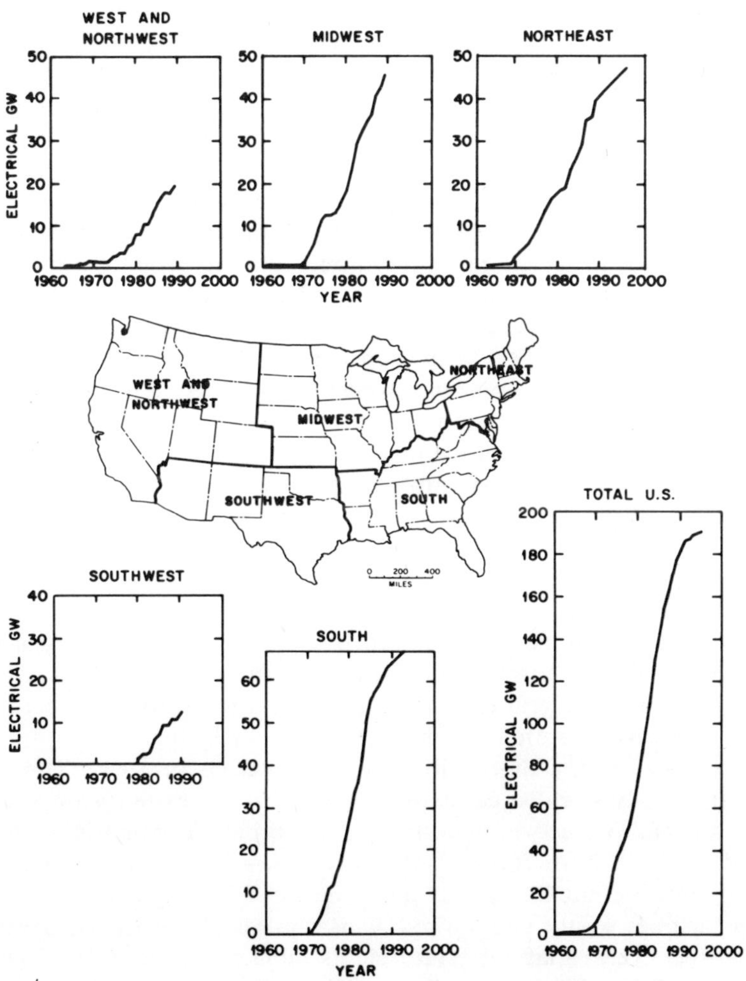


FIGURE 1. Electrical capacity of nuclear power stations operable, under constuction, or on order, June 30, 1978.

supply fresh fuel and to dispose of spent fuel discharged from the reactors. Adequate capability has been developed in the U.S. for preparation of fuel from virgin uranium ore, but there are essentially no facilities for handling spent commercial power reactor fuel. Indeed, present national policies and regulations limit the options now available for disposition of spent fuel. It has been decided that reprocessing of spent fuel from commercial power reactors will not be permitted at the present time because of the possibility that separated plutonium could be diverted for illicit use as weapons. The very fact that spent fuel contains fissile plutonium formed by nuclear reactions during irradiation as well as residual $^{235}U$ at a concentration greater than that in natural uranium is the reason reprocessing has been proposed. The objective of reprocessing is to recover these values as reactor fuel. The uranium would be sent back through isotopic enrichment to increase the $^{235}U$ concentration needed for LWR reactor fuel. The plutonium could be stored for future use as fuel for a fast breeder reactor, which the present U.S. administration has also delayed, in part because of weapons proliferation potential, or could be used as mixed $UO_2$-$PuO_2$ to fuel LWRs. Since national policy prohibits reprocessing, spent fuel must be stored. The U.S. Department of Energy is considering temporary storage in away-from-reactor (AFR) facilities and long-term storage (both retrievable and nonretrievable) in geologically stable repositories.

In a study for the Allied Gulf Nuclear Service Plant,[5] it was determined that for 50 nuclear reactors operating over a 20-year period, spent fuel reprocessing with recycle of the plutonium and uranium reduces the demand for $U_3O_8$ by nearly 11 million lb/year or 37%. Furthermore, separative work requirements for enrichment of $^{235}U$ are reduced by about 1.5 million separative work units (SWU) per year or approximately 25%.

The pictorial representation of the major steps of the nuclear fuel cycle for LWRs shown in Figure 2 gives some indication of the broad industrial base needed to support nuclear power stations. Each step requires a substantial installation. Supply of fuel starts with mining of the uranium followed by milling to produce a crude concentrate called yellowcake. The yellowcake is sent to a conversion plant where it is refined and converted to pure uranium hexafluoride ($UF_6$). At the enrichment plant, gaseous $UF_6$ is processed to enhance the $^{235}U$ to the concentration needed for LWR fuel. Enriched $UF_6$ is sent to a fuel preparation plant where it is converted to uranium dioxide and pressed into pellets. Fuel fabrication, which may be carried out in the same facility as the fuel preparation or may be done in a separate plant, involves cladding the pellets in Zircaloy tubing followed by assembly of the tubes into fuel elements. The finished fuel elements are then ready for use in the reactors. When spent fuel is discharged from the reactor, it is stored for a period of time in a pool of water at the reactor site. The water serves both as shielding for gamma radiation and as a heat sink to remove fission product decay heat.

Under the present national policy, spent fuel would not be sent to a reprocessing plant as shown in Figure 2, but would be sent to an AFR storage site and eventually to a long-term storage federal repository — when a repository has been established. Even if the reprocessing option becomes available, spent fuel will always be stored at least 160 days to allow the shorter-lived fission products to decay. The fuel will then be shipped in heavily shielded casks to a chemical reprocessing plant where chemical separations are made to recover plutonium and uranium. The plutonium may be stored as indicated in Figure 2 for future use — perhaps as fuel for fast breeder reactors. If future regulations permit the use of plutonium in LWRs, mixed oxide (MOX) fuel fabrication facilities would be required. This represents a major significant addition to industrial fuel cycle capability, since plants fabricating fuel from slightly enriched uranium cannot be readily adapted for handling plutonium. Additional containment

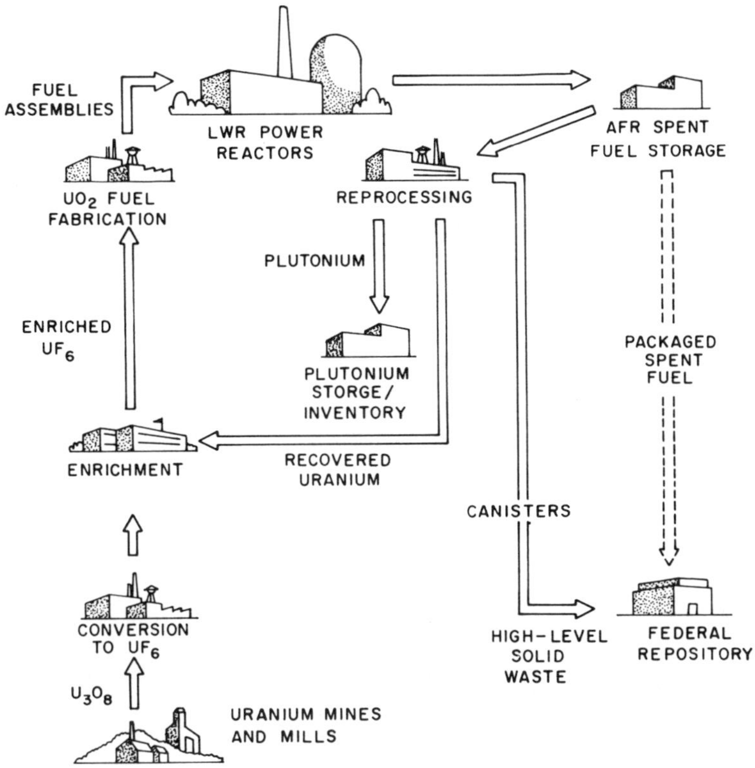


FIGURE 2. LWR nuclear fuel cycle without plutonium recycle.

and handling techniques are needed to protect personnel from the more toxic plutonium, and strict safeguards are needed to prevent unauthorized diversion of the plutonium. Uranium product from reprocessing in the form of pure uranyl nitrate solution would be converted to $UF_6$, probably at the reprocessing plant, and sent back for isotopic enrichment. Fission products would be converted to chemically inert solids and stored on site until a federal repository is ready to accept them for permanent storage. Although the major amounts of radioactivity are contained in the fuel reprocessing waste, effluents from all parts of the fuel cycle must receive adequate treatment and management to limit release of radioactivity to the environment. The constraints and requirements on release of radioactivity are still evolving and impose continually changing technological requirements. A series of reports on segments of the nuclear fuel cycle was prepared by Oak Ridge National Laboratory for use in establishing guides concerning limits of release.[6-12]

Another vital part of the fuel cycle is transportation of the radioactive materials to and from industrial sites which may be widely spread throughout the U.S. Table 2, taken from WASH-1174-74, The Nuclear Industry 1974,[13] summarizes the flow of major materials in the nuclear fuel cycle.

Even though this broad industrial support is needed for the nuclear power industry, and substantial capital investments are required to provide for all of the fuel cycle facilities, in a mature nuclear economy with many reactors, investments for all the supporting facilities are estimated to be only about 15% of the capital needed for the reactors.[14]

### Table 2
### MATERIAL FLOW IN THE NUCLEAR FUEL CYCLE[13]

| Material | Point to point movements | Packaging |
|---|---|---|
| Uranium ore | Uranium mines to uranium mills | None |
| Uranium ore concentrates | Uranium mill to feed preparation plant | Drums |
| Normal $UF_6$ | Feed preparation plant to enriching (gaseous diffusion) plant | Pressurized cylinders |
| Enriched $UF_6$ | Enriching plant to materials processing and/or fuel fabrication plant | Pressurized cylinders and protective packages |
| Enriched uranium oxide powder or pellets | Materials processing plant to fuel fabrication plant | Drums and birdcages with protective packages |
| New fabricated fuel assemblies | Fuel fabrication plant to nuclear power reactors | Special fuel element shipping containers |
| Low-level radioactive wastes | All facilities to commercial burial sites | Drums and shielded protection packages |
| Irradiated fuel | Nuclear power reactor to AFR storage or chemical reprocessing plant | Shielded casks |
| Uranyl nitrate | Chemical reprocessing plant to fuel fabrication plant or to $UF_6$ feed preparation plant | Tank trucks, tank cars, and protective packages |
| Plutonium oxide | Chemical reprocessing plant to fuel fabrication plant | Protective packages and special plutonium containers |
| Fission product radioisotopes | Chemical reprocessing plant to Government or industry users | Shielded casks |
| High-level radioactive solid wastes | Chemical reprocessing plant to Federal repository | Shielded casks |

## III. URANIUM MINING AND MILLING

Uranium ore is the ultimate source of fuel for light water reactors. The U.S. has significant amounts of uranium ore with most deposits of economic importance found in flat-lying sedimentary rock located in three major districts: (1) the Wyoming basins, (2) the Colorado plateau, and (3) the Texas gulf coast. The average assay of ore in the early 1950s was around 0.3% $U_3O_8$. The grade of ore since that time has been dropping and during 1976 and 1977 the average grade was slightly above 0.15% $U_3O_8$. On January 1, 1978, the U.S. Department of Energy[15] estimated that uranium ore reserves in the forward cost category of \$50/lb of $U_3O_8$ contained approximately 890,000 tons of $U_3O_8$.* In addition to these reserves, potential resource estimates in the \$50/lb cost category includes another 2.9 million tons of $U_3O_8$, described as probable and possible. A list of principal uranium minerals and representative occurrences of these ores in the U.S. is given in Table 3.

U.S. mining and milling activities have fluctuated widely over the past 30 years. In 1948, about 100 tons of $U_3O_8$ was recoverd from 38,000 tons of ore. By 1960, the annual production of $U_3O_8$ had increased to more than 17,000 tons from about 8 mil-

* All reserves with forward costs of no more than \$50/lb of $U_3O_8$ are included. Only estimated operating costs and those capital costs not yet incurred are used to calculate the forward cost. Since it does not include profit, cost of money, taxes, and already incurred expenditures (such as property acquisition, exploration, mine development, and mill construction), forward cost is not representative of market price.

### Table 3
### PRINCIPAL URANIUM MINERALS AND THEIR REPRESENTATIVE OCCURRENCES IN THE U.S.

| Mineral | Composition | Some occurrences |
|---|---|---|
| Uraninite (Pitchblende) | Uranium oxides | Colorado, Utah, New Mexico |
| Coffinite | Uranium silicate | Utah, Wyoming |
| Carnotite | Potassium uranium vanadate | Colorado, Utah, Arizona, New Mexico |
| Tyuyamunite | Calcium uranium vanadate | New Mexico |
| Torbernite | Copper uranium phosphate | Colorado, Utah |
| Carbonate-fluorapatite | Calcium fluorophosphate containing uranium | Central Florida, Idaho, Utah, Wyoming |

lion tons of ore. It remained at this level through 1962, but by 1967 the annual production had dropped to 9125 tons of $U_3O_8$. Since that time, production has again increased and was 14,940 tons in 1977 from more than 10 million tons of ore (see Table 4). On January 1, 1977, there were 36 open pit mines and 251 underground mines. Half of the uranium produced in 1977 came from underground mines, 46% from open pit mines, and 4% from miscellaneous sources including heap leaching, mine water, solution mining, and low-grade stock piles. Although today's U.S. production is adequate and ore reserves are sufficient, it has been predicted that uranium will be in short supply sometime between 1990 and 2000.* The government has taken two steps[2] to help relieve this impending shortage: (1) regulation changes to allow increased use of uranium from foreign countries and (2) comprehensive evaluation of domestic uranium resources. The use of imported uranium supplied by a domestic customer for enrichment has been restricted, but with the new regulations, purchases of foreign uranium started again in 1975. Comprehensive evaluation of domestic uranium resources and identification of favorable areas for exploration are the major goals of the National Uranium Resources Evaluation program (NURE) which began in 1974. Three classes of potential uranium resources (probable, possible, and speculative) are considered in order to accommodate potential estimates in a variety of geological environments with varying degrees of reliability.

## A. Uranium Requirements

The amount of $U_3O_8$ needed to supply fuel for a 1000-MWe LWR (and hence the mining, milling, conversion, and enrichment requirements) is dependent upon whether or not uranium and/or plutonium from spent fuel is recycled and upon the $^{235}U$ assay in tails from the isotopic enrichment plant. Estimated needs also vary somewhat with the details of the model used to make the calculations. For one model, the annual requirements averaged over the entire lifetime of a 1000-MWe LWR operating at a capacity factor of 75% are given in Table 5. It can be seen that the $U_3O_8$ requirements are quite sensitive to these variables. For example, from 20 to 24% more uranium is required if the enrichment plant tails assay is 0.30% $^{235}U$ instead of 0.20% $^{235}U$. When the enrichment plant is operated at a tails assay of 0.20% $^{235}U$, uranium requirements are decreased by 20% by recycling uranium from spent reactor fuel, and if both uranium and plutonium are recycled, uranium requirements are decreased about 34%.

* The lifetime of a reactor is assumed to be 30 years. In the period 1990 to 2000 there will be enough reactors built so that the uranium required to operate them for 30 years will result in a commitment of uranium in an amount that may lead to a shortage. That is, there may not be enough uranium to satisfy the 30-year lifetime needs of additional reactors.

**Table 4**
**SUMMARY OF URANIUM MINING AND MILLING IN THE U.S.[15-17]**

| Calendar year | Ore mined (1,000 short tons) | $U_3O_8$ milled (short tons) |
|---|---|---|
| 1960 | 7,970 | 17,766 |
| 1965 | 4,386 | 10,341 |
| 1970 | 6,324 | 12,905 |
| 1971 | 6,279 | 12,250 |
| 1972 | 6,418 | 12,900 |
| 1973 | 6,537 | 13,235 |
| 1974 | 7,027 | 11,528 |
| 1975 | 7,057 | 11,600 |
| 1976 | 8,608 | 12,700 |
| 1977 | 10,325 | 14,940 |

*Note:* As a result of processing lag and accumulation of small ore piles, ore mined in a calendear year is not all milled in the same year.

**Table 5**
**ANNUAL $U_3O_8$ REQUIREMENTS, IN METRIC TONS, FOR A MODEL 1000-MWe LWR**

| | Enrichment tails assay (% $^{235}U$) | | |
|---|---|---|---|
| | 0.30 | 0.25 | 0.20 |
| No fuel recycle | 248 | 225 | 206 |
| Uranium recycle | 205 | 182 | 162 |
| Uranium and plutonium recycle | 170 | 151 | 136 |

The average annual requirement for a model 1000-MWe LWR is 225 metric tons of $U_3O_8$ if the enrichment plant operates at a tails assay of 0.25% $^{235}U$ and there is no recycle of uranium or plutonium from spent fuel. At these conditions, the 1977 mining and milling rate ($\sim$14,940 short tons of $U_3O_8$ per year) is sufficient to support about 60 1000-MWe LWRs. In July of 1978, there were 69 power reactors in the U.S. having a total electrical generating capacity of 51,420 MWe.[18] However, $U_3O_8$ production requirement is not a function of today's electrical generating capacity, but is a function of the generating capacity several years in the future. This is a result of the long lead times needed to bring new facilities on line and to produce the uranium and fabricate it into reactor fuel elements. Annual $U_3O_8$ requirements to produce the predicted 560 billion kWh from nuclear power stations in 1985 are about 19,000 metric tons (21,000 short tons) and for 920 to 940 billion kWh in 1990 are 31,500 to 32,200 metric tons (34,600 to 35,400 short tons). This is more than a 130% increase in requirements which must be met at least 5 years before 1990.

### B. Uranium Mining Methods

Typical methods normally used to recover other metallic ores are also used for mining uranium. Some modifications are made to fit the particular characteristics of uranium deposits in sandstone, from which most of the domestic supply is now derived. Two basic mining methods have been widely used: (1) open pit mining, which has provided about 55% of the ore, and (2) underground excavation, which has provided about 45%. In recent years, solution mining has been developed and used for recovering uranium from a few ore bodies. This method offers several economic and environmental impact advantages; however, not all ores are amenable to solution mining.

Factors which affect the selection of uranium mining methods are (1) size, shape, depth, and slope of the ore body; (2) physical and mechanical properties of the deposit and rock around it; (3) hydrology of the surrounding area; (4) ore grade, production rate, and comparative mining costs; and (5) environmental factors.

*1. Open Pit Mining*

Open pit or surface mining is used when the ore body lies under a relatively friable overburden and at depths of no more than about 100 m. Single bench mining (single level of operation) is used for shallow deposits. For deeper deposits, multiple bench mining is used with the sides of the open pit stepped for stability against slides. The overburden that must be removed from an open pit mine may be 30 to 50 times the amount of ore recovered.

A model open pit mine producing 1600 metric tons of 0.2% $U_3O_8$ ore per day for 300 days/year will supply about 960 metric tons of $U_3O_8$ per year or enough for 5.3 1000-MWe LWRs if enrichment assay is 0.25% $^{235}U$ and uranium from spent fuel is recycled (4.3 1000-MWe LWRs if neither uranium or plutonium is recycled). With an overburden-to-ore ratio of 30:1, about 9.5 million $yd^3$/year of overburden must be moved. The removed overburden is stored temporarily in huge piles that may contain as much as 19 million metric tons and cover up to 250 acres. As needed, this overburden containing small amounts of uranium and its daughters is used to back-fill mined out areas. The active open pit may be approximately 120 acres.

Mine drainage water containing several curies of radioactivity (uranium and daughters) might be as much as 650 million gal/year. This water is pumped from the open pit into seepage ponds where the radioactive materials are returned to the ground by ion exchange as the waste water seeps through the soil.

There are small amounts of dust particles containing traces of uranium and its daughters produced as the ore body is exposed and broken up during mining. Somewhat larger amounts of radon-222 gas, also a uranium daughter, escape from the broken ore. In high concentrations, the alpha-emitting daughter products are a health hazard and on long exposure can be conducive to lung cancer. However, in open pit mining, this gas is rapidly diluted to acceptable levels. Since $^{222}Rn$ and its daughters are short-lived, discharge to the atmosphere produces only local effects on the radiation level in the air.

A material balance flowsheet for open pit mining and subsequent milling of uranium ore to meet the annual requirements (182 metric tons of $U_3O_8$) for a model 1000-MWe LWR are given in Figure 3.

*2. Underground Excavation Mining*

Underground excavation has been used extensively to recover uranium ore bodies that are covered with more than about 120 m of overburden or that lie under rock strata that cannot be readily broken and removed. Surface installations required for an underground mine include service buildings, a head frame with a truck-loading facility, a mine waste pile, and a settling pond for treating mine water pumped from underground sumps. Only a few acres are required for the service buildings, but the

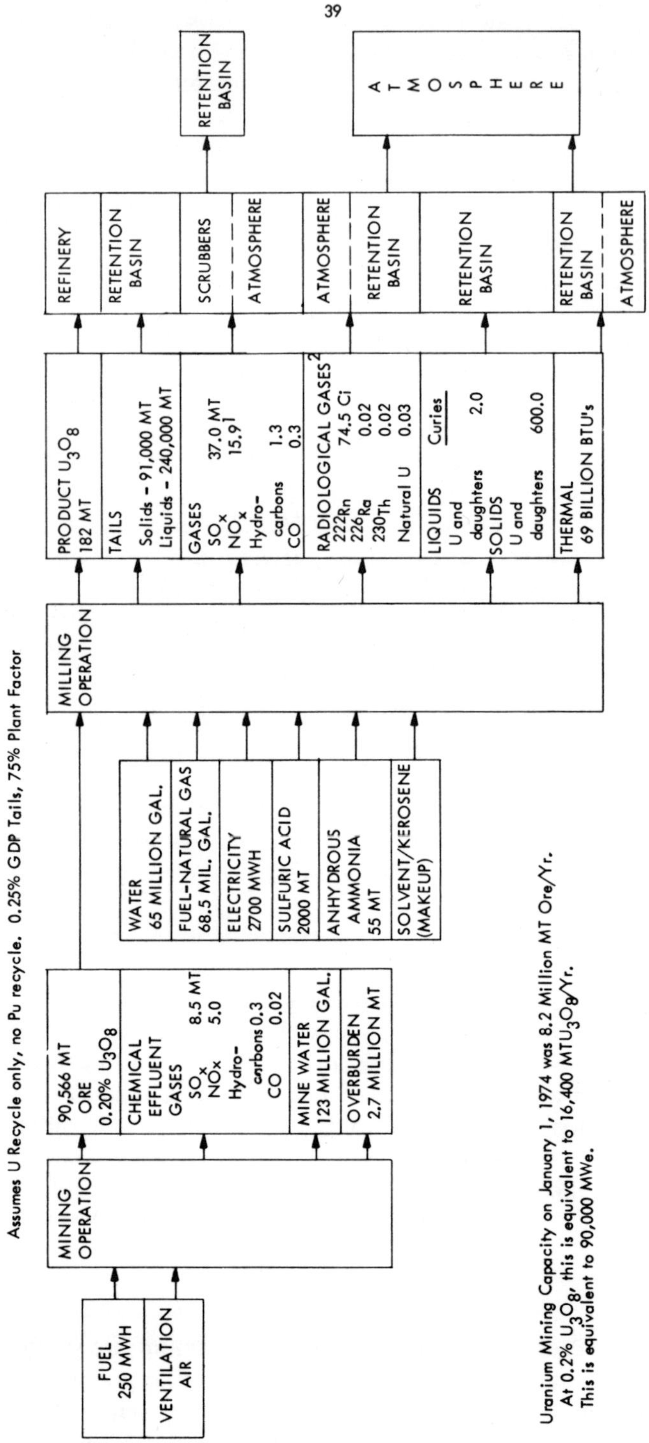


[1]Forty percent from natural gas use.
[2]Including airborne particulates.

FIGURE 3. Uranium milling material balance flowsheet normalized to one model 1000-MWe LWR annual fuel requirement (182 metric tons of $U_3O_8$).[19]

volume of the mine waste pile might be equal to the volume of the ore removed. The underground mine itself might range up to distances of over a mile.

Underground mines must be well ventilated to keep the concentrations of radon gas and its daughter products at acceptably low levels. The required volume of ventilating air is much larger than that for other mines. The exhaust air is discharged to the atmosphere where it is rapidly diluted to acceptable radiation levels.

*3. Solution Mining*

Solution mining[20-22] is the *in situ* leaching of ore bodies to recover valuable minerals by selective dissolution. It is commonly accomplished by injecting leach solutions through wells into a permeable ore body and then, after it has passed through the ore, recovering the leach liquor from other wells for further processing. Under certain conditions, this method is suitable for some uranium ore bodies that cannot be developed economically by conventional mining methods. Thus, it has the effect of expanding the uranium resources. Solution mining also has a much less significant impact on the environment than conventional methods since it requires a minimum of surface installations and eliminates the need for handling large tonnages of ore and disposing of large volumes of waste. Many of the hazards associated with underground excavations are eliminated in solution mining. These significant reductions in personnel hazards, waste disposal, and land rehabilitation costs are important factors in reducing overall mining costs. Furthermore, the significantly smaller number of processing steps required results in decreased energy, operating costs, labor requirements, capital investment, and lead times. It thus becomes possible in some cases to economically mine low-grade ore bodies containing as little as 100,000 lb of $U_3O_8$.

Uranium ore bodies must meet several special criteria to make them amenable to solution mining. The uranium must be contained within a nearly horizontal layer of a permeable host deposit located below the water table and lying between two layers of impermeable shale. Generally, solution mining in the U.S. is limited to the Wyoming and Texas roll front-type deposits. The majority of amenable deposits consists of uraninite and minor coffinite precipitated as minor interstitial components of a sandstone host. It has been estimated[20] that solution mining can probably be used for 30 to 50% of the western sandstone deposits; however, it doesn't work well with most deposits in the Colorado plateau or with high-content vanadium ores.

Solution mining of uranium may be carried out with either an acid or an alkaline leach solution. Current technology has centered primarily around an alkaline process that permits complete recycle of the leach solution. In this process, a solution of sodium carbonate and sodium bicarbonate is injected into the ore body where the uranium is leached in a two-stage process. Tetravalent uranium is oxidized to the hexavalent state, which is then complexed by carbonate to form the soluble uranyl tricarbonate anion, $UO_2(CO_3)_3^{-4}$. The uranium-rich leach solution is then pumped from the ground, uranium is recovered by anion exchange,[23] and the leach solution is regenerated and reused.

Since the carbonate leach solution is quite specific for uranium, most of the radioactive daughters of uranium remain in the ore body. The radiation levels of the solutions result almost entirely from minute quantities of $^{226}Ra$ carried along with calcium dissolved from the ore, but these concentration levels can be controlled by ion exchange so that most of the radium remains in the ground and in some cases even the small amount leached is returned to and deposited in the ore body when the leach solution is recycled.

The quantity and nature of wastes produced by solution mining vary greatly with the process. Large quantities of wastes may be produced if sulfuric acid is used as a leach solution and recycling cannot be practiced because of impurity build-up. With the alkaline process in which the leach solution is recycled, wastes may vary from 1 to 2 lb for each pound of $U_3O_8$ produced, or about 1 to 2 tons of tailings for every 1500 tons of ore treated. These tailings consist of small quantities of degenerated ion-exchange resin and solid impurities dissolved from the ore, primarily calcium. It may be desirable to precipitate the uranium and ship it from the mining plant as yellow cake slurry to avoid the need for expensive equipment to control low-level airborne dust during drying and packaging. This can also eliminate the need for waste dumps and tailings ponds at the mine site. These final operations of drying and packaging with precise quality control, product weighing, and assaying can more economically be carried out at a bulk facility large enough to handle yellow cake slurry from several mines.

Heap leaching[24] is a modification of solution mining in which a large pile of ore, either confined or in an open pile, is leached to recover the uranium. It is accomplished by passing leaching solution down through a heap, usually on a slope, with troughs under the heap leading toward the downhill side where the leach solution is collected for additional processing. These piles are usually 300 to 400 ft along each side and 20 to 30 ft high. After leaching, the piles are generally abandoned and new piles are constructed next to them. All heap leaching in the U.S. uses sulfuric acid to dissolve uranium from typical sandstone ore.

### C. Uranium Recovery from Wet-Process Phosphoric Acid

Phosphoric acid produced in the U.S. is a potential source of significant amounts of uranium. Phosphate deposits contain varying amounts of trace quantities of uranium. The U.S. Department of Energy estimates that potential uranium reserves in phosphate deposits are 4 million tons.[25] Primary interest in uranium from phosphate is in two tri-state areas: Wyoming, Montana, and Utah in the West and central Florida and the Carolinas in the South. So far, most activity has centered around the Florida phosphates, with uranium content of about 100 to 200 ppm. During treatment of the phosphate deposits to produce phosphoric acid, the uranium is dissolved while most of the radioactive daughters remain in the solid residue. Uranium is an undesirable radioactive contaminant in the phosphoric acid. So far, only a token amount of uranium ($\sim$19 tons $U_3O_8$) has been recovered as by-product from phosphoric acid. However, it is estimated that wet-process phosphoric acid produced in the U.S. in 1976 contained about 3000 tons of dissolved $U_3O_8$.[26] The amount is projected to increase to about 8000 tons/year by the year 2000. Since total uranium production in the U.S. was only about 13,500 tons in 1976, it can be seen that phosphoric acid as a source of uranium offers an attractive potential.

Two different processes[27] were used initially for recovering uranium from Phosphate Rock. One method involved the extraction of uranium into nonyl pyrophosphoric acid from crude phosphoric acid. The other method involved reduction of uranium in a monosodium phosphate solution to precipitate uranous phosphate. In the early 1970s, improved solvent extraction processes were developed at Oak Ridge National Laboratory.[26,28,29]

Extractants that have been used to recover uranium from sulfate leach solutions of western ores are not satisfactory for recovering uranium from phosphoric acid, since the uranium is complexed by the phosphate, and extraction factors are not great enough. A synergistic extractant combination of di(2-ethylhexyl)phosphoric acid and trioctylphosphine oxide (DEPA-TOPO) can effectively extract hexavalent uranium, and a mixture of mono- and dioctylphenylphosphoric acid (OPAP) can effectively extract tetravalent uranium. Two extraction processes have been developed based on this information.

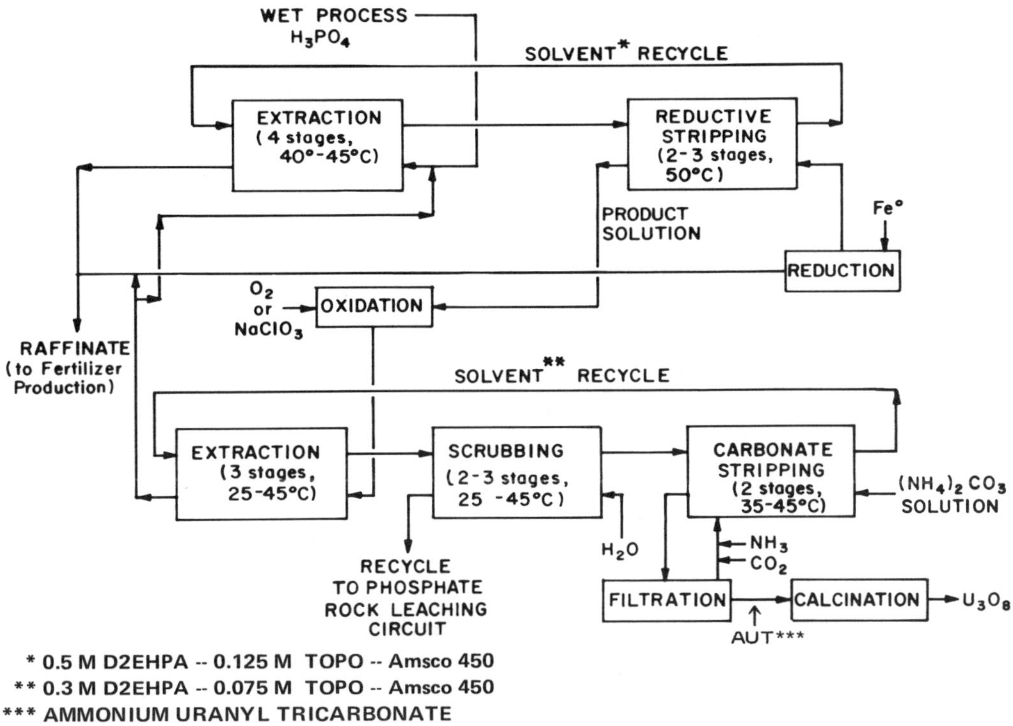


FIGURE 4. Flowsheet for recovery of uranium from phosphoric acid by extraction with di(2-ethylhexyl)phosphoric acid plus trioctylphosphine oxide (DEPA-TOPO) in an aliphatic diluent.

The DEPA-TOPO process is presented in Figure 4. In the first cycle, hexavalent uranium in the acid from pretreatment is extracted in four to five stages with 0.5 *M* DEPA-0.125 *M* TOPO in an aliphatic diluent (such as refined kerosene). Uranium is stripped from the solvent in three stages by contacting it with phosphoric acid containing ferrous iron in order to reduce the uranium to its less extractable tetravalent state. It is convenient to use a small volume of the raffinate from the extraction system since it is suitably concentrated in $H_3PO_4$ and, after reduction with iron metal, contains a suitable concentration of $Fe^{2+}$. Fluoride present in the acid catalyzes the reduction. Strip solutions containing about 12 g of uranium per liter are readily obtained. The solutions which are a factor of 70 or so richer in uranium than the original acid are amenable to treatment in a second cycle using the same extractant.

In the second cycle, uranium in oxidized feed from the first cycle is extracted with DEPA-TOPO at conditions that allow high uranium loading of the solvent that minimizes extraction of impurities such as iron and aluminum. The solvent is scrubbed with water to remove small amounts of extracted phosphoric acid and then stripped with ammonium carbonate solution at conditions that precipitate uranium continuously as ammonium uranyl tricarbonate. The carbonate is then filtered and calcined to produce a crude $U_3O_8$ product.

An alternate process uses a different initial cycle to concentrate uranium in a small stream of phosphoric acid coupled with the same DEPA-TOPO second cycle. In this alternate process, uranium in its tetravalent state is extracted into 0.3 *M* solution of OPAP. As a result, there is no need to oxidize organic impurities normally contained in the raw phosphoric acid. Another advantage is that effective extraction can be accomplished from more concentrated phosphoric acid than is possible with DEPA-TOPO. Uranium is stripped from the OPAP by contacting it with 10 $M$ $H_3PO_4$ containing an oxidant such as $NaClO_3$ to convert the uranium to its less extractable hexavalent state. A convenient source of strip solutions is 50 to 54% $P_2O_5$ product acid from the evaporator. The strip solution can contain up to 15 to 20 g uranium per liter. After dilution to 6 $M$ $H_3PO_4$ it can be fed directly into the second cycle as shown in Figure 5.

### D. Uranium Milling

Milling operations are carried out to extract uranium from ores that normally contain between 0.10 and 0.30% $U_3O_8$ to produce a semirefined concentrate that is referred to as "yellowcake". This product may be any of several impure compounds, such as $U_3O_8$, ammonium diuranate, or sodium diuranate, and usually contains the equivalent of 75 to 85% $U_3O_8$. It is packed as a dry solid in 55-gal drums, such as that shown in Figure 6, and shipped to refining and chemical conversion plants where pure $UF_6$ is produced.

Most of the domestic uranium mills are located in the western U.S. adjoint or very near to the mines, so the large tonnages of ore can be trucked relatively short distances over privately owned or leased land. As of January 1978, there were 18 uranium mills[15] in the U.S. with a total capacity of 39,210 tons of ore per day or about 21,000 tons of $U_3O_8$ per year. A list of these plants, their location, and the nominal daily capacity is given in Table 6. These mills vary in size from 400 to 7000 tons of ore per day, but the majority have capacities of about 1000 to 3000 tons/day. As the grade of the ore drops, mills must process larger amounts of ore in order to maintain a profitable production rate of yellowcake.

Processing methods used by the different mills vary widely because of the different compositions and characteristics of the ores handled. There are nearly as many different processes as there are mills. However, basic treatments at all the mills are somewhat similar. The sequence of these steps is given schematically[9] in Figure 7. Ores are blended and crushed to pass through a ½-in. screen. The crushed ore is then wet ground in a rod or ball mill and is transferred to leaching tanks where the ore is contacted with an appropriate solution containing an oxidizing agent to solubilize the uranium. The residual solids are separated, washed with water, and slurried to an impoundment area for disposal. Product liquor is treated, usually by solvent extraction or ion exchange to produce a semipure concentrated uranium solution. The uranium is then precipitated with ammonia or caustic, and the slurry is thickened and filtered or centrifuged to recover the uranium-bearing solids from the waste liquid which is also sent to the tailings impoundment area. The solids are dried and may be calcined and pulverized to produce yellowcake which is drummed and shipped to a refinery.

Although each mill has its own particular processing method, the processes in use at the active mills fall into three general types:[8] (1) sulfuric acid leach followed by solvent extraction to purify the uranium (10,100 tons/day), (2) sulfuric acid leach followed by ion exchange to purify the uranium (9100 tons/day), and (3) sodium carbonate leach with no additional uranium purification step (3350 tons/day). A survey[8] of ten mills made in 1973 shows that a variety of chemical reagents is used and that there is a considerable difference in the amounts of these reagents needed for the different mills (Table 7). A discussion of specific processes used at a number of mills is presented by Merritt.[30] Approximately 95% of the uranium in the ore is recovered as yellowcake.

#### *1. Acid Leach Mills*

Most of the U.S. uranium ores are suited to acid leaching, and about 80% of all

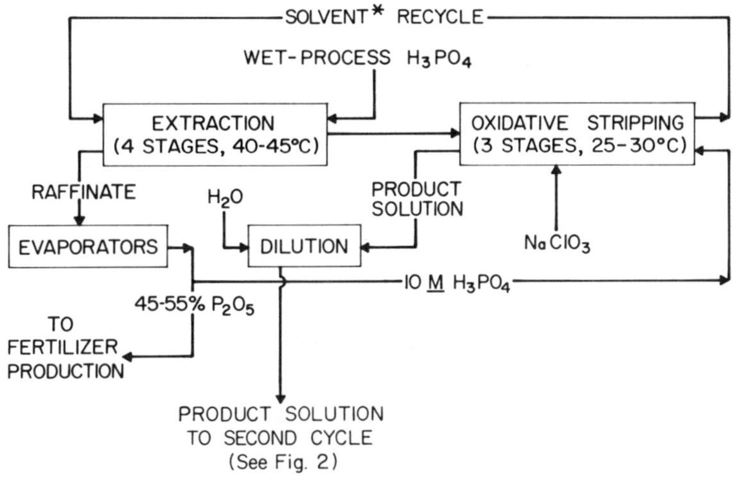


FIGURE 5. Flowsheet for recovery of uranium from phosphoric acid by extraction with a mixture of mono- and dioctylphenylphosphoric acid (OPAP) in an aliphatic diluent.

ores are treated by this method. Sulfuric acid is commonly used because it is cheaper than other acids. Uranium minerals containing carbonates, e.g., $CaCO_3$ and $MgCO_3$, consume acid in direct proportion to their amounts. If ore contains more than about 15% $CaCO_3$ by weight, it is likely that acid consumption and costs will be too high, and carbonate leaching would probably be used.

A schematic flowsheet of a model acid-leach mill which uses solvent extraction for uranium purification is given in Figure 8; the material balance flowsheet normalized to the annual requirements for one model 1000-MWe LWR was given in Figure 3.

The current trend is to increase use of this type process for treating most uranium ores in the U.S. Ore is brought from the mines in trucks, dumped, and passed through a primary crushing circuit, where it is crushed to about ½ in. and screened. All oversized material is recycled. The crushed ore is conveyed on endless rubber belts to elevated storage bins. All equipment handling dry ore is vented through dust collectors, and exhaust air is discharged through a short stack on the roof.

The ore is then ground to minus 28 mesh in rod mills as a slurry containing about 65% solids, and the fine ore is sent to the leach circuit. Acid leaching is carried out differently at different mills and with different ores; however, it is usual to employ a number of tanks arranged for series flow that provide a total residence time of about 7 or 8 hr. Sulfuric acid containing an oxidant, usually manganese dioxide or sodium chlorate, is added continuously, and propeller or turbine-type agitators are used to mix the slurry for efficient leaching. The oxidant plays the important and necessary role of oxidizing uranium in those ores containing tetravalent uranium and in preventing the reduction of uranium by other ore constituents or by metallic iron introduced during ore grinding.

Slurry from the leach tanks passes through hydroclones to remove the coarse sand fraction, which is then washed countercurrently in a series of about six classifiers. Leach solution containing the dissolved uranium is recovered by sending the hydroclone and classifier overflows through a series of six thickeners to which flocculants are added to promote settling of the slimes, which are washed countercurrently with fresh water. The washed sands, which constitute about 70% of the ore processes, and slimes, which are the other 30%, are pumped as slurries to the tailings pond.

FIGURE 6. Photograph showing shipping method for yellowcake.

## Table 6
## URANIUM PRODUCTION MILLS OPERATED IN THE U.S. AS OF JANUARY 1, 1978[15]

| Company | Location | Nominal capacity (tons ore per day) |
|---|---|---|
| Anaconda Company | Grants, N.M. | 6,000 |
| Atlas Corporation | Moab, Utah | 1,100 |
| Conoco-Pioneer | Falls City, Tex. | 2,900 |
| Cotter Corporation | Canon City, Colo. | 450 |
| Dawn Mining Company | Ford, Wash. | 400 |
| Exxon Company, USA | Powder River Basin, Wyo. | 3,000 |
| Federal-American Partners | Gas Hills, Wyo. | 950 |
| Kerr-McGee Nuclear Corp. | Grants, N.M. | 7,000 |
| Lucky Mc Uranium Corp. | Gas Hills, Wyo. | 1,650 |
| Lucky Mc Uranium Corp. | Shirley Basin, Wyo. | 1,800 |
| Rio Algom Corporation | La Sal, Utah | 700 |
| Rocky Mt. Energy | Powder River Basin, Wyo. | 1,000 |
| Sohio-Reserve Oil | Cebolleta, N.M. | 1,660 |
| Union Carbide Corporation | Uravan, Colo. | 1,300 |
| Union Carbide Corporation | Gas Hills, Wyo. | 1,200 |
| United Nuclear Corporation | Church Rock, N.M. | 3,000 |
| United Nuclear-Homestake Partners | Grants, N.M. | 3,400 |
| Western Nuclear, Inc. | Jeffrey City, Wyo. | 1,700 |
| Total | | 39,210 |

When solvent extraction is used, the uranium is recovered by countercurrent contact in four extraction stages with a long-chain tertiary amine, such as Alamine® 336 which is primarily tricapryl amine, dissolved in kerosene. The uranium is then removed from the extractant by backwashing or stripping it in four countercurrent stages with a solution of ammonium sulfate. The extractant is rather expensive, so it is essential that it be recycled back to the extraction circuit.

Uranium is precipitated from the ammonium sulfate solution by the addition of gaseous ammonia, concentrated and washed in thickeners, and collected on filters. The dried product, or yellowcake, is packaged in 55-gal steel drums and shipped to a refinery. Most of the uranium daughters, which are in secular equilibrium in the ore, remain in the ore tailings. The uranium product contains about 5% of the Th ($1.4 \times 10^{-2}$ $\mu$Ci/g $U_3O_8$) and 0.2% of the Ra ($5.5 \times 10^{-4}$ $\mu$Ci/g $U_3O_8$). Since no other radionuclides are present in significant amounts, no shielding is required for the shipping drums.

Ion exchange is used in some mills to purify uranium contained in the sulfuric acid leach.[24] This is accomplished by contacting the leach solution with coarse particles of an organic, solid ion-exchange resin that will selectively sorb uranium. Since uranium is present in the leach solution as an anionic uranyl sulfate complex, strong base anion-exchange resin is used. Ion-exchange processes are frequently carried out as a series of semicontinuous operations. Typical vessels used to contain the resin are rubber-lined steel pressure tanks with dished ends and may be 6 to 9 ft in diameter and 12 to 15 ft high. A support of sized crushed rock or a perforated plate is placed in the bottom of the tank. The bed of resin fills only about half of the tank to allow room for bed expansion during backwashing, which is an operation during regeneration of the resin in which solutions are passed up through the resin bed to expand it and to flush out entrained solids. Plants using fixed-bed ion exchange usually have three or four columns. It is common practice to use at least two columns in series during sorption of

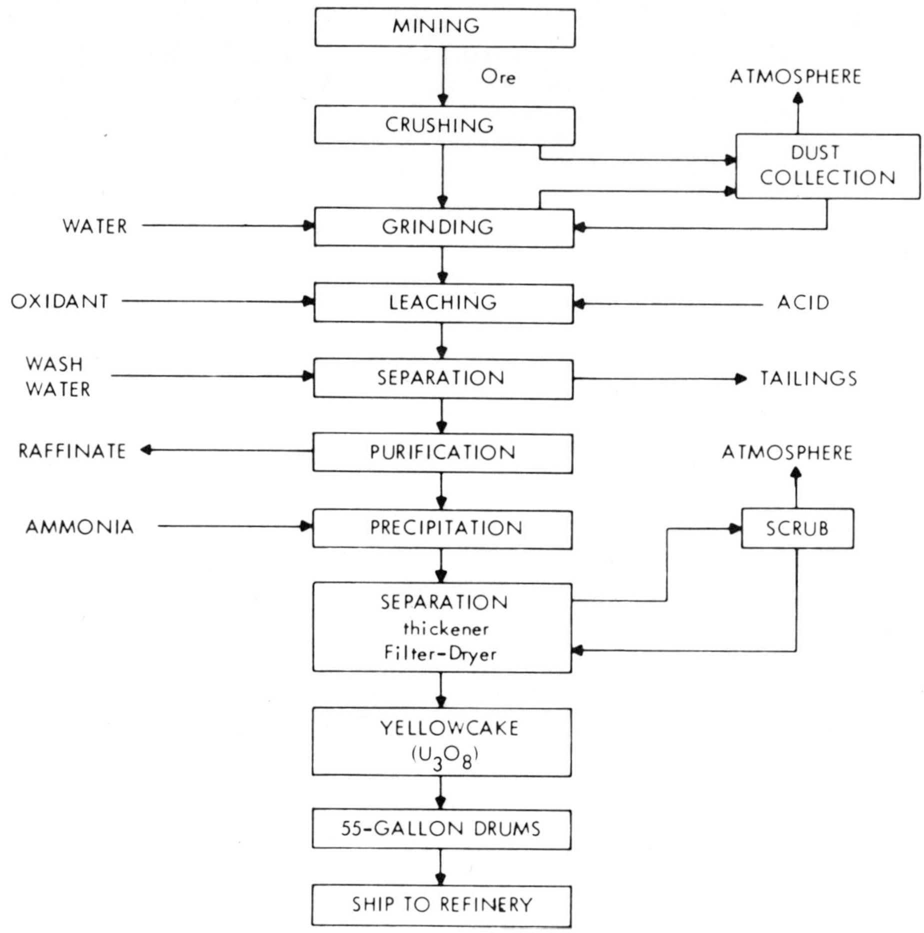


FIGURE 7. Simplified flow diagram of uranium milling process steps.

the uranium so that the first column can be loaded to uranium saturation before excessive uranium breaks through in solution from the second column. Upon saturation, the first column is put into the elution cycle, the second column becomes the first, and the third column, which has been regenerating, becomes the second column. This sequence of operation is repeated each time a column becomes loaded with uranium.

*2. Alkaline-Leach Mills*

Although only a small fraction of the uranium ore is treated by alkaline leaching, this method plays an important role in recovering uranium from ores that are not amenable to acid leaching. Leaching of ores with a solution containing sodium carbonate and sodium bicarbonate has become of more importance in recent years.

A schematic flowsheet[8] of a model alkaline-leach uranium mill is given in Figure 9. In many cases it is not necessary to further purify uranium at the mill after it has been leached, since the carbonate leach is quite selective and does not dissolve much of the other ore constituents. The head end operations for an alkaline-leach mill are similar to those for an acid mill; however, the ground ore usually must be much finer (from 25 to 80% minus 200 mesh). To accomplish this, a wet-grinding system consisting of a ball mill operated in closed circuit with a classifier is used, the grinding is done at

## Table 7
## CHEMICAL USAGE IN URANIUM MILLS, SPRING 1973

Chemical usage (lb/ton of ore processed)

| | Mill | $H_2SO_4$ | $Na_2ClO_3$ | $MnO_2$ | $NH_3$ | NaOH | MgO | CaO | NaCl | $NH_4NO_3$ | $Na_2CO_3$ | Amine[a] | Alcohol[b] | Kerosene | Flocculent | Ion-exchange resin | Grinding rods |
|---|---|---|---|---|---|---|---|---|---|---|---|---|---|---|---|---|---|
| Solvent | 1 | 60 | 2.0 | — | 1.7 | — | — | — | — | — | — | 0.03 | 0.08 | 0.08 | 0.13 | — | 0.5 |
| Extraction | 3 | 85 | 2.4 | — | 2.8 | — | — | — | — | — | — | 0.04 | 0.04 | 1.19 | 0.19 | — | Yes |
| | 4 | 59 | — | 7.8 | 0.75 | — | 1.4 | 4.4 | 38 | — | — | — | — | — | — | 0.32 | 0.5 |
| Ion | 5 | 92 | 2.0 | — | 8.5 | — | — | — | — | — | — | 0.03 | 0.03 | 0.30 | — | 0.34 | Yes |
| Exchange | 6 | 60 | 5.0 | — | 2.5 | 0.2 | — | — | — | — | — | 0.008 | 0.02 | 0.42 | — | 0.34 | Yes |
| | 7 | 47 | 1.5 | — | 4.4 | — | — | — | — | — | — | 0.02 | 0.02 | 0.31 | — | 0.04 | Yes |
| | 8 | 165 | — | 4 | 2 | 1.2 | — | 12 | — | 8 | — | — | — | — | Yes | Yes | Yes |
| Alkaline | 9 | — | — | — | — | 15 | — | — | — | — | 12 | — | — | — | — | — | Yes |
| Circuit | 10 | 3.3 | — | — | — | 76 | — | — | — | — | 6 | — | — | — | 1.8 | — | Yes |

[a] Alamine 336 and Adogen 364 are used interchangeably.

[b] Isodecanol and tridecanol are used interchangeably although isodecanol is more common.

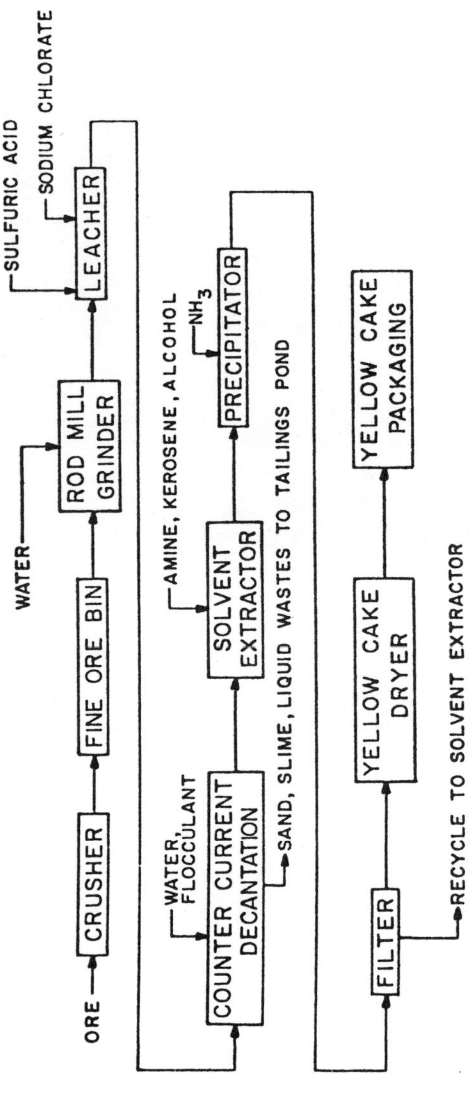


FIGURE 8. Flowsheet for a model acid leach-solvent extraction uranium mill.

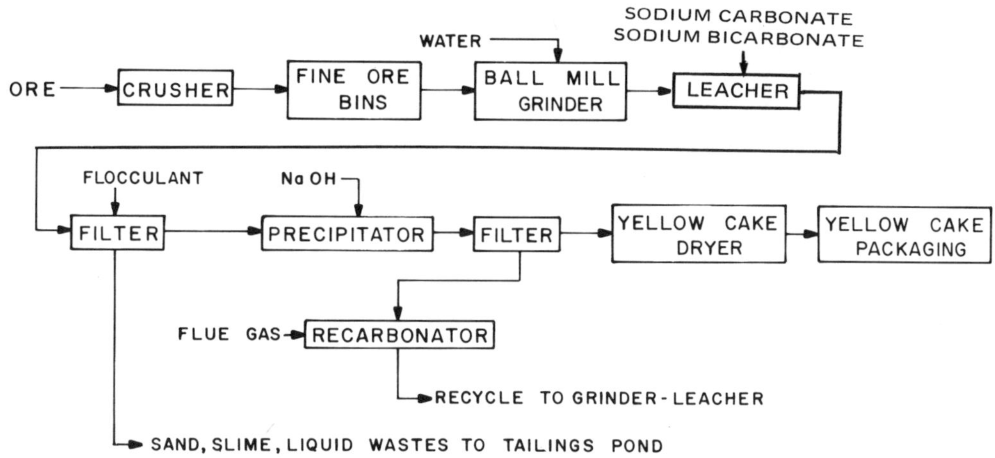


FIGURE 9. Flowsheet for a model alkaline leach uranium mill.

65% solids in a sodium carbonate-sodium bicarbonate solution. Effective leaching can be accomplished in two stages consisting of a 5-hr leach at 65 psig pressure and 200°F, followed by an 18-hr leach at atmospheric pressure and 185°F. The leaching is frequently carried out in Pachuca tanks, which are large tanks, 40 to 60 ft deep, with conical bottoms (60° included angle) and cylindrical bodies having an overall height-to-diameter ratio of about 2.5:1. Steam may be added through nozzles and compressed air may be added through a central lift or bottom jet. This air not only serves to agitate the contents of the tank but also is an effective oxidant. After leaching, solids are separated and washed free of uranium by several stages of countercurrent filtration. The solids, which consists of a 50-50 mixture of sands and slimes, are repulped with fresh water and pumped to the tailings pond.

When no additional purification is required, uranium is recovered from the leach solution by the addition of sodium hydroxide, which forms insoluble sodium diuranate (yellowcake). The solid is filtered, washed, dried, and packaged in 55-gal steel drums for shipment. If product specifications require low sodium content, the sodium diuranate is dissolved in sulfuric acid and reprecipitated with hydrogen peroxide or ammonia.

Because the chemicals are expensive, alkaline mills have always practiced the recycle of leach solution, with only a small bleed stream sent to waste to prevent build-up of impurities. Excess caustic in the filtrate is reacted with flue gas (largely carbon dioxide) to convert it to leach solution containing the required ratio of sodium carbonate and sodium bicarbonate.

Some of the early mills[24,30] treated the carbonate leach solution by ion exchange to purify and recover the uranium. All of the alkaline leach mills now in operation precipitate sodium diuranate directly from the leach solution.

## IV. REFINING AND CHEMICAL CONVERSION

Yellowcake produced at uranium mills is a concentrated product containing the equivalent of 75 to 85% uranium oxide; however, it is still far from pure. The composition of a typical yellowcake is given in Table 8. Refining and conversion operations are carried out to remove these impurities and to convert the uranium to the volatile hexafluoride required as feed for the uranium isotope-enrichment plant.

**Table 8**
**TYPICAL YELLOWCAKE FEED TO A $UF_6$ CONVERSION PLANT[12]**

| Constituent of feed | Concentration (wt %) | Quantity fed (metric tons/year) |
|---|---|---|
| Uranium (U) | 73.53 (wt %, U basis) | 10,000 |
| Impurities | | |
| Ammonium ($NH_4^+$) | 3.09 | 420 |
| Sodium (Na) | 2.41 | 329 |
| Silica ($SiO_2$) | 1.2 | 163 |
| Sulfate ($SO_4^{2-}$) | 2.94 | 400 |
| Arsenic (As) | 0.06 | 8 |
| Boron (B) | 0.003 | 0.4 |
| Calcium (Ca) | 0.19 | 26 |
| Carbonate ($CO_3^{2-}$) | 0.31 | 42 |
| Chloride, bromide, iodide ($Cl^-$, $Br^-$, $I^-$) calculated as $Cl^-$ | 0.07 | 10 |
| Fluoride ($F^-$) | 0.01 | 1 |
| Iron (Fe) | 0.38 | 52 |
| Molybdenum (Mo) | 0.10 | 14 |
| Phosphate ($PO_4^{3-}$) | 0.26 | 35 |
| Potassium (K) | 0.13 | 18 |
| Vanadium (V) | 0.12 | 16 |
| Water ($H_2O$) | 1.91 | 260 |
| Extractable organics | 0.05 | 7 |
| Nitric acid-insoluble uranium | 0.01 | 1 |
| Principal radionuclides | | (Ci/year) |
| $U_{nat}$[a] | | 3333 |
| $^{234}Th$ | | 3333 |
| $^{234m}Pa$[b] | | 3333 |
| $^{230}Th$ | | 141.7 |
| $^{226}Ra$ | | 15.67 |
| $^{222}Rn$ | | 15.67 |

[a] One curie of natural uranium ($U_{nat}$) is defined in 10 CFR 20 as the sum of $3.7 \times 10^{10}$ dis/sec from $^{238}U$, $3.7 \times 10^{10}$ dis/sec from $^{234}U$, and $9 \times 10^8$ dis/sec from $^{235}U$; i.e., 1 Ci of $U_{nat}$ consists of 1 Ci of $^{238}U$, 1 Ci of $^{234}U$, and $2.43 \times 10^{-2}$ Ci of $^{235}U$; it is also equivalent to 3000 kg of natural uranium.

[b] Metastable $^{234m}Pa$, $t_{1/2}$ = 1.18 min.

Two different processes[13] for refining yellowcake and converting it to uranium hexafluoride are used commercially in the U.S. The Allied Chemical Company plant at Metropolis, Ill., with an annual conversion capacity of about 13,000 metric tons of uranium, uses the "dry hydrofluor process", and the Kerr-McGee Nuclear Plant at Sequoyah, Okla., with an annual conversion capacity of about 10,500 metric tons of uranium, uses the "wet solvent process". Kerr-McGee increased capacity from the original 4500 metric tons/year in mid-1978 and added the capability of accepting up to two thirds of its total throughput in slurry form.[31] An additional charge of ~60¢/lb will be made for processing slurries; however, this will provide economic advantages in many cases. This seems to be especially true of uranium producers using solution mining, heap leaching, and recovery from wet phosphoric acid, since their basic product is a slurry. Since they can ship slurry, there will be no need for these producers to add the expensive equipment required to dry and package yellowcake and the associated dust control equipment required for environmental and personnel protection.

Both plants are designed for virtually complete recovery of uranium, total utilization of elemental fluorine, and high utilization of the other main reactants (hydrogen, anhydrous hydrogen fluoride, ammonia, and nitric acid).

There are major facilities for converting ore concentrates to uranium hexafluoride in Canada, the United Kingdom, and France. Plants in these countries supplied 34% of the U.S. uranium hexafluoride feed to the gaseous diffusion plants for toll enriching during fiscal year 1974.

### A. Hydrofluor Process

The dry hydrofluor process treats the mill concentrates directly through reduction, hydrofluorination, and fluorination in a series of fluidized-bed reactors. The crude uranium hexafluoride is then purified by fractional distillation. Impurities associated with the uranium concentrates are eliminated as solid constituents in the fluorination tower ash or as volatile compounds. The majority of the volatile impurities are removed from the off-gases by caustic scrubbers. A simplified flow diagram of the hydrofluor process is shown in Figure 10.

Uranium concentrate in 55-gal drums is received at the conversion plant, weighed, sampled, and stored for subsequent processing. In some cases, concentrates are treated with ammonium sulfate solution to remove sodium, which causes plugging during subsequent hydrofluorination. The concentrates are then dried, roasted, and sized for the fluidized-bed reactors. A fluidized-bed reduction unit is shown schematically in Figure 11. Uranium trioxide and $U_3O_8$ are reduced in this unit with $H_2$ at 540 to 620°C.

$$UO_3(s) + H_2(g) \rightarrow UO_2(s) + H_2O(g)$$

Hydrogen is supplied at 100% excess by cracking ammonia at 870°C.

$$2NH_3 \rightarrow N_2 + 3H_2$$

The reduction reaction is exothermic; however, additional heat input is required to compensate for heat losses. Close temperature control is required to make certain that essentially complete reduction is accomplished and that sintering of the particle surfaces does not occur. Unreduced oxide reacts with anhydrous hydrogen fluoride to form $UO_2F_2$, which consumes more elemental fluorine for conversion to $UF_6$ than does $UF_4$, evolves more heat during fluorination, and does not fluorinate as well, thus resulting in more ash that must be recycled.

Both single-stage and two-stage reductions are used, and nitrogen is sometimes added to the cracked ammonia to maintain the fluidizing velocity of the bed.

The presence of sulfate serves as a chemical promoter and increases productivity in both the reduction step and the hydrofluorination step. With optimum operation, the reduced product is a highly reactive uranium dioxide which is readily converted to uranium tetrafluoride with only 5 to 10% excess hydrogen fluoride.

The reactive uranium dioxide can be hydrofluorinated with anhydrous HF at 350 to 540°C in either a stirred bed or a fluidized bed. Schematic diagrams of typical equipment are shown in Figure 12. The basic reaction during hydrofluorination is

$$UO_2(s) + 4HF(g) \rightleftarrows UF_4(s) + 2H_2O(g)$$

Conversion efficiency is 98.0 to 98.5%, and the overall uranium yield for the hydrofluorination step is 99.99%. The presence of large amounts of sodium may result in fusion of the particles, since sodium fluoride forms a relatively low-melting compound

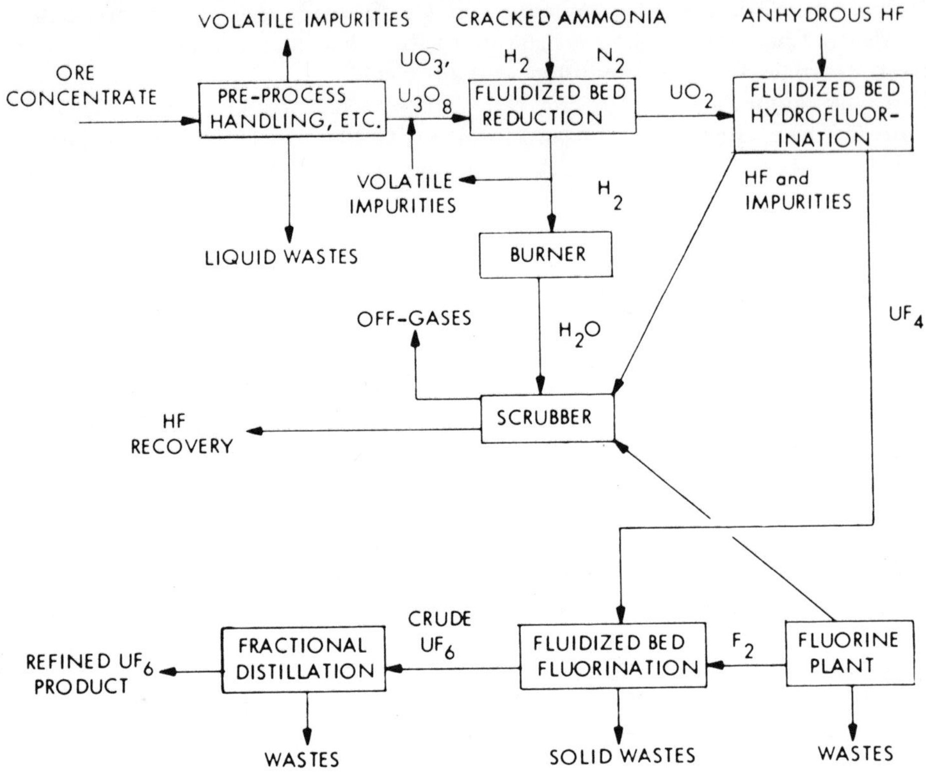


FIGURE 10. Flowsheet of $UF_6$ production from $U_3O_8$ by the "dry hydrofluor" process.

with $UF_4$ ($7NaF \cdot 6UF_4$ with a melting point of ∿675°C compared to 960°C for $UF_4$). Since particle sintering restricts diffusion of HF to unconverted $UO_2$ and may even result in a plastic mass that plugs the bed, concentrates that contain large amounts of sodium are pretreated with ammonium sulfate solutions to remove sodium.

Uranium tetrafluoride still containing most of the original impurities in the yellowcake is fed to a fluidized-bed fluorinator (Figure 13) in which uranium hexafluoride is produced by the highly exothermic reaction with fluorine,

$$UF_4(s) + F_2(g) \rightarrow UF_6(g)$$

Operating temperature is controlled at 425 to 540°C, and from 80 to 90% of the fluorine is utilized. Heat removal is the limiting factor in reactor design, and unless precautions are taken the rate of heat generation can exceed the rate of heat transfer across the fluidized bed. Therefore, $CaF_2$ particles are used as the fluidized media and $UF_4$ is fed to the reactor at a controlled rate to limit the reaction rate. The uranium hexafluoride produced along with several volatile impurities, i.e., $VF_5$, $VOF_3$, and $MoF_6$, are volatilized from the reactor, filtered, and collected as a solid in cold traps. The off-gas is recycled or treated for recovery of the excess fluorine. During fluorination, some of the fluidized solids are removed from the bottom of the bed to limit the build-up of nonvolatile impurities, such as sodium, that form low-melting compounds with $UF_6$ and may cause caking and to avoid the build-up of nonvolatile radioactive daughter products of uranium. These solids plus the filter fines, referred to as fluorination ash, are drummed and stored for at least 6 months to allow $^{234}Th$ and $^{234m}Pa$ to decay.

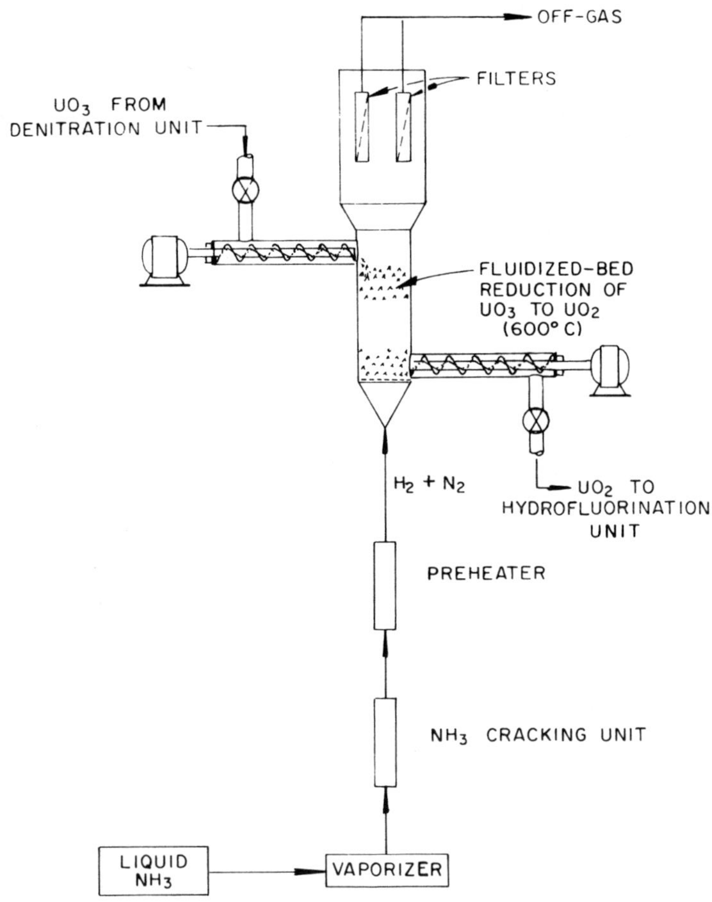


FIGURE 11. Schematic diagram of a typical fluidized bed reduction unit.

The ash is then leached with sodium carbonate to recover uranium. The remainder of the ash, largely $CaF_2$ sludge, becomes a solid, low-level, non-TRU waste (Chapter 6).

After fluorination, the uranium hexafluoride which has been collected in the cold trap is melted and transferred to the distillation feed tanks where it is maintained in a molten state. Two monel bubble cap columns are used to purify the uranium hexafluoride by fractional distillation. The crude $UF_6$ is vaporized into the first column which separates $UF_6$ (sublimation point 56°C, triple point 64°C) from the more volatile $VF_5$ (bp 48°C), $MoF_6$ (bp 35°C), and traces of $SiF_4$, $CF_4$, $SF_6$, etc. Liquid $UF_6$ containing the high-boiling impurities flows to the second column. Vanadium oxyfluoride, with a sublimation point of 110°C, has a solubility of only 0.7 wt % in $UF_6$ at the operating conditions and limits the separation attainable when present in large amounts. During the fractionation, high-purity $UF_6$ is collected in cold traps. Later, it is melted and drained into a 10- or 14-ton shipping cylinder (see photograph in Figure 14) for transfer to the isotope-enrichment plant. Still bottoms consist principally of an unidentified molybdenum compound (probably an oxyfluoride), $VOF_3$, a small amount of $UF_6$, and traces of particulates that have passed through the filters. Still tops from the first column and bottoms from the second column are stored without further treatment, since the values they contain do not justify recovery.

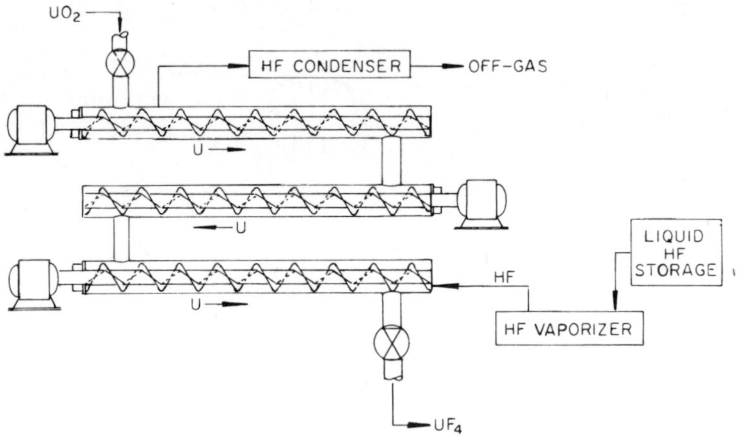


Stirred-Bed Hydrofluorination System.

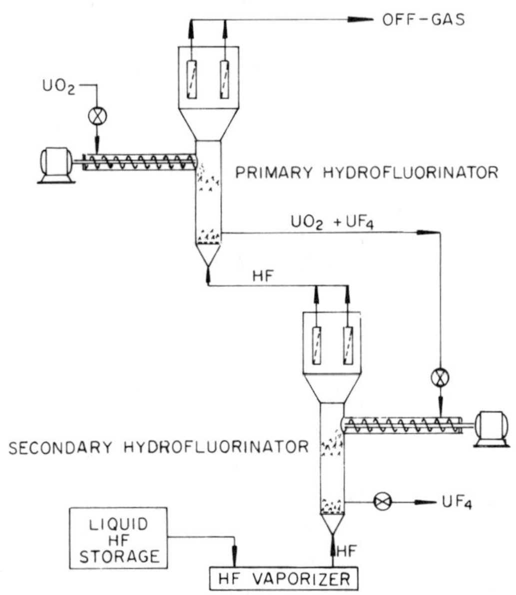


Fluidized-Bed Hydrofluorination System.

FIGURE 12. Schematic diagrams of hydrofluorination systems.

Several types of filters are used to remove particulates from the off-gas streams. Bag filters are used to recover uranium dusts during the initial handling operations. These efficiently remove essentially all of the particles down to 1 μm in size. Sintered stainless steel filters are used for off-gas from the reduction step, and sintered nickel or monel filters are used for off-gas from the fluorination step. Sintered metal filters with a nominal 10-μm pore size will remove 98% of the 0.7-μm particles. Off-gas from the hydrofluorination step is passed through porous carbon filters, which remove >99.999% of the uranium dust. High Efficiency Particulate Air (HEPA) filters are used as backup filters for all off-gas streams, just as they are used in other parts of the nuclear fuel cycle.

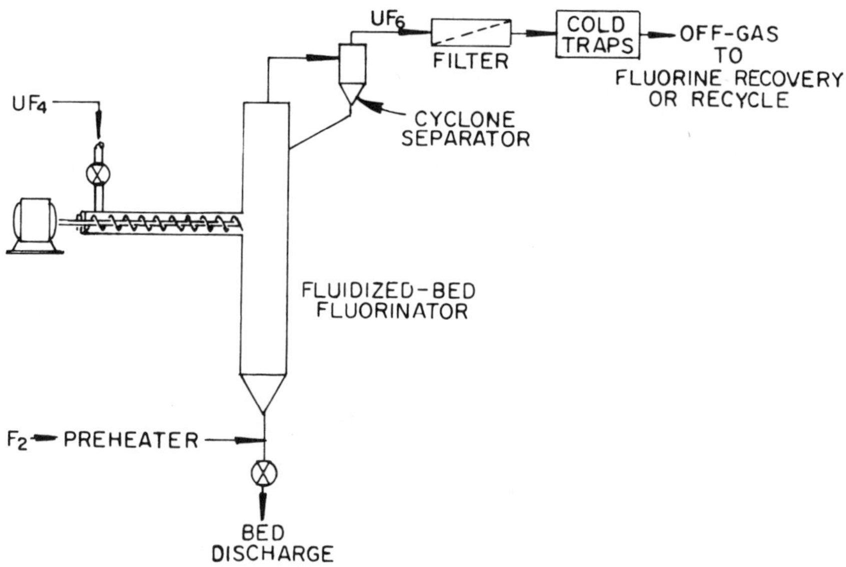


FIGURE 13. Schematic diagram of a fluidized bed for production of $UF_6$ from $UF_4$.

## B. Solvent Extraction Process

The solvent extraction process (sometimes referred to as the wet process) treats the uranium mill concentrates to prepare a high-purity uranium feed prior to reduction, hydrofluorination, and fluorination steps. Therefore, most of the impurities are contained in aqueous waste streams from the solvent extraction treatment, and there is no need for the fractional distillation step following fluorination since the $UF_6$ is already of acceptable purity. A simplified flow diagram of the solvent extraction process is shown in Figure 15.

The receiving, weighing, sampling, and storage operations are essentially the same as for the hydrofluor process. In the wet process, the concentrate is then dissolved by digesting it in hot nitric acid. The solution of uranyl nitrate is then passed countercurrent to a solution of tributyl phosphate (TBP) in an organic diluent such as hexane. Uranyl nitrate extraction into the TBP-hexane solution is highly selective, and virtually all of the impurities are left in the aqueous raffinate stream. The uranyl nitrate is then back-extracted or stripped into a dilute nitric acid solution. This solution is pumped to a steam-heated reboiler where it is evaporated to form uranyl nitrate hexahydrate $[UO_2(NO_3)\cdot 6H_2O]$. The $UO_2(NO_3)\cdot 6H_2O$ is then calcined, or denitrated, to $UO_3$ in a bed fluidized by superheated steam at a temperature of about 315°C. This pure $UO_3$ is then treated through reduction, hydrofluorination, and fluorination in a series of fluidized beds to produce $UF_6$. These processing steps are essentially the same as those for the hydrofluor process.

## C. $UF_6$ from Reprocessed Uranium

Since no commercial reprocessing plant is in operation, there is no recycled uranium to convert to $UF_6$. However, the process steps necessary for this conversion are essentially the same as those for the solvent extraction process for converting uranium mill concentrates to $UF_6$. The uranium product from reprocessing plants is a quite pure solution of uranyl nitrate. There are traces of transuranium elements and fission products resulting from reactor irradiation. Selective sorption of trace impurities on packed

FIGURE 14. Typical $UF_6$ shipping containers for natural uranium. (The near container holds 14 tons and the far container holds 10 tons.)

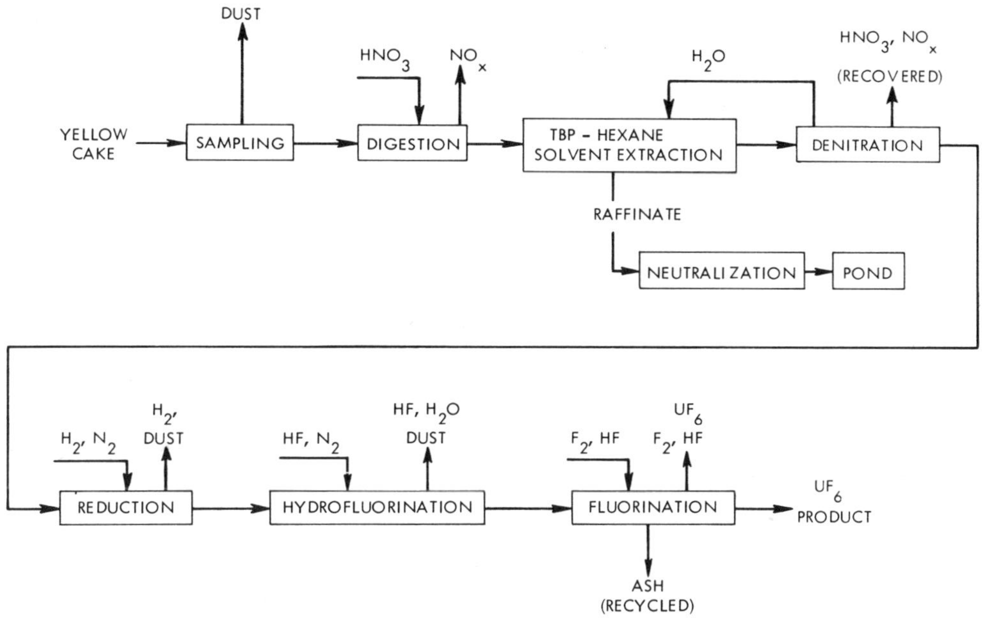


FIGURE 15. Wet process for conversion of yellowcake to $UF_6$.

beds of sodium fluoride is required to meet the rigid specifications for $UF_6$ feed to diffusion plants given in Table 9.

## V. URANIUM ENRICHMENT

Naturally occurring uranium is a mixture of fixed composition containing uranium-238, uranium-235, and uranium-234. Of these isotopes, only uranium-235, which constitutes 0.711 wt % of natural uranium, is fissile and can be used as fuel for nuclear reactors. When a fissile nucleus, such as uranium-235, captures a thermal (or slow) neutron, it normally splits into two fission fragments of unequal mass and releases two or three fast neutrons and a large amount of energy. In order to sustain a nuclear chain reaction, the fast neutrons must be slowed down (moderated) by numerous collisions with light-atomic weight nuclei and then be available to produce additional fission reactions. Since natural uranium contains only 0.711 wt % uranium-235, it can be used only in those reactors constructed of materials that do not absorb an appreciable fraction of the neutrons. Therefore, either highly purified graphite or heavy water ($D_2O$), which have very small capture cross-sections for neutrons, must be used for moderators. Most of the power reactors in the U.S. use ordinary water, and since there is appreciable parasitic capture of neutrons by the light hydrogen isotope, these reactors must use uranium fuel containing 2 to 4% uranium-235 to counteract these neutron losses.

Treatment of uranium to enrich the $^{235}U$ content is a key step in the LWR fuel cycle and is also a dominant cost in that cycle. A wide variety of methods[32] for uranium isotope separation was considered during World War II. These early research efforts were concentrated on electromagnetic processes, gaseous diffusion, thermal diffusion, and gas centrifugation. Large-scale enrichment plants were built using the first three processes, but centrifuge technology was not sufficiently advanced at that time. Gas-

Table 9
**SPECIFICATIONS FOR $UF_6$ DELIVERED TO DIFFUSION PLANTS**

| Value[a] | Numerical value |
|---|---|
| Maximum vapor pressure of filled container at 200°F in pounds per square inch, absolute | 75 |
| Minimum weight percent of $UF_6$ in material | 99.5 |
| Maximum mol percent of hydrocarbons, chlorocarbons, and partially substituted halohydrocarbons | 0.01 |
| Maximum number of parts of elements indicated per million parts of total uranium | |
| Antimony | 1 |
| Bromine | 5 |
| Chlorine | 100 |
| Niobium | 1 |
| Phosphorus | 50 |
| Ruthenium | 1 |
| Silicon | 100 |
| Tantalum | 1 |
| Titanium | 1 |
| Total number of parts of elements forming nonvolatile fluorides (having a vapor pressure of one atmosphere or less at 300°C) per million parts of total uranium, e.g., aluminum, barium, bismuth, cadmium, calcium, chromium, copper, iron, lead, lithium, magnesium, manganese, nickel, potassium, silver, sodium, strontium, thorium, tin, zinc, and zirconium | 300 |
| Maximum number of parts of elements or isotopes indicated per million parts of uranium | |
| Chromium | 1500 |
| Molybdenum | 200 |
| Tungsten | 200 |
| Vanadium | 200 |
| Uranium-233 | 500 |
| Uranium-232 | 0.110 |
| Maximum thermal neutron absorption of total impurity elements as equivalent parts of boron per million parts of total uranium | 8 |
| Maximum total of gamma activity due to fission products and U-237 as percent of gamma activity of aged natural U and as measured in a high-pressure ionization chamber (Drawing D-AWM-8798 of Nuclear Division, Union Carbide Corporation) | 20 |
| Maximum beta activity due to fission products as percent of beta activity of aged natural uranium | 10 |
| Maximum alpha activity from all transuranic elements in disintegrations per minute per gram of total uranium | 1500 |

[a] All specification analyses on $UF_6$ shall be performed on samples removed in the liquid state from each cylinder while its contents are liquid and homogeneous.

eous diffusion proved to be the most economical of the three large plants, so it was chosen as the method for major use, and today it is the only uranium isotope separation method in wide use. Continued technical developments over the last 20 years have led to increased interest in the use of gas centrifuges, and they will assume a prominent role in future enrichment plants. Development of advanced methods for uranium enrichment is also receiving much attention in the U.S. Extensive research and development is being conducted on atomic vapor laser isotope separation, molecular laser isotope separation, and the plasma separation process. These methods have the potential of attaining very high separation in a single stage, thus making it feasible to recover essentially all of the fissile U-235 in an enriched form. Most of the work on these programs is classified.

Both the gaseous diffusion and the centrifuge processes use uranium hexafluoride since it is the only known usable uranium compound that exists as a gas at suitably low temperatures. It sublimes at 56°C, and its triple point if 64°C. Therefore, the enrichment plants must be operated at elevated temperatures to prevent formation of solid $UF_6$ in the system. Both methods are physical processes based upon the small mass difference between $^{235}UF_6$ and $^{238}UF_6$ molecules.

**A. Enrichment Capacity in the U.S.**[13,32,33]

During World War II, the gaseous diffusion process was developed and a government-owned plant was built at Oak Ridge, Tenn. to produce large quantities of enriched uranium-235 for military purposes. A second large gaseous diffusion plant at Paducah, Ky. began operation in 1951, and a third plant at Portsmouth, Ohio has been producing enriched uranium since 1955. Today, these plants are operated as an integrated complex and are used almost exclusively to produce enriched uranium for nuclear power reactor fuel and are a significant source of such fuel for the western world. Enrichment capability exists in Europe and new facilities are under construction.

When uranium is separated isotopically, two uranium streams are produced: the product stream enriched in $^{235}U$ and a waste stream, referred to as "tails", that is depleted in $^{235}U$. The amount of work required for isotope separation is a function of the quantity of uranium treated and the $^{235}U$ content of the feed (0.711% for natural uranium), product, and tails. The separative work unit (SWU) is a measure of the effort expended to separate a specified amount of uranium of given assay into two components, an enriched product and depleted tails. Common practice is to refer to a kilogram separative work unit simply as a separative work unit or as a SWU, and production capacities of enrichment plants are usually given in SWUs. Uranium is enriched in the government-owned gaseous diffusion plants according to a "toll enrichment" arrangement. Under this arrangement, the customer supplies a quantity of uranium feed and receives in return a smaller amount of uranium enriched to the requested $^{235}U$ assay; the customer may also claim the tails. For this service, the U.S. Department of Energy (DOE) makes a service charge, or toll, based upon the amount of separative work required.

The three gaseous diffusion plants have an annual base capacity of about 17.2 million SWUs. As a result of research and development carried out over the last 30 years, improved technology is now capable of providing higher process efficiency. A cascade improvement program (CIP) is underway to equip the plants with more efficient diffusion barriers and with modified equipment that results in lower pressure losses. Completion of CIP in 1981 will increase plant efficiency by 32% and will provide an additional 5.5 million SWU/year with no increase in the current power requirements of 6065 MW. At the same time the CIP modifications are being made, the equipment is also being uprated as part of the Cascade Uprating Program (CUP) to permit power levels and process pressures to be increased safely. The improved diffusion stages will then be capable of using more power efficiently to provide an additional 4.6 million SWU/year. The three gaseous diffusion plants operating at an overall power consumption of 7380 MWe will then have an annual capacity of 27.3 million SWU. The number of 1000-MWe LWRs for which those plants can provide fuel is dependent upon the tails assay and whether or not uranium and plutonium from spent reactor fuel are recycled. If the tails assays is 0.25% $^{235}U$ and only the uranium is recycled, a model 1000-MWe LWR requires an average of 116,000 SWU/year over its lifetime.[19] For these conditions, the uprated plants can supply enough enriched uranium for nuclear power plants with 235 GWE capacity. Although this appears to be an ample supply to satisfy U.S. needs, it must be remembered that from ¼ to ⅓ of the enriched uranium will be used in foreign reactors. It is also important to note that disproportionately larger amounts of enriched uranium are required during initial start-up of the reactors

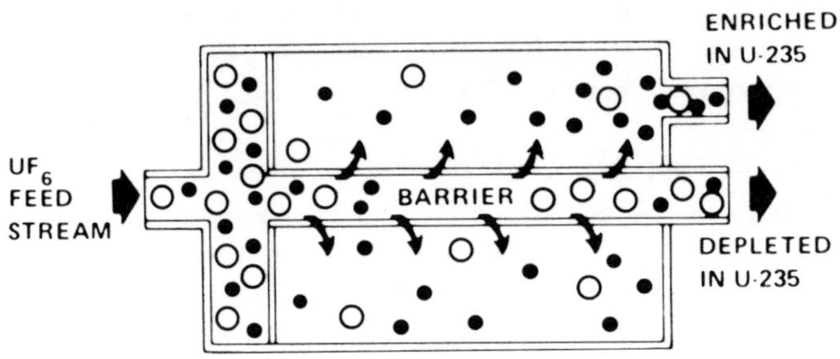


FIGURE 16. Schematic diagram of a gaseous diffusion stage. (Since the lighter $^{235}UF_6$ molecules have a slightly higher average velocity than the $^{238}UF_6$ molecules, the portion of gas effusing through the barrier wall is slightly enriched in $^{235}U$. The average molecular speeds of the two types of molecules differ by only about 0.4%.)

since the entire fuel core must be supplied at start-up compared to an annual replacement of about one third of the core to sustain an operating reactor.

In order to provide for future needs, the U.S. government is committed to further expansion of uranium enrichment capacity through construction of the Gas Centrifuge Enrichment Plant by the DOE. Plans are to construct this expansion in capacity adjacent to the gaseous diffusion plant at Portsmouth, Ohio. The design is for an 8.8 million SWU/year addition in eight separate buildings. However, because of the nature of the centrifuge equipment, it is feasible and economical to construct smaller units so the addition will be constructed in stages. Plans are to complete the first two buildings and have them in full production of 2.2 million SWU/year by 1988. Construction of the additional capacity will be scheduled in response to demands for enrichment services.

## B. Gaseous Diffusion

The gaseous diffusion process[32] is a physical separation method based upon the difference in rates at which gaseous $^{235}UF_6$ and $^{238}UF_6$ effuse through a thin barrier containing many millions of uniform pores commensurate in size with the mean free path of the gaseous molecules (see Figure 16). In a gas made up of molecules containing different isotopic species, a molecule containing the light isotope will, on the average, have a higher velocity than one that contains the heavy isotope and therefore will come in contact with the walls of the containment vessel more frequently. If these walls are a barrier with pores of the correct size, the lighter molecules have a greater probability of hitting and passing through the barrier. The maximum theoretical separation that can be achieved is proportional to the ratio of the average velocities or to the square roots of their masses. In the case of $^{235}UF_6$ and $^{238}UF_6$ molecules, the difference in mass is very small, and the maximum theoretical separation is only 1.0043. This effect can be multiplied to give the desired enrichment by making a cascade consisting of a large number of stages. For example, about 1200 stages[34] are required in the gaseous diffusion plant to enrich natural uranium feed to about 4% $^{235}U$ in the product with a tails assay of about 0.3% $^{235}U$.

The operation of a gaseous diffusion cascade is shown schematically in Figure 17. About one half of the $UF_6$ fed to a stage is allowed to effuse through the barrier and is slightly enriched in $^{235}UF_6$. This stream is advanced to the next stage toward the product end of the cascade, and the other half of the $UF_6$, slightly depleted in $^{235}UF_6$, is rejected downward to the next stage toward the tails end of the cascade. Thus, the feed to a stage in the cascade is a combination of the reject stream from the upper

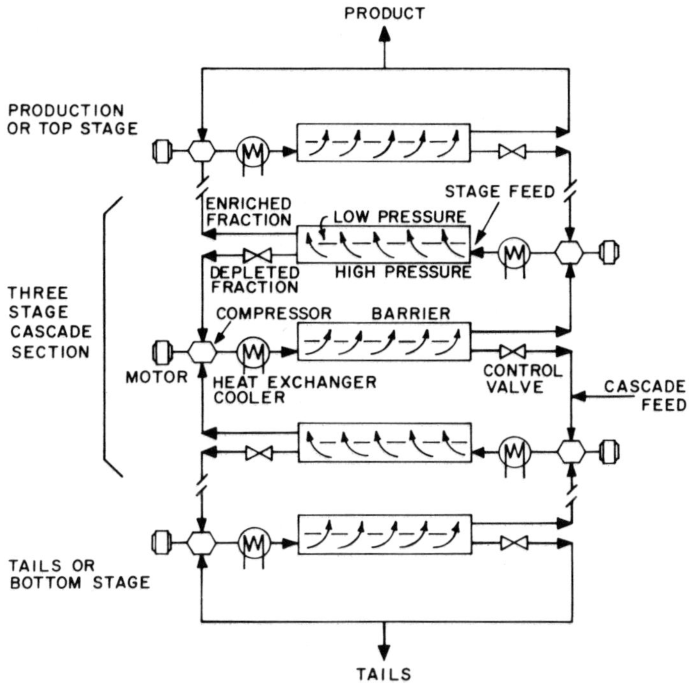


FIGURE 17. Cascade of gaseous diffusion stages.

stage and the enriched stream from the lower stage. In an "ideal" cascade, stage flows are adjusted so the $^{235}U$ assays of two such streams are the same. Pure uranium hexafluoride is received from the conversion plant as a solid in 12- or 14-ton shipping cylinders (see Figure 14). These cylinders are placed in a feed station autoclave and heated with steam to about 110°C to vaporize the $UF_6$ which is fed as gas into the central part of the cascade. Enriched product is withdrawn by condensing it at low temperature into product cylinders constructed of steel. The depleted tails stream is withdrawn from the bottom of the cascade in 14-ton steel cylinders by compressing and cooling the gas to a liquid. Motor driven compressors, which are used to move the $UF_6$ streams from stage to stage in the cascade, are the only major dynamic equipment. Except for the compressor drive motor, all process equipment, piping, compressors, and valves are kept within insulated enclosures that maintain the required warm ambient air temperature.

Since the $UF_6$ is heated during compression, an elaborate system is required to cool the $UF_6$. An evaporative coolant, $CClF_2$-$CClF_2$, is used as the primary coolant in a closed recirculating system. It is transferred as a liquid to a heat exchanger at each stage where it removes heat as it vaporizes. The gaseous $CClF_2$-$CClF_2$ is then transported to a water-cooled condenser where heat is removed to condense the primary coolant for recirculation. Heat is ultimately discharged to the environment by means of large cooling towers. Electric power and cooling water are the major utilities needed for a gaseous diffusion plant. A new 8.75-million-SWU/year plant using the latest technology would require 2400 MWe, with over 90% of this for running the gas compressors. About 250,000 gal of cooling water per minute would be required to remove waste heat.

Gaseous diffusion plants have been operating satisfactorily for over 30 years. The

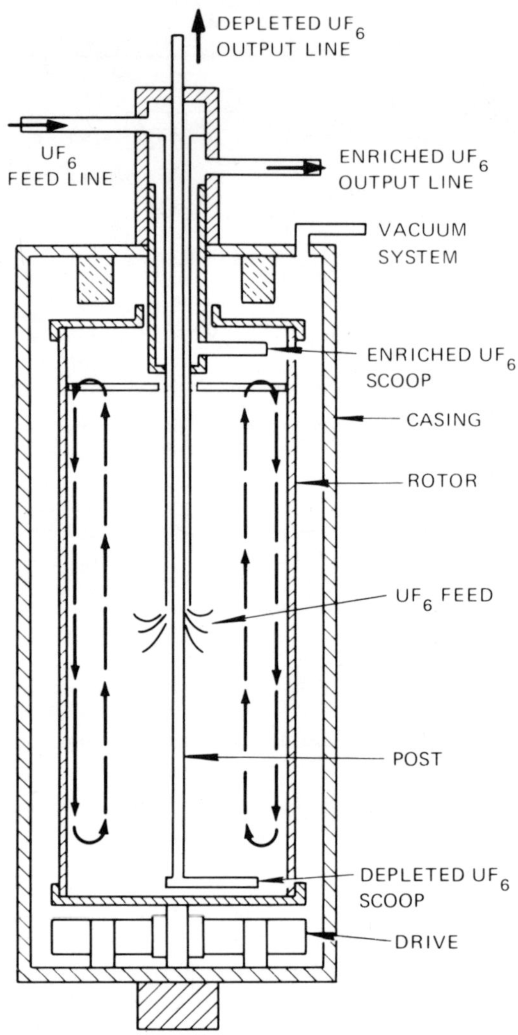


FIGURE 18. Schematic diagram of a gas centrifuge.

operation is very reliable and requires only a moderate amount of maintenance and a small staff.

## C. Gas Centrifuge

The gas centrifuge process,[35] like gaseous diffusion, is a physical separation method based upon the difference in mass of the $^{235}UF_6$ and $^{238}UF_6$ gas molecules. The process is carried out by spinning $UF_6$ gas in a high-speed centrifuge inside an evacuated casing (Figure 18). As the gas accelerates to nearly the speed of the rotor, centrifugal force causes the heavier $^{238}UF_6$ molecules to move preferentially to the rotor wall and the gas at the axis becomes slightly enriched in $^{235}UF$. This partial separation is enhanced by an axial countercurrent flow of gas within the rotor. Uranium hexafluoride feed is introduced near the middle of the rotor, and enriched and depleted gas streams are removed from near the ends of the rotor. The degree of enrichment in a single centrifuge is not only dependent upon the mass difference of the isotopes, but also on the length of the rotor, and most importantly upon the speed of rotation. In large, high-

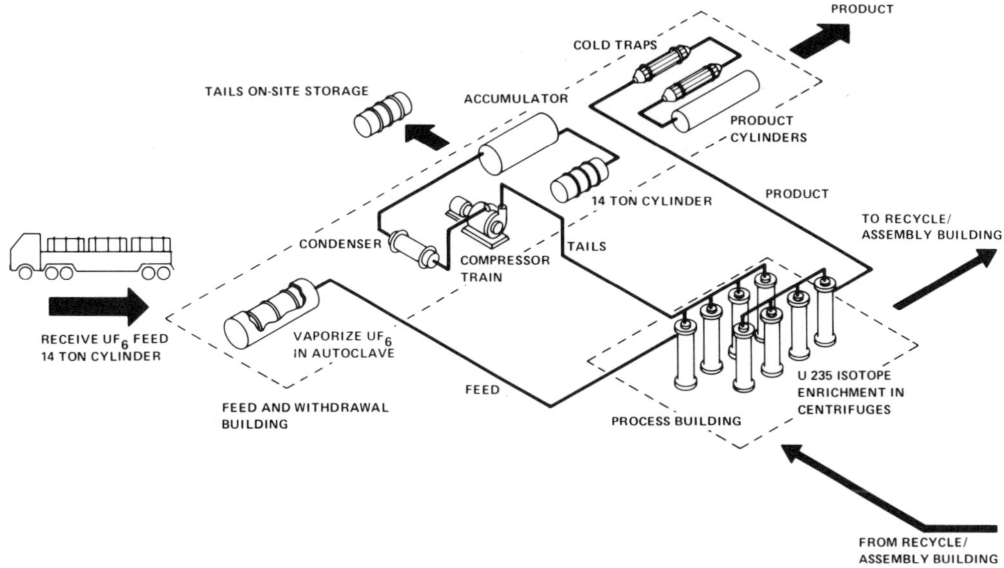


FIGURE 19. Schematic diagram of a centrifuge enrichment facility.

speed machines the degree of enrichment is many times that attainable in a single gaseous diffusion stage. Because of this greater separation, significantly fewer stages are required in a centrifuge cascade than for a gaseous diffusion plant to achieve the same enrichment. Only a small amount of gas can flow through a centrifuge, so a large number of machines must be connected in parallel for each stage to provide a large enough flow for a large-capacity plant.

A gas centrifuge plant requires only about 4% of the power needed for a diffusion plant of the same capacity. Whereas an 8.8-million-SWU/year diffusion plant requires about 2700 MWe, a centrifuge plant of the same size needs only 100 MWe. Figure 19 shows a schematic diagram of a centrifuge enrichment plant. Figure 20 is a photograph of advanced gas centrifuge equipment undergoing performance testing in a small cascade. Uranium hexafluoride received from conversion plants in 14-ton cylinders is fed into the cascade by heating to vaporize the $UF_6$. Waste $UF_6$ depleted in uranium-235 is taken from the bottom of the cascade by condensing the gas into 14-ton cylinders for storage. Enriched product is taken from the top of the cascade and collected in smaller steel cylinders with a capacity of 2½ tons. These cylinders are placed in a protective carrier for shipment to fuel fabrication plants.

Although centrifuge enrichment plants require considerably fewer stages, much less electrical power, and less cooling water than a gaseous diffusion plant, they will require a larger operating staff and much maintenance.

## VI. FUEL FABRICATION

Enriched $UF_6$ containing 2 to 4% $^{235}U$ received from the enrichment plant in steel shipping cylinders is converted into pellets of $UO_2$ at commercial nuclear fuel fabrication plants. The pellets, which are about ½ in. in diameter and ½ in. long, are loaded into metal tubes that are then fitted with end caps and welded. The metal cladding allows heat to flow from the fuel to the cooling water, but protects the $UO_2$ from corrosion by the water and prevents contamination of the coolant with radioactive

FIGURE 20. Gas centrifuges installed at the Component Test Facility, Oak Ridge, Tenn.

fission products. Some of the early reactors used stainless steel cladding, but most of the LWR fuel is now clad in Zircaloy, which is predominantly zirconium containing small amounts of tin, nickel, chromium, and iron. Zircaloy is corrosion resistant and has a low capture cross-section for neutrons. Zirconium, as it is found in nature, always contains hafnium which is very similar in many chemical properties but has a very high capture cross-section for neutrons. In order to prepare zirconium acceptable for use in nuclear reactors, it must be specially processed to remove the hafnium to very low levels.

Clad tubes of $UO_2$ are assembled into fuel element bundles that have sufficient rig-

FIGURE 21. Pressurized water reactor fuel elements fabricated by Westinghouse. (Photograph courtesy of Westinghouse.)

idity and open area between tubes to allow unrestricted flow of cooling water. A number of completed pressurized water reactor fuel elements fabricated by Westinghouse are shown in the photograph of the storage room in Figure 21. Fuel elements are placed in protective cases prior to transporting them, usually by truck, to the reactors (Figure 22).

Fabrication of fuel elements from slightly enriched uranium for LWR reactors is a commercial industry with nine licensed plants in the U.S. These are listed in Table 10. The capacity is sufficient to fabricate about 3000 tons of fuel per year, or enough to supply fuel for about 90 1000-MWe reactors. Two of the fuel fabrication facilities simply prepare $UO_2$ powder or pellets which are then sent to other plants for assembly into completed fuel elements. All four of the companies that manufacture LWRs, Gen-

FIGURE 22. Protective cases for shipment of fuel elements to reactors.

**Table 10**
**LICENSED, COMMERCIAL LWR FUEL FABRICATION PLANTS**

| Company | Plant location | Plant feed material | Plant product |
|---|---|---|---|
| Babcock & Wilcox | Lynchburg, Va. | $UO_2$ powder | Fuel assemblies |
| Babcock & Wilcox | Apollo, Pa. | $UF_6$ | Fuel assemblies |
| Combustion Engineering | Hematite, Mo. | $UF_6$ | $UO_2$ powder or pellets |
| Combustion Engineering | Windsor, Conn. | $UO_2$ powder | Fuel assemblies |
| Exxon Nuclear | Richland, Wash. | $UF_6$ | Fuel assemblies |
| General Electric | Wilmington, N.C. | $UF_6$ | Fuel assemblies |
| Kerr-McGee | Cimarron, Okla. | $UF_6$ | Fuel assemblies |
| Nuclear Fuel Services | Erwin, Tenn. | $UF_6$ | $UO_2$ powder or pellets |
| Westinghouse | Columbia, S.C. | $UF_6$ | Fuel assemblies |

eral Electric, Westinghouse, Babcock and Wilcox, and Combustion Engineering, also have facilities for fabricating completed fuel elements.

### A. Preparation of Slightly Enriched $UO_2$ Fuel

Low-enrichment uranium is only moderately toxic and its radioactivity involves little penetrating radiation. Therefore, no shielding is required to protect the workers in fuel fabrication plants. Most operations are conducted in hoods, vented vessels, and enclosures from which the exhaust is filtered or otherwise treated to prevent discharge of uranium to the environment. Dry $UO_2$ powder is usually handled in chemical fume hoods or in ventilated boxes with small openings through which personnel can work. Ventilation is arranged so air flow is into the boxes and discharges through filtered

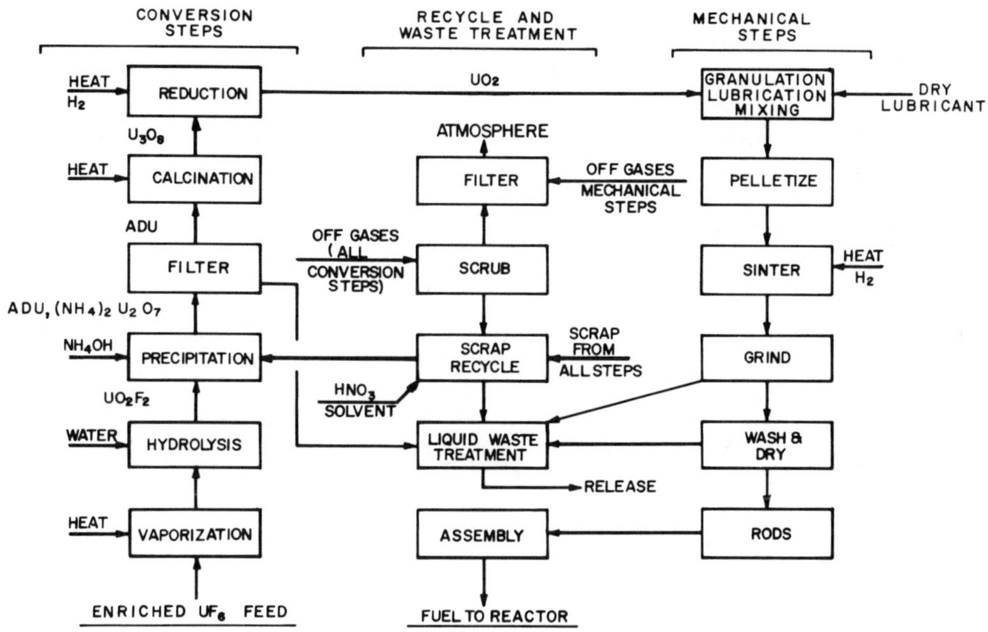


FIGURE 23. Simplified ADU process flow diagram for LWR fuel fabrication.

exhaust. Pellet forming operations are similar to those in normal ceramics industrial practice, except that special precautions are taken to prevent the spread of uranium contamination to the work area or atmosphere.

Three different commercial processes are in use for the preparation of $UO_2$ powder for pellet pressing. These are the ammonium diuranate process (ADU), the direct conversion process (DC), and the ammonium uranyl carbonate process (AUC). The ADU process is the most widely used.

### *1. Ammonium Diuranate (ADU) Process*

A simplified flow diagram of the ADU process[7,9] is shown in Figure 23. The cylinder of enriched $UF_6$ is heated electrically or with steam to sublime $UF_6$, which is piped into a column or tank where it is reacted with water to produce a solution of $UO_2F_2$ containing HF.

$$UF_6(g) + 2H_2O \rightarrow UO_2F_2 + 4HF$$

The uranyl fluoride solution is then treated with ammonium hydroxide to precipitate ammonium diuranate quantitatively. The precipitate is collected by centrifugation or filtration and washed. The ADU filter cake is calcined in air to produce $U_3O_8$, and subsequently treated with hydrogen (usually cracked ammonia) to prepare $UO_2$ powder.

The principal steps to produce reactor grade $UO_2$ pellets consist of: (1) a pretreatment of $UO_2$ powder by comminution, compaction, and granulation to a controlled size distribution; (2) pelletizing; (3) sintering in $H_2$ to adjust the oxygen-to-uranium ratio; (4) grinding the pellets to exact size specifications; and (5) final washing and drying. The pellets are carefully inspected, and those meeting the specifications of size, density, surface condition, and chemical purity are loaded into Zircaloy tubes. The tubes are fitted with end caps and welded. Fuel rods are then assembled into completed fuel element assemblies.

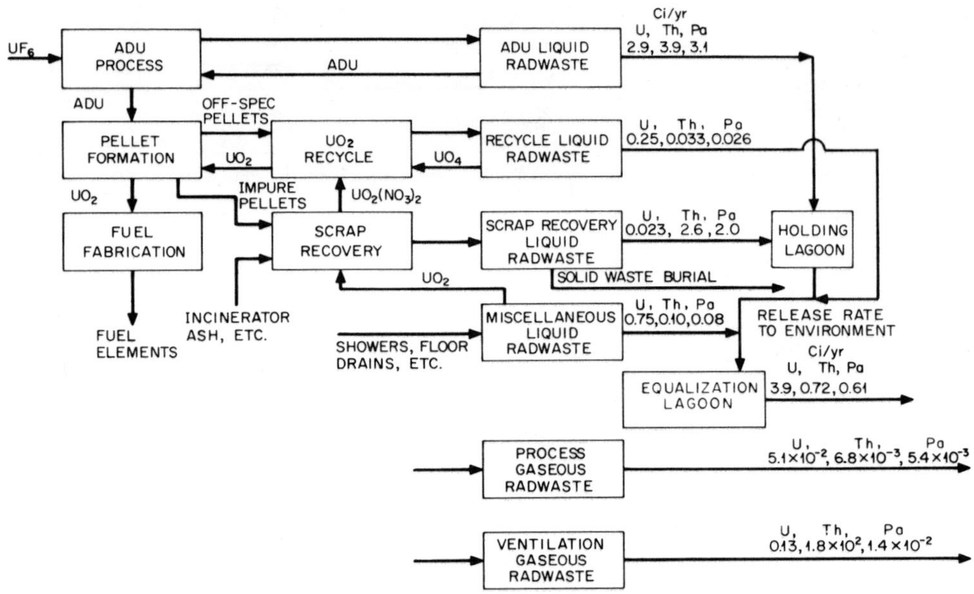


FIGURE 24. Process and radwaste systems for a model ADU plant.

It is estimated that 10 to 15% of the uranium ends up as scrap, most of which can be recycled directly into the fabrication line. A percent or two of the uranium must be processed through a purification cycle consisting of scrap dissolution in nitric acid, uranium purification by solvent extraction, and uranyl nitrate conversion to $UO_2$ by the ADU process.

An analysis[7] was made of radioactivity releases from various process streams of a 1500 metric tons of uranium per year "model" fuel fabrication plant based on the ADU process. Several different cases were studied with case 1 as the base case. This is the lowest cost case studied, and, although not a specific plant, the model is based upon technology used in some current facilities. The model process and radwaste systems for the base case are shown in Figure 24.

### 2. *Direct Conversion (DC) Process*

In the DC process,[7] $UF_6$ gas is sublimed from the shipping cylinder into a fluidized bed where it is reacted with steam. This can be represented by the following reaction:

$$UF_6(g) + 2H_2O(g) \rightarrow UO_2F_2(s) + 2HF(g)$$

Off-gas from the fluidized bed containing HF is filtered through a sintered nickel filter. The solid $UO_2F_2$ is treated in a second fluidized bed with cracked ammonia and steam,

$$UO_2F_2(s) + H_2(g) \rightarrow UO_2(s) + 2HF(g)$$

Most of the $UO_2F_2$ is converted to $UO_2$, but the product is treated in a third fluidized bed, again with steam and cracked ammonia, to complete the reaction. The resulting $UO_2$ powder is spherical with an average diameter of $\sim$120 $\mu$m. Off-gas from all the fluidized beds is filtered through sintered nickel filters to prevent solids entrainment and the filtered gas is treated to remove HF. The HF is treated with lime to produce a $CaF_2$ sludge which is stored on site in waste ponds.

Subsequent steps for pellet fabrication are similar to those for $UO_2$ powder prepared by the UNH method.

*3. Ammonium Uranyl Carbonate (AUC) Process*

In the AUC process, $UF_6$ gas along with $CO_2$ and $NH_4$ gas is injected into demineralized water to form a crystalline precipitate of ammonium uranyl carbonate. The following equation represents the overall reaction:

$$UF_6(g) + 10NH_3(g) + 3CO_2(g) + 5H_2O(\ell) \rightarrow (NH_4)_4\,UO_2(CO_3)_3\downarrow + 6NH_4F$$

The crystals of AUC are about 60 $\mu$m in size. They are filtered from the ammonium fluoride solution, washed with an ammonium carbonate solution and then with methanol. Crystalline AUC is converted to $UO_2$ powder by reacting it in a fluidized bed with steam and cracked ammonia. The $UO_2$ powder is then pressed into pellets in a similar manner to that used for $UO_2$ powder prepared by the UNH method.

### B. Preparation of Mixed $UO_2$-$PuO_2$

Although there are no large commericial facilities, there are several small-scale pilot plants that have the capability of preparing fuel containing plutonium. Pellets made from uranium and plutonium dioxides are called mixed oxide fuel and are often referred to as MOX. It has been estimated that the total capacity in the U.S. is about 100 tons of MOX fuel per year.[13]

Fabrication of MOX fuel includes essentially the same process steps as those for enriched uranium fuel.[10] The principal difference is that $PuO_2$ powder is prepared separately and then mechanically blended with $UO_2$ powder prior to pelletizing. In this case, the $UO_2$ is made from natural instead of enriched uranium. The most commonly used method for preparing $PuO_2$ is based upon the precipitation of plutonium oxalate from a nitric acid solution of tetravalent plutonium (see Figure 25). The plutonium oxalate is calcined in air to produce $PuO_2$, which is then crushed and screened to size before blending with $UO_2$. Figure 26 is a flow diagram showing the subsequent steps for preparation of MOX fuel rods.

Fabrication and processing of MOX fuel must be carried out in equipment specially designed for handling plutonium. These facilities must provide for complete containment so traces of plutonium cannot enter the working areas occupied by personnel. The usual practice is to carry out all processing in equipment inside of glove box enclosures. When transfers from one operation to another are required, plutonium-containing materials are handled in sealed containers. After the MOX pellets are sealed inside a fuel rod and the outside of the rod has been decontaminated of traces of plutonium, the MOX fuel rods can be handled outside of an enclosure. Waste treatment and disposal methods are much more restrictive than those required for a fabrication facility handling only low-enrichment $UO_2$. An analysis of a model MOX fabrication plant, waste treatment methods, and potential impact on the environment has been made.[10]

## VII. CHARACTERISTICS OF LWRS

The first nuclear reactors were constructed during the early 1940s to produce plutonium for use in weapons. Fuel for these reactors consisted of uranium metal clad in aluminum. Several countries, notably France and England, extended the technology developed for these production reactors and built graphite-moderated, gas-cooled reactors for the generation of electrical power. Efficiency was poor because temperatures

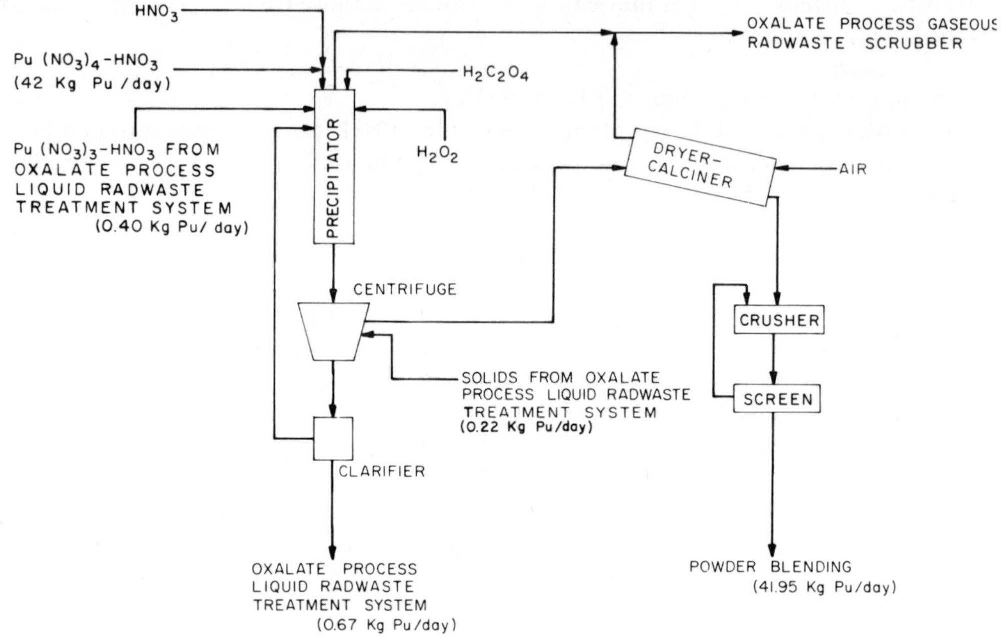


FIGURE 25. Oxalate process for production of plutonium oxide.

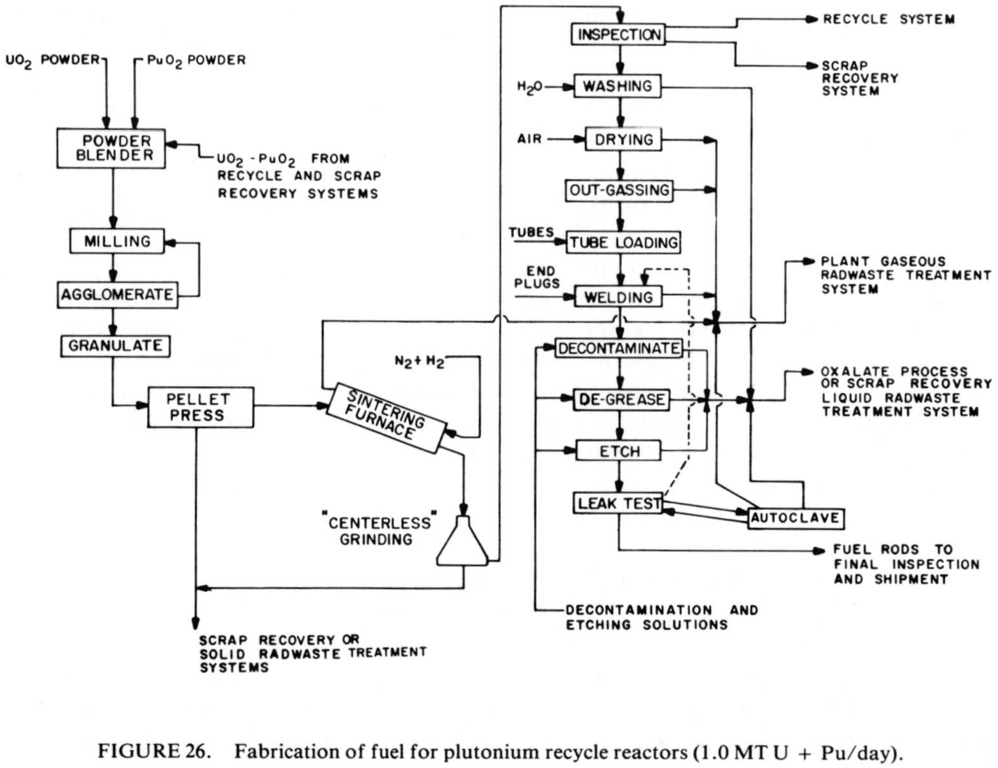


FIGURE 26. Fabrication of fuel for plutonium recycle reactors (1.0 MT U + Pu/day).

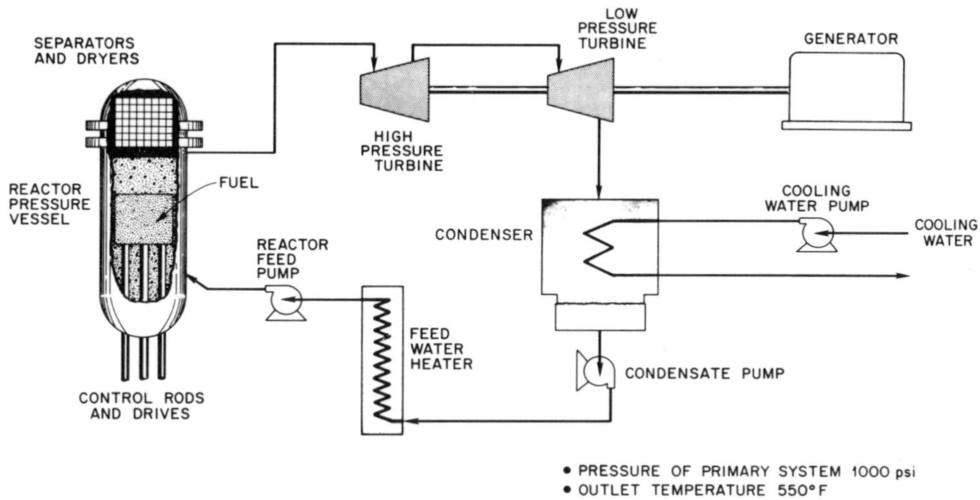


FIGURE 27. Boiling water reactor power plant.

were kept low to prevent melting of the uranium metal clad in either aluminum or magnesium. Advances have been made, especially in fuel composition and coolant technology, so that the latest gas-cooled reactors can operate at the more efficient, higher temperatures.

There are two major varieties of LWRs: pressurized water reactors (PWRs) and boiling water reactors (BWRs). Details in design and engineering approaches for the nuclear steam supply systems of PWRs differ considerably from plant to plant, since they are supplied by three vendors, Westinghouse, Babcock and Wilcox, and Combustion Engineering. Even though BWRs are offered only by General Electric, design details are tailored to meet specific requirements of the utilities. As a result, there is also considerable variation in BWR steam supply systems. Nuclear fuel for both PWRs and BWRs consists of uranium dioxide pellets fabricated from uranium enriched to contain 2 to 4% $^{235}U$ and encased in Zircaloy tubing. Both reactors use ordinary water ($H_2O$) as a moderator to slow down fission neutrons and as the medium for heat transfer from the reactor core. Hence, the general name of light water reactors.

BWRs use a direct power cycle in which steam is generated within the reactor vessel at about 1000 lb/in.$^2$ A schematic diagram[13] showing the concept of a BWR power plant is given in Figure 27. Steam from the reactor core passes through separators and driers and then is sent directly to the turbine system for generating electrical power. Spent steam is condensed and recycled to the reactor core in a closed system to minimize release of radionuclides.

Pressurized water reactors are also cooled and moderated with ordinary water; however, pressure in the reactor core and primary coolant system is sufficiently high ($\sim$2250 lb/in.$^2$) to prevent boiling. This concept is shown schematically[13] in Figure 28. The primary coolant circulates from the reactor core through heat exchangers where heat is transferred to a secondary coolant (also ordinary water) which is at a pressure of about 1000 lb/in.$^2$ Steam generated on the low pressure side of the heat exchanger passes through separators and driers and then to the turbine-generator unit. Primary coolant in the closed loop through the reactor core never enters the steam turbine. Therefore, radioactive nuclides from leaking fuel elements or that are produced by neutron activation do not enter the steam turbine. Most PWRs contain a soluble neutron-absorbing component, such as boric acid. By adjusting its concentration, fine control of the reactivity of the reactor can be accomplished.

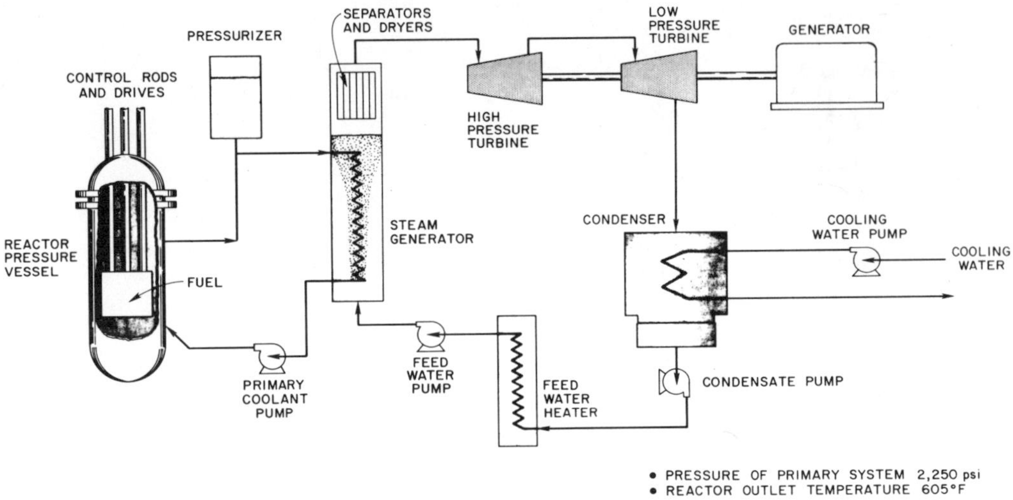


FIGURE 28. Pressurized water reactor power plant.

Some characteristics of typical fuel for boiling water and pressurized water reactors are given in Table 11. Thermal efficiencies are around 33%. However, the specific operating power and the irradiation level of spent fuel for BWRs is considerably lower than for PWRs. The total amount of fuel in the reactor core varies from one reactor to another. A 1000-MWe BWR requires about 120 to 190 metric tons of fuel and a PWR of the same electrical capacity requires only 80 to 120 metric tons.

Uranium in spent fuel from LWRs has a $^{235}U$ content of 0.8 to 0.9%, slightly greater than the 0.711% in natural uranium. It also contains fissile plutonium that grows in during irradiation by neutron capture of $^{238}U$ which subsequently transforms to $^{239}Pu$. The objective of reprocessing spent fuel is to recover these fissile values for recycle as nuclear fuel, thus decreasing the amount of virgin uranium required to sustain electrical generation by nuclear reactors.

Nuclear reactors are constructed with several barriers to minimize the amount of radioactive fission products and other radionuclides that enter the environment. The bulk of the radioactivity is retained inside the fuel. Spent fuel removed from the reactor core is stored in water-filled pools at the reactor. Eventually, this fuel will be shipped from the reactors in heavily shielded carriers to away-from-reactor (AFR) storage facilities when they have been provided and licensed. Two spent fuel storage concepts for AFR storage are being considered: (1) storage in water-cooled pools and (2) storage in air-cooled vaults. In addition, consideration is also being given to retrievable spent fuel storage in geologic repositories. An AFR water-cooled pool would be quite similar to the pools at the reactors but would have larger storage capacity. It would consist of water-filled cells in heavily reinforced concrete lined with stainless steel. Auxiliary equipment would be supplied to control water quality and to cool the water circulated from the pool to remove fission product heat from the large amount of spent fuel.

If reprocessing is allowed, spent fuel will be shipped to licensed facilities for recovering the fissile values it contains. A photograph of the carrier used to ship spent fuel to the Nuclear Fuels Services Reprocessing Plant in West Valley, N.Y. is shown in Figure 29. A sketch for a conceptual spent fuel cask capable of carrying 10 PWR or 25 BWR fuel assemblies is shown in Figure 30.

Other wastes[37] generated at nuclear power reactors receive special treatment to con-

**Table 11**
**COMPARISON OF SOME TYPICAL BWR AND PWR FUEL CHARACTERISTICS**[36]

| Characteristic | BWR | PWR |
|---|---|---|
| Thermal efficiency (%) | 34 | 33 |
| Specific power ($MW_{th}$/MTU) | 28 | 38 |
| Initial core, average | | |
| Irradiation level ($MWD_{th}$/MTU) | 17,000 | 22,600 |
| Fresh fuel assay (wt % $^{235}U$) | 2.03 | 2.26 |
| Spent fuel assay (wt % $^{235}U$) | 0.86 | 0.74 |
| Fissile Pu recovered (kg/MTU) | 4.8 | 5.8 |
| Replacement loadings (annual rate at steady state and 75% plant factor) | | |
| Irradiation level ($MWD_{th}$/MTU) | 27,500 | 32,600 |
| Fresh fuel assay (wt % $^{235}U$) | 2.73 | 3.21 |
| Spent fuel assay (wt % $^{235}U$) | 0.84 | 0.90 |
| Fissile Pu recovered (kg/MTU) | 5.9 | 7.0 |

fine radioactivity. Figure 31 indicates typical systems for waste treatment at PWRs and Figure 32 shows typical systems at BWRs. Solid radioactive wastes are packaged and sent to licensed low-level waste burial grounds. A recent survey has shown the BWRs consistently shipped a larger total volume of solid radwaste than PWRs for each thermal megawatt-hour produced. The average volume of radwaste from PWRs was $3.1 \times 10^{-5}$ $m^3$/MWh(t) and from BWRs was $6.4 \times 10^{-5}$ $m^3$/MWh(t). The specific activities of these wastes averaged 1.0 Ci/$m^3$ for PWRs and 1.6 Ci/$m^3$ for BWRs.

## VIII. SPENT FUEL REPROCESSING

Fully irradiated fuel discharged from light water reactors will contain from 7.4 to 9.0 kg of $^{235}U$ and from 4.8 to 7.0 kg of fissile plutonium in each metric ton of spent fuel (see Table 11). A metric ton of the fuel also contains from 17 to 33 kg of fission products, many of which have large neutron capture cross-sections and thus poison the nuclear reaction. By reprocessing the fuel, the fissile values could be recovered for reuse as nuclear fuel, thus decreasing the amount of virgin uranium needed to sustain the nuclear power industry. The basic technology proposed for commercial fuel reprocessing plants is that which was developed by the U.S. Atomic Energy Commission during the 1940s and early 1950s for use at the Hanford and Savannah River plants to recover plutonium. This technology is centered around the Purex®[38] solvent extraction process, which isolates pure streams of plutonium nitrate and uranium nitrate. Most of the fission products, the actinides (neptunium, americium, and curium), and small amounts of plutonium and uranium remain in a nitric acid waste stream, referred to as high-level TRU waste.

### A. Commercial Reprocessing Plants in the U.S.

Three commercial plants for reprocessing LWR fuel have been constructed in the U.S.;[13] however, none is now operating and only one of the three has actually processed spent fuel. The only commercial fuel reprocessing has been carried out at the Nuclear Fuel Services Plant in West Valley, N.Y.[13] This plant, with a nominal 1 metric ton/day capacity, started operation in April 1966 and had processed about 620 metric tons of fuel by 1972, when it was shut down for modification and enlargement. Because of a number of new regulations that went into effect since the plant was first constructed, the plant was never relicensed. Extensive building and facility changes at great cost would have been required to meet the new regulations.

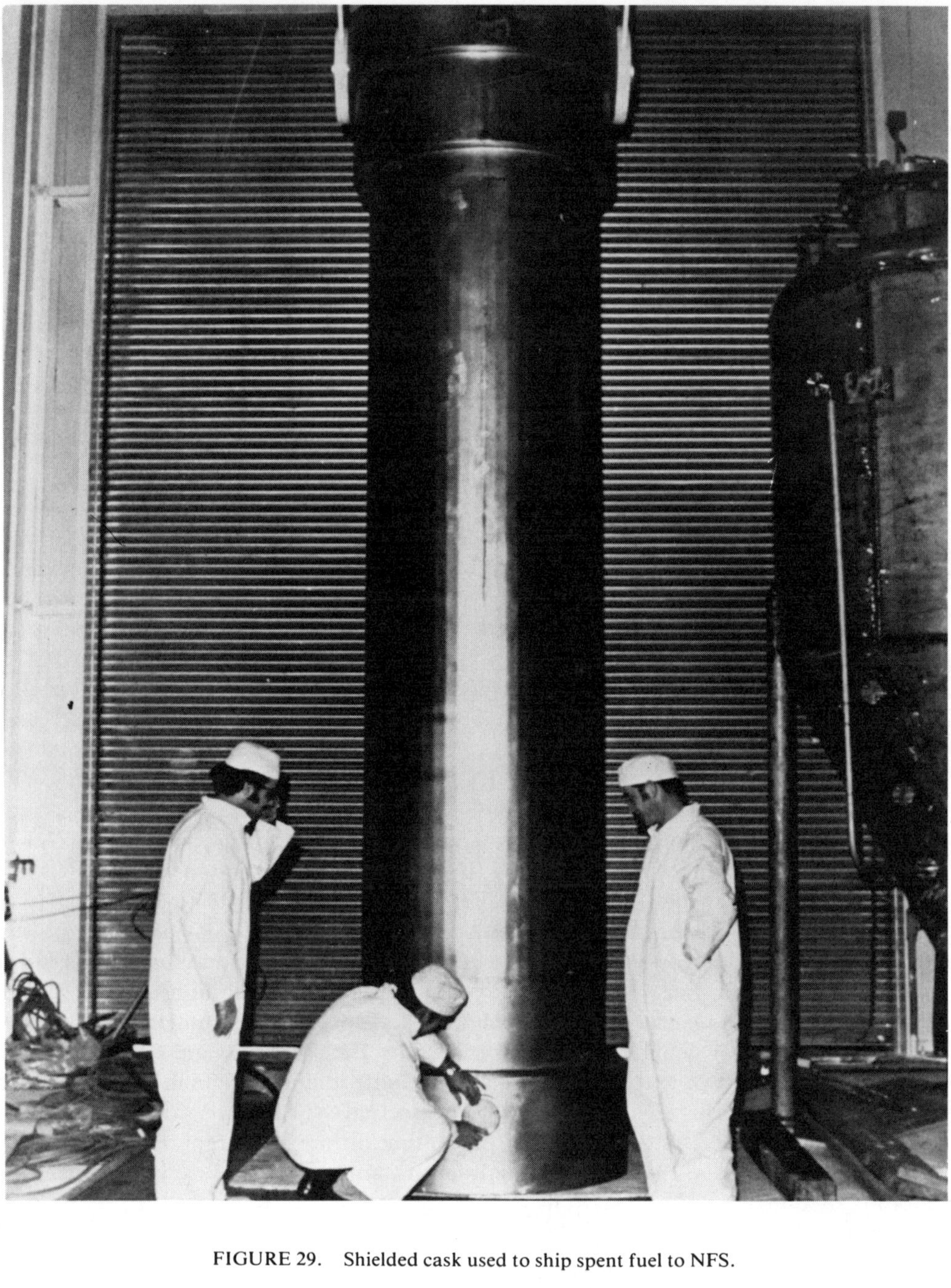

FIGURE 29. Shielded cask used to ship spent fuel to NFS.

The General Electric Company constructed the Midwest Fuel Recovery Plant at Morris, Ill. This facility was to have a nominal 1 metric ton/day capacity and was to use an advanced processing method based upon an initial solvent extraction cycle followed by fluoride volatility for final uranium purification and anion exchange for final plutonium purification. During pretesting, it was determined that the direct maintenance concept used for this facility was not feasible because of the excessive radioactivity levels that would be encountered. The decision was made in 1974 not to spend the additional funds necessary to correct this deficiency. Although the facility will not

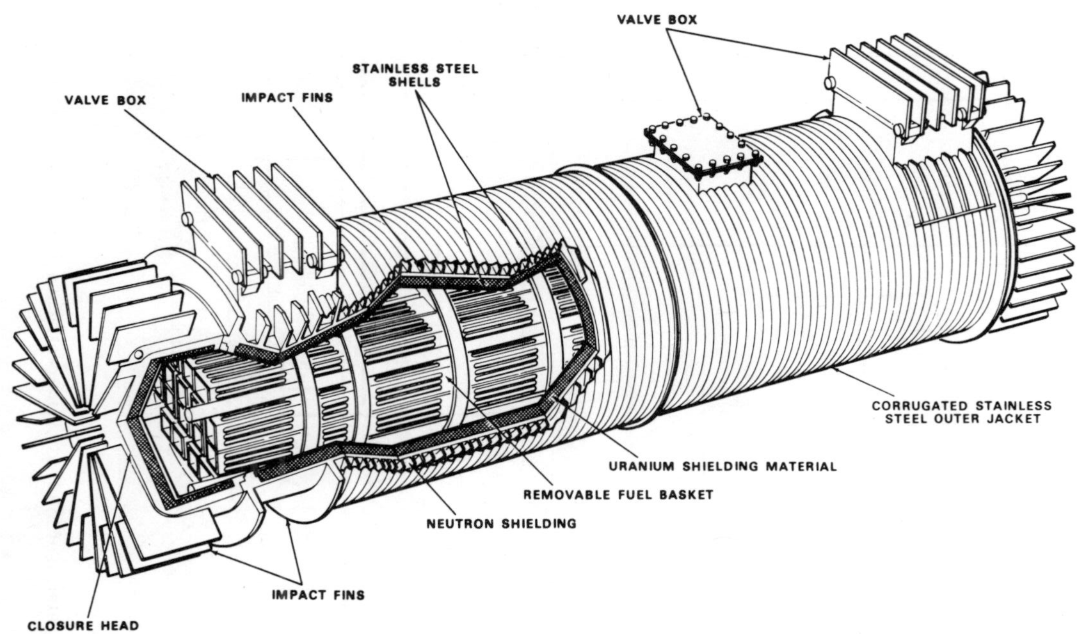


FIGURE 30. Concept of an advanced cask for shipping spent fuel.

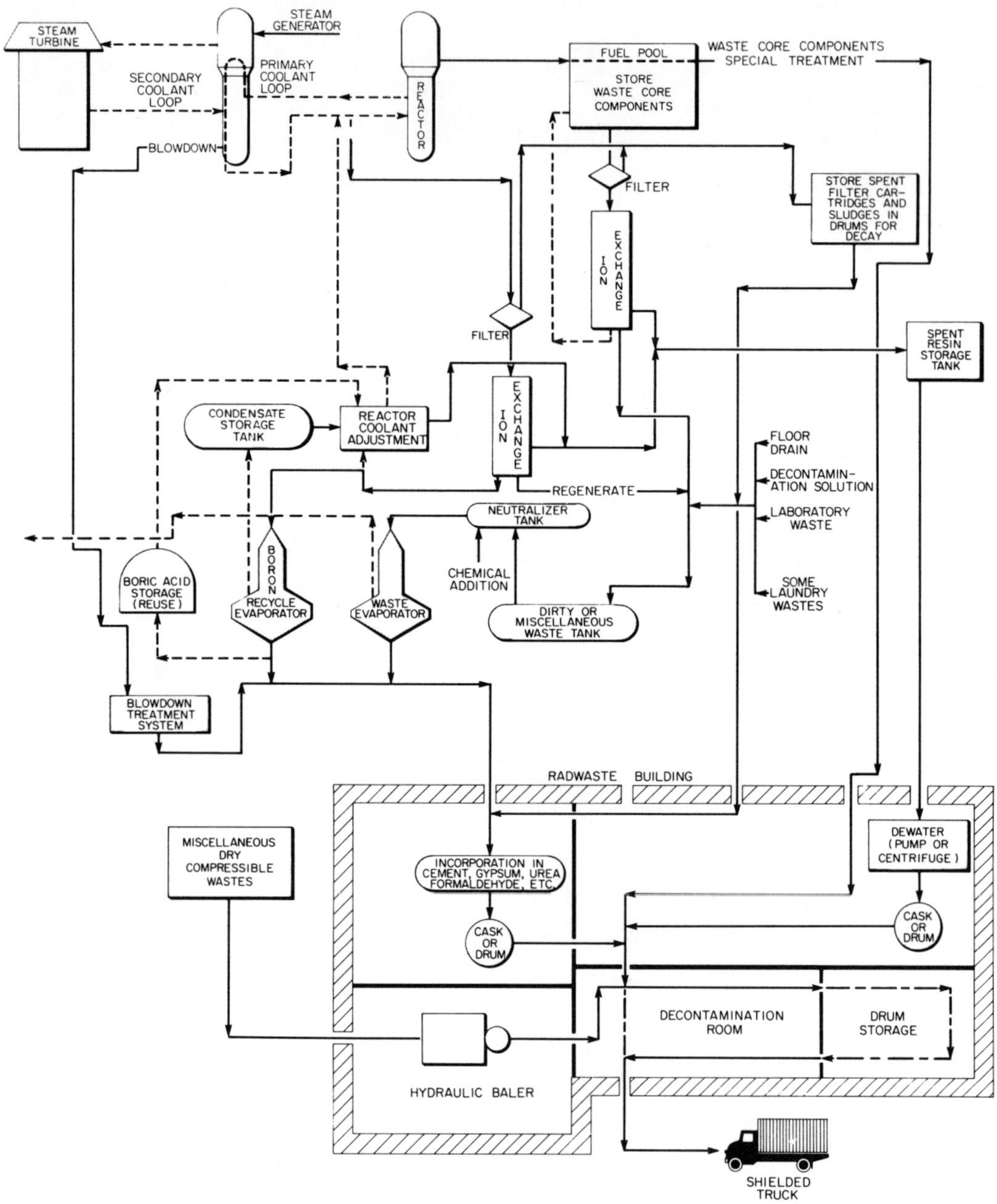


FIGURE 31. Typical system for treatment of solid radioactive wastes at a pressurized water reactor.

be used for fuel reprocessing, it does have a water pool capable of accepting spent fuel and temporarily storing it.

In 1971, construction began on the Barnwell Nuclear Fuel Plant at Barnwell, S.C. This facility is owned by Allied-General Nuclear Services (AGNS), which is a partnership between Allied Chemical and the General Atomic Company. The nominal capacity is 5 metric tons/day or 1500 metric tons/year. Construction of the main plant was completed. However, design and construction of some vital ancillary operations, including facilities for plutonium conversion to $PuO_2$ and waste solidification, were de-

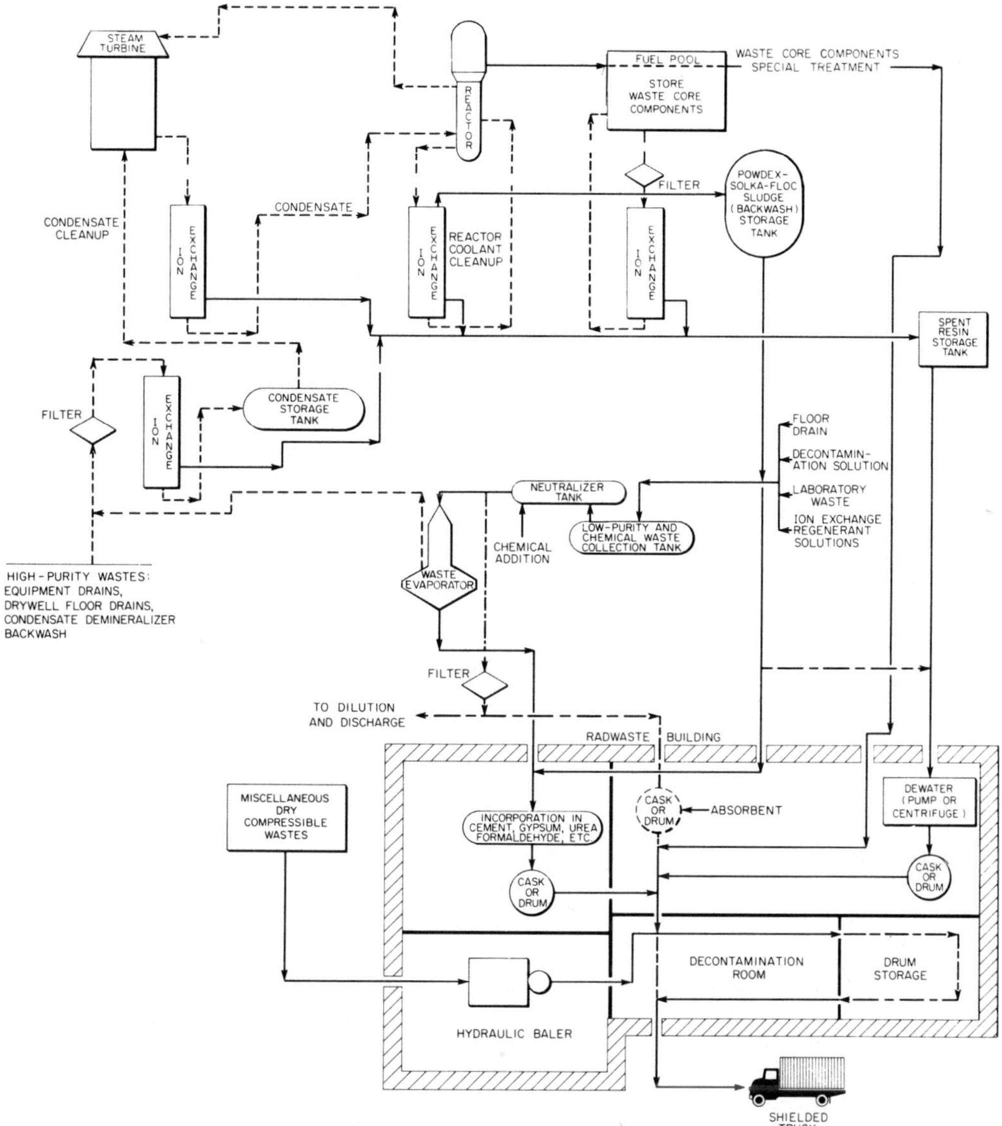


FIGURE 32. Typical system for treatment of solid radioactive wastes at a boiling water reactor.

layed pending issuance of federal regulations. The decision not to permit reprocessing of spent LWR fuel in the U.S. has deferred completion of the facility. The facility does have capability of receiving and temporarily storing spent LWR fuel.

### B. Reprocessing Plant Operations

A nuclear fuel reprocessing plant requires a massive building which may be 200 ft wide by 115 ft high by 300 ft long.[11] Because of the high radiation levels from fission products in the spent fuel, the main operations must be housed in a heavily reinforced concrete structure with walls from 3 to 5½ ft thick for shielding from gamma radiation. This structure is normally subdivided into smaller enclosures, or cells, that are isolated from each other by shielding and ventilation systems appropriately designed for the process function, radioactivity level, and maintenance. Figures 33 and 34 are cross sections of a "typical" fuel reprocessing plant used for a cost-benefit study to determine the cost and effectiveness of radioactive waste treatment systems.[11] This "typical" plant is based upon information from the Barnwell Nuclear Fuel Plant Environmental Report.[39]

A pictorial representation of operations at a fuel reprocessing plant is shown in Figure 35. Spent fuel is received in heavily shielded casks and stored in water-filled pools. Storage capacity would normally be about one fourth of the annual reprocessing capacity. As processing begins, the ends and shrouds of the fuel element that do not contain fuel are cut away and packaged for disposal. These are radioactive because of neutron activation. The fuel elements are then sheared to approximately 1-in. lengths in order to expose the oxide fuel for subsequent dissolution. This is accomplished by means of a hydraulic shear which exerts up to 320 tons of force on the shear blade. The sheared fuel may then be exposed to air at high temperature in a process called voloxidation, to remove tritium as tritiated water vapor. Trapping and control of tritium and other volatile radioactive fission products released during these head-end steps are discussed in Chapter 5.

Following voloxidation, the fuel is transported to a dissolver where it is treated with nitric acid. The fuel is selectively dissolved leaving the cladding hulls; after the dissolver solution is removed, the hulls are leached with nitric acid, washed, dried, and packaged. The hulls contain neutron activation products and trace amounts of fission products, uranium, and plutonium, and along with other structural components of the fuel comprise the fuel assembly structural material wastes (FASM). The hulls, packaged in steel containers, may be kept temporarily at the reprocessing plant but would eventually be sent to a federal repository for final storage.

The nitric acid solution of spent fuel is treated by the Purex® process, which is based upon solvent extraction with tributyl phosphate (TBP) to recover purified uranium and plutonium. Several extraction cycles are needed to completely decontaminate uranium from plutonium and fission products, and several cycles are also needed to attain adequate purification of the plutonium. Details of reprocessing chemistry are discussed in Chapter 4.

The purified uranium product is in the form of a uranyl nitrate solution, which may be converted to $UF_6$ (see Section IV. C.) for recycle to the enrichment plant. The plutonium product is also a purified nitrate solution. It must be converted to a solid (probably $PuO_2$) for shipment.

Aqueous waste from the first extraction cycle contains essentially all of the radioactive fission products; the transuranium elements neptunium, americium, and curium; and small amounts of residual uranium and plutonium. This high activity level waste, along with lower activity level wastes from other processing cycles, may be stored temporarily as acid solution in large underground tanks. The waste must eventually be treated to convert the dissolved radioactive materials to an inert solid package in steel containers. The packaged waste must then be shipped to a federal repository for final disposal. Although no such repository now exists, it is unlikely that fuel reprocessing plants will be licensed before waste repositories have been established. A discussion of waste handling and disposition is given in Chapter 6.

## ACKNOWLEDGMENTS

Even though it is not possible to give adequate recognition, the author is grateful to all those that gave assistance in preparation of this chapter. An attempt was made to give credit through references to previously published work. However, in some cases

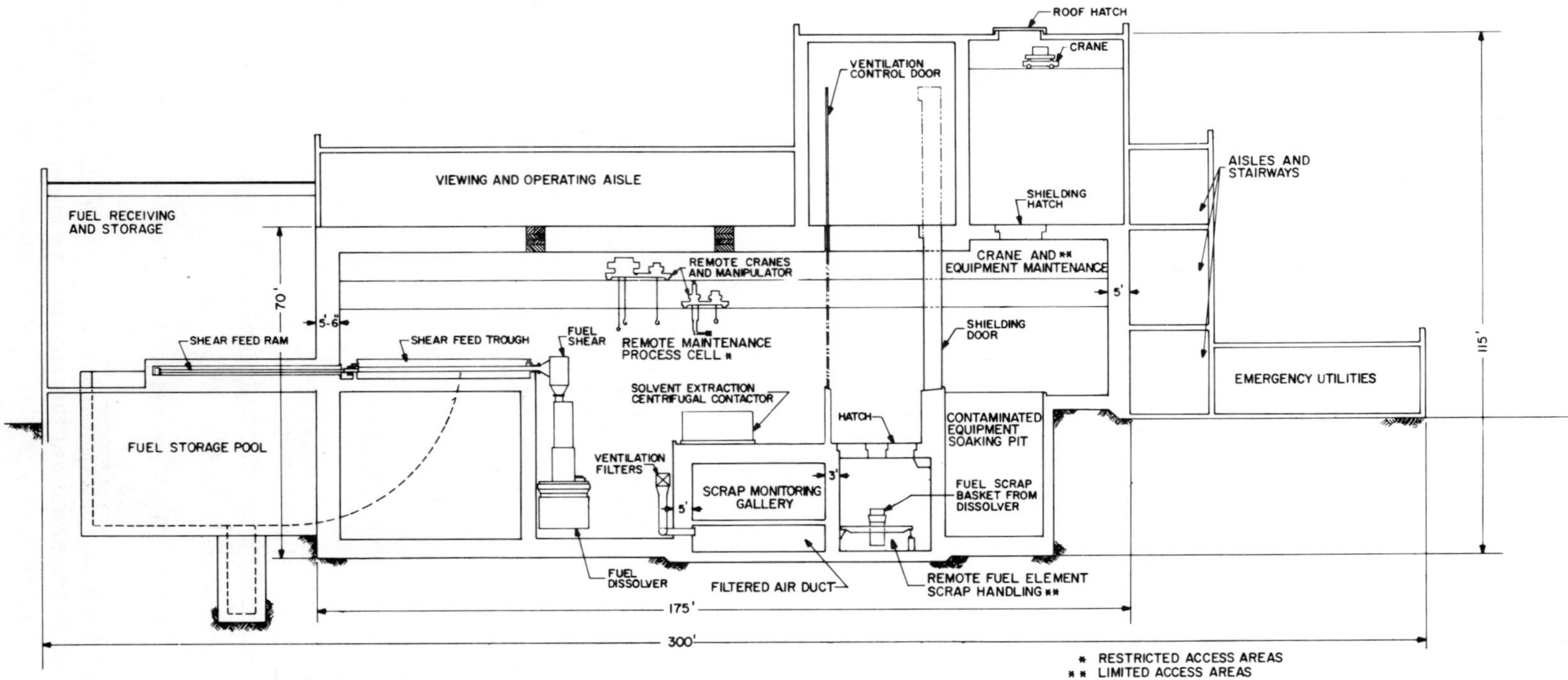


FIGURE 33. Longitudinal cross section of a typical fuel reprocessing plant.

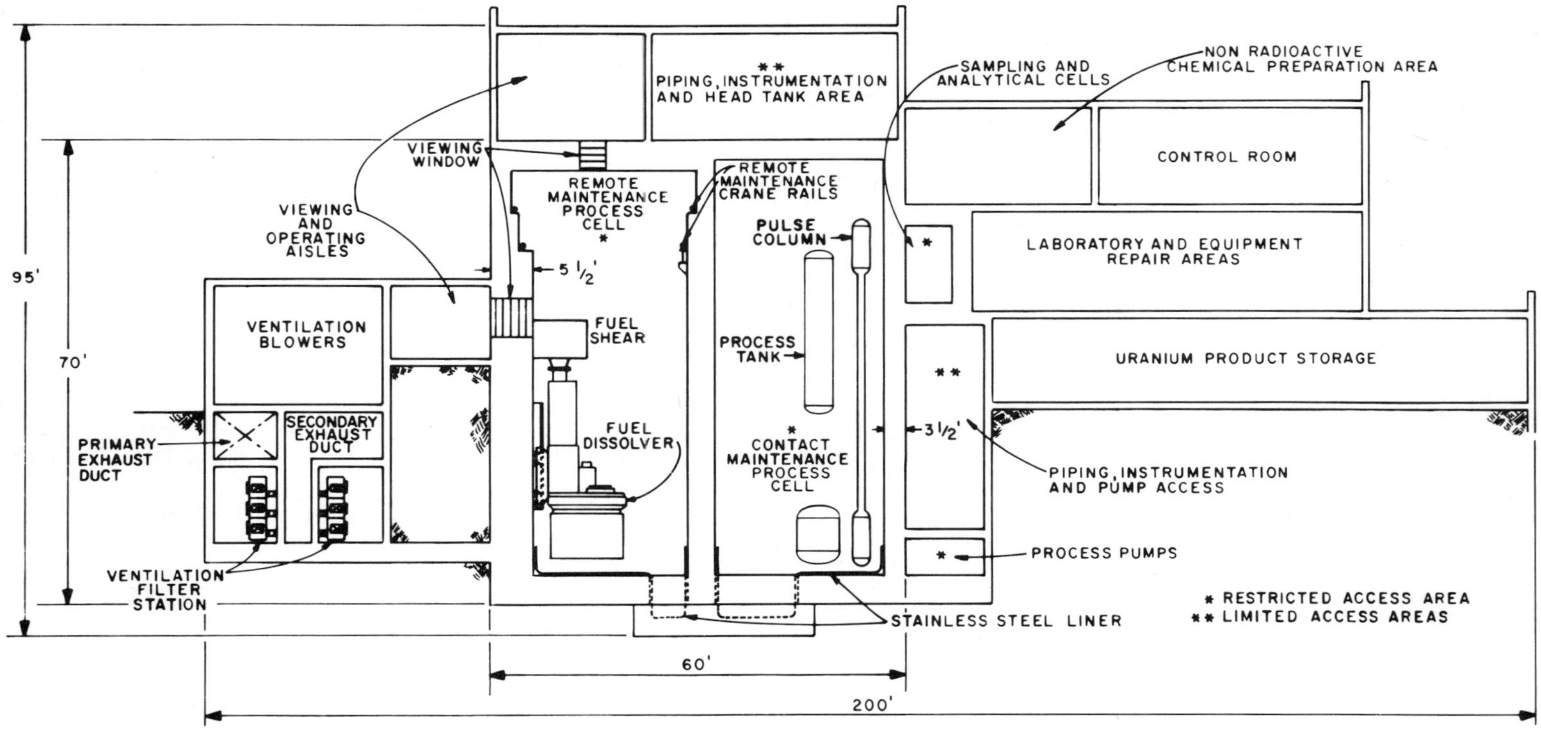


FIGURE 34. Cross section of a typical fuel reprocessing plant.

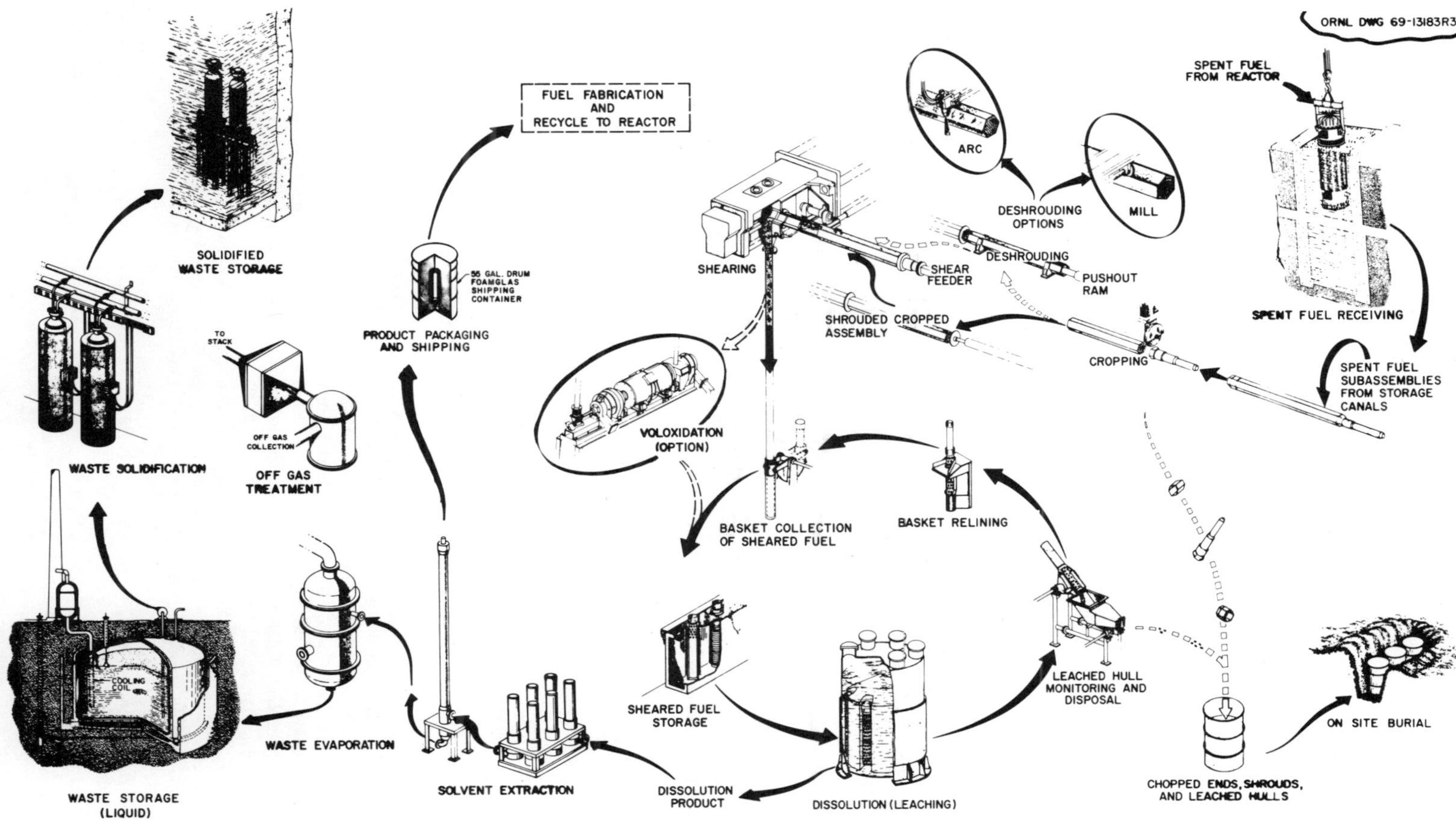


FIGURE 35. Pictorial representation of spent nuclear reactor fuel reprocessing.

this was not possible. Special thanks are due to Dr. R. G. Wymer and Dr. H. W. Godbee of the Chemical Technology Division, Oak Ridge National Laboratory. Each of them had prepared surveys of the nuclear fuel cycle for oral presentations. Much of the information presented here is based upon draft copies of their surveys. Fred Hurst, of the Chemistry Division, Oak Ridge National Laboratory, supplied a draft copy of a report on uranium recovery from phosphoric acid prepared for oral presentation as well as reprints for his published work.

## REFERENCES

1. *Statistical Abstract of the United States, 1977,* 98th annu. ed., Bureau of the Census, U.S. Department of Commerce, Washington, D.C., 1977, 591.
2. **Walske, C.,** "Nuclear power is well on its way to becoming second only to coal . . . " *Nucleonics Week,* 19(16), 9, 1978.
3. "National electricity generation by fuel type," *Inside D.O.E.,* 13, May 15, 1978.
4. World list of nuclear power plants, *Nucl. News (Hinsdale, Ill.),* 21(10), 67, 1978.
5. "AGNS study shows $10-million/reactor/year benefit for fuel recycle," *Nucleonics Week,* 17(16), 1976.
6. **Finney, B. C., Blanco, R. E., Dahlman, R. C., Kitts, F. G., and Witherspoon, J. P.,** Correlation of Radioactive Waste Treatment Costs and the Environmental Impact of Waste Effluents in the Nuclear Fuel Cycle for Use in Establishing "as Low as Practicable" Guides — Nuclear Fuel Reprocessing, Rep. ORNL-TM-4901, Oak Ridge National Laboratory, Oak Ridge, Tenn., May 1975.
7. **Pechin, W. H., Blanco, R. E., Dahlman, R. C., Finney, B. C., Lindauer, R. B., and Witherspoon, J. P.,** Correlation of Radioactive Waste Treatment Costs and the Environmental Impact of Waste Effluents in the Nuclear Fuel Cycle for Use in Establishing "as Low as Practicable" Guides — Fabrication of Light-Water Reactor Fuel from Enriched Uranium Dioxide, Rep. ORNL-TM-4902, Oak Ridge National Laboratory, Oak Ridge, Tenn., May 1975.
8. **Sears, M. B., Blanco, R. E., Dahlman, R. C., Hill, G. S., Ryon, A. D., and Witherspoon, J. P.,** Correlation of Radioactive Waste Treatment Costs and the Environmental Impact of Waste Effluents in the Nuclear Fuel Cycle for Use in Establishing "as Low as Practicable" Guides — Milling of Uranium Ores, ORNL-TM-4903, Vol. 1, Oak Ridge National Laboratory, Oak Ridge, Tenn., May 1975.
9. **Ryon, A. D. and Blanco, R. E.,** Correlation of Radioactive Waste Treatment Costs and the Environmental Impact of Waste Effluents in the Nuclear Fuel Cycle for Use in Establishing "as Low as Practicable" Guides — Appendix A. Preparation of Cost Estimates for Volume 1, Milling of Uranium Ores, ORNL-TM-4903, Vol. 2, Oak Ridge National Laboratory, Oak Ridge, Tenn., May 1975.
10. **Groenier, W. S., Blanco, R. E., Dahlman, R. C., Finney, B. C., Kibbey, A. H., and Witherspoon, J. P.,** Correlation of Radioactive Waste Treatment Costs and the Environmental Impact of Waste Effluents in the Nuclear Fuel Cycle for Use in Establishing "as Low as Practicable" Guides — Fabrication of Light-Water Reactor Fuels Containing Plutonium, ORNL-TM-4904, Oak Ridge National Laboratory, Oak Ridge, Tenn., May 1975.
11. **Finney, B. C., Blanco, R. E., Dahlman, R. C., Hill, G. S., Kitts, F. G., Moore, R. E., and Witherspoon, J. P.,** Correlation of Radioactive Waste Treatment Costs and the Environmental Impact of Waste Effluents in the Nuclear Fuel Cycle — Reprocessing Light-Water Reactor Fuel, ORNL/NUREG/TM-6, Oak Ridge National Laboratory, Oak Ridge, Tenn., January 1977.
12. **Sears, M. B., Blanco, R. E., Finney, B. C., Hill, G. S., Moore, R. E., and Witherspoon, J. P.,** Correlation of Radioactive Waste Treatment Costs and the Environmental Impact of Waste Effluents in the Nuclear Fuel Cycle — Conversion of Yellowcake to Uranium Hexafluoride. I. The Fluorination-Fractionation Process, ORNL/NUREG/TM-7, Oak Ridge National Laboratory, Oak Ridge, Tenn., September 1977.
13. The Nuclear Industry 1974, Rep. WASH-1174-74, U.S. Atomic Energy Commission, Washington, D.C., 1975.
14. The Fuel Cycle Task Force, Nuclear Fuel Cycle, Rep. ERDA-33, U.S. Energy and Research Development Administration, Washington, D.C., March 1975.
15. Statistical Data of the Uranium Industry, Rep. GJO-100(78), U.S. Department of Energy, Grand Junction Office, Grand Junction, Colo., January 1978.

16. Commodity Data Summaries 1977, Bureau of Mines, U.S. Department of the Interior, Washington, D.C., 1977, 182.
17. Staff of the Bureau of Mines, Metals, Minerals, and Fuels, *Miner. Yearb.*, 1,1327, 1976.
18. "Nuclear electricity generation in June," *Nucleonics Week*, 19(31), 16, 1978.
19. Environmental Survey of the Uranium Fuel Cycle, Rep. WASH-1248, U.S. Atomic Energy Commission, Washington, D.C., April 1974.
20. **Hunkin, G. G.**, The environmental impact of solution mining for uranium, *Min. Congr. J.*, 61(10), 24, 1975.
21. Solution mining — can it speed development of uranium reserves?, *The Mining Record*, Denver, Colo., June 23, 1976.
22. **White, L.**, *In-situ* leaching opens new uranium reserves in Texas, *Eng. Min. J.*, 176(7), 73, 1975.
23. **Merritt, R. C.**, *The Extractive Metallurgy of Uranium*, Colorado School of Mines Research Institute, Golden, 1971, 138.
24. **Merritt, R. C.**, *The Extractive Metallurgy of Uranium*, Colorado School of Mines Research Institute, Golden, 1971, 108.
25. "Earth sciences to gauge potential of getting uranium from phosphate," *Nuclear Fuel*, 3(20), 5, 1978.
26. **Hurst, F. J.**, Recovery of uranium from wet process phosphoric acid by solvent extraction, *Trans. Soc. Min. Eng. AIME*, 262, 240, 1977.
27. **Clegg, J. W. and Foley, D. D.**, *Uranium Ore Processing*, Addison-Wesley, Reading, Mass., 1958, 372.
28. **Hurst, F. J., Arnold, W. D., and Ryon, A. D.**, Recovering uranium from wet-process phosphoric acid, *Chem. Eng.*, 84, 56, 1977.
29. **Hurst, F. J., Arnold, W. D., and Ryon, A. D.**, Progress and problems in recovering uranium from wet-process phosphoric acid, in *Proc. 26th Annu. Fertilizer Industry Roundtable*, Spillman, A., Ed., Fertilizer Industry Roundtable, Glen Arm, Md., October 1976, 100.
30. **Merritt, R. C.**, *The Extractive Metallurgy of Uranium*, Colorado School of Mines Research Institute, Golden, 1971, 379.
31. "Kerr-McGee engineers a quiet coup in uranium hexafluoride conversion," *Nuclear Fuel*, 3(17), 1, 1978.
32. United States Gaseous Diffusion Program for Uranium Enrichment, Prepared for the U.S. Department of Energy by Nuclear Division Union Carbide Corporation, 1978.
33. United States Uranium Enrichment Services, U.S. Department of Energy, Washington, D.C., 1978.
34. AEC Gaseous Diffusion Plant Operations, Rep. ORO-684, U.S. Atomic Energy Commission, Washington, D.C., January 1972.
35. United States Gas Centrifuge Program for Uranium Enrichment, Rev. 1, prepared for the U.S. Department of Energy by Nuclear Division, Union Carbide Corporation, 1978.
36. Nuclear Power Growth 1974—2000, Rep. WASH-1139(74), U.S. Atomic Energy Commission, Washington, D.C., February 1974.
37. **Wymer, R. G. and Blomeke, J. O.**, Radioactive Wastes: Sources, Treatment, and Disposal, presented at Energy Sources for the Future Symp., CONF-750733, Oak Ridge National Laboratory, Oak Ridge, Tenn., 1975, 134.
38. **Ferguson, D. E.**, Chemical reprocessing of nuclear fuel, in *24th Int. Congr. Pure and Applied Chemistry*, Vol. 6, Butterworth, London, 1974.
39. Barnwell Nuclear Fuel Plant — Environmental Report, Docket No. 50-332, Allied-Gulf Nuclear Services, Barnwell, S.C., November 1971.

Chapter 3

# CHEMICAL ASPECTS OF LWR FUEL REPROCESSING

**D. O. Campbell***

## TABLE OF CONTENTS



* Chemical Technology Division, Oak Ridge National Laboratory, operated by Union Carbide Corporation under Contract W-7405-eng-26 with the U.S. Department of Energy.

# I. INTRODUCTION

The original purpose of nuclear fuel reprocessing was to recover the plutonium produced in the fuel of the plutonium production reactors. More recently, a second goal that has been recognized is to place the radioactive by-products into a form suitable for safe long-term storage. There have been proposals that certain other elements, fission products, and transuranium isotopes, are also of value, but the undeveloped market and high cost have justified such recovery operations only on a very limited scale.

Following normal exposure in a light water reactor (LWR) fueled initially with low-enriched uranium (about 3%) as $UO_2$, typical of many power reactors now operating, the residual uranium is comparable to natural uranium in its content of fissile mass 235, or about 0.7 to 0.8%. In addition, plutonium is produced in the fuel by neutron capture by $^{238}U$; some is fissioned within the reactor, but the remainder is discharged from the reactor and constitutes about 1% of the spent fuel. In broad terms, if the fuel is reprocessed and the fissile content is recycled to new reactor fuel, the requirements for natural uranium and for the separative work required to enrich that uranium are reduced by some 35% and 25%, respectively, compared to the mode of operation with no recycle.[1] The economic incentive for reprocessing, then, is this saving in uranium mining and enrichment.

Reprocessing was the central activity early in the nuclear age, and its more recent application to power reactors followed naturally from that experience. It is well known that the initial purpose was to produce plutonium for weapons use, and reprocessing was an essential operation. Later on, electric power was derived from graphite-moderated gas-cooled reactors, especially in Europe. These operated to a burnup of only a few thousand MWd per tonne of fuel, and the spent fuel contained a large fraction of the fissile content initially present. Accordingly, it was necessary to recover and recycle the fuel to utilize a substantial fraction of the energy it contained. Presently, LWRs operate with a fuel burnup in the vicinity of 30,000 MWd/t, which corresponds to utilization of the larger portion of the initial fissile content. Although additional fissile material is produced during irradiation (plutonium), the spent fuel contains less than half as much fissile material as did the initial fuel.

The purpose of this chapter is to summarize the operations normally planned and utilized in LWR fuel reprocessing, with emphasis on the underlying chemistry. We are dealing here with systems containing over 40 chemical elements, some with a rich variety of chemical properties resulting from oxidation-reduction and complexation reactions, and with extremely high radiation intensities. Further, organic solvents are used to separate selectively uranium and plutonium from all other constituents. It is clear that a substantial amount of chemistry is involved. Finally, the processing is accomplished using countercurrent multistage processes that are extremely powerful separations methods. By taking advantage of both the varied chemical properties of the individual elements and the multistage separations methods, remarkable versatility and performance have been achieved.

## A. Process Requirements

In all countries in which the capability has been developed, reprocessing has centered on the recovery and purification of uranium and plutonium from irradiated fuel. In a few cases uranium was not recovered, but was left with the waste for subsequent recovery, if desired. The requirements can be stated qualitatively: uranium and plutonium should be as free as possible from chemical impurities and radioactivity; losses of these elements should be as small as possible; the impact on the environment from the entire operation, now and for all time to come, should be minimized; and the laws of economics are applicable in the final analysis. It is clear that trade-offs must be made, as they must in any industrial operation, but in this case they are receiving more attention than is normal.

The process goals can be stated somewhat more meaningfully by drawing on past experience. Losses of uranium and plutonium have typically been in the range of a few tenths of 1% to 1%. A recovery more quantitative cannot be justified on economic grounds alone, but smaller losses may be necessary for environmental reasons. Product purity must be high enough that the elements can be utilized for the intended purposes, which are, for uranium, reenrichment to increase the $^{235}U$ content for fuel recycle, and for plutonium, fabrication into new fuel. The purity requirements are more stringent for uranium, in many cases, because impurities may interfere with the enrichment process and may themselves be enriched. In any case, purity specifications that have been established over the years have generally been met in practice. The radioactivity of the uranium and plutonium is typically reduced by a factor (DF) in the range of $10^7$ to $10^8$, which has permitted hands-on operations with the products. However, this is one area where changes have been made from time to time; such changes always have been in the direction of specifying higher purity. Plutonium may become an exception, with a lower DF being acceptable because the radiation inherent in recycled plutonium is significant and may require remote handling, independently of other impurities present.

The whole subject of wastes, or process effluents, is another matter. This problem has largely been postponed in the past, and it is not resolved today. The wastes contain a large amount of radioactivity, and the solutions have simply been concentrated and stored in large tanks, as liquids or slurries, as a temporary expedient. The approaches under consideration now involve the removal of all water and acid, which can then be recycled, leaving the permanent solids and radioactivity which can be converted to one of several insoluble solid forms (calcine, glass, etc.) for permanent storage in a stable geologic formation (see Chapter 6).

Since the waste resulting from fuel reprocessing is highly radioactive, it is a very expensive material to handle. There is strong incentive to minimize the volume by avoiding addition of materials which result in waste solids (metal ions, salts, etc.). Furthermore, volatile materials (ammonium compounds, halides, sulfate, etc.) that would interfere with acid evaporation and materials corrosive to the evaporator (chlorides, fluorides, sulfate) should be avoided. These constraints have had increasing influence the last few years as attention has focused on modifications of process steps that were satisfactory from the chemical point of view but contributed to these waste problems. Examples are the use of ferrous sulfamate for plutonium valence control, fluoride to aid dissolution of spent fuel, and sodium carbonate scrubbing for purifying the solvent used in fuel reprocessing. Alternatives for all of these are under study.

The physical characteristics of a typical pressurized water reactor (PWR) fuel assembly are summarized in Table 1, and the design is indicated in Figure 1. A boiling water reactor (BWR) fuel assembly is generally similar except that fuel rods are somewhat larger in diameter and there are fewer rods per assembly, such as a 7 × 7 array instead of 15 × 15. The chemical composition of the fuel, both before irradiation (discharge) and after irradiation followed by the indicated cooling times (90 days to 10 years), is summarized in Table 2 for fission products and in Table 3 for the actinide elements. The corresponding values for decay energy (watts) are presented in Tables 4 and 5, and the radioactive disintegration rate (curies) in Tables 6 and 7. These data, which are derived from ORIGEN computer calculations,[2] offer reasonable approximations of the composition of both BWR and PWR fuel. An additional source of radioactivity is activation of stable elements in the cladding, structural components, and impurities in the system; this depends on the actual materials and compositions exposed to neutrons.

With a plutonium recycle fuel there is, of course, an increase in plutonium and transplutonium actinides, and also a significant increase in a few important fission products on the edge of the peaks of the fission yield curve, most notably ruthenium, silver, and iodine. In process solutions containing 1 *M* uranium a number of fission products are present at concentrations in the range of 0.001 to 0.01 *M*, and radiation levels are the order of $10^3$ Ci/ℓ.

The major generic steps in aqueous reprocessing of spent LWR fuel are outlined in Figure 2. The fuel subassembly, after storage for a period of time, is broken down and fuel rods are sheared into short lengths to expose the oxide fuel, which is then dissolved in nitric acid. After clarification, accountability measurements, and composition adjustments, the solution enters a series of solvent extraction operations in which the uranium and plutonium are separated, purified, and recovered. Three basic processes are utilized, each including extraction of the desired metal from the aqueous into an organic phase, followed by stripping of the metal from the organic back into another aqueous phase. These include: (1) a "codecontamination" cycle in which uranium and plutonium are extracted and stripped together, (2) a "partition" cycle in which uranium and plutonium are extracted together but stripped separately (partitioned), and (3) "purification" cycles in which either uranium or plutonium, after partitioning, is separately purified further. The arrangement varies somewhat from plant to plant, and this example is typical of the French plant at La Hague. An alternative is to partition in the first cycle instead of the second, thereby eliminating the codecontamination cycle, but to compensate for the attendant loss of decontamination by utilizing two purification cycles instead of one. The purified products are converted to suitable chemical and physical forms for recycle within the overall nuclear economy, and the wastes are stored temporarily as liquids prior to conversion to a solid and permanent waste management.

## II. HISTORICAL DEVELOPMENT OF THE CHEMISTRY OF AQUEOUS REPROCESSING METHODS

### A. Introduction

Reprocessing methods have been developed to satisfy the specific requirements of the nuclear industry at the time. The work started with the Manhattan Project, a crash program to develop weapons during World War II; recovery of plutonium as rapidly as possible was the overriding goal. Subsequently, nuclear reactors developed along several lines. The graphite-moderated gas-cooled reactors using natural or slightly enriched uranium for fuel were a logical extension of the "production reactors", and power reactors based on these systems were built in several countries, notably France and Britain. These early production and power reactors used uranium metal fuel clad with aluminum or magnesium, and they were limited in temperature. More recently, development has turned to increasing the temperature by using helium cooling (instead of carbon dioxide) and much more complex fuel structures, but these will not be considered in this text.

In the U.S. a different reactor evolution occurred, based on development of PWRs for ship propulsion (primarily submarines), and the LWR power reactors derived from this. Water served as both coolant and moderator. Although alloys of uranium with various other metals (zirconium, molybdenum, etc.) were intensively investigated in the 1950s, and stainless steel cladding was used in some cases, the LWR fuel that finally

### Table 1
### PHYSICAL CHARACTERISTICS OF A REFERENCE PWR FUEL ASSEMBLY

| Property | PWR |
|---|---|
| Overall assembly length, m | 4.059 |
| Cross section, cm | 21.4 × 21.4 |
| Fuel element length, m | 3.851 |
| Active fuel height, m | 3.658 |
| Fuel element OD, cm | 1.07 |
| Fuel element array | 15 × 15 |
| Assembly total weight, kg | 657.9 |
| Uranium per assembly, kg | 461.4 |
| $UO_2$ per assembly, kg | 523.4 |
| Zircaloy per assembly, kg | 108.4 |
| Hardware per assembly, kg | 26.1 |
| Total metal per assembly, kg | 134.5 |
| Nominal volume per assembly, $m^3$ | 0.186 |

evolved was low-enrichment uranium oxide clad in Zircaloy, a zirconium alloy. A variation was the BWR, which was similar in many respects but allowed boiling to provide heat transfer by means of steam rather than pressurized water. Reactors rather similar to these LWRs were also developed in the U.S.S.R. In Canada, heavy water-moderated reactors using natural uranium were developed (CANDU). Here, too, the cladding was Zircaloy, chosen because of corrosion considerations and neutron economy.

There was, at the same time, a rapid development in research and test reactors which differed substantially from power reactors; they generally used highly enriched uranium in the form of a cermet or alloy, clad with aluminum, and operated at a temperature too low for a useful power cycle. At present, the focus throughout the world is on the liquid metal fast breeder reactor (LMFBR), using a metal-clad ceramic fuel and sodium metal for the coolant. Fast reactors utilize a much higher fissile isotope (plutonium) concentration than thermal power reactors. There has been interest in several other reactor types, including fluid-fuel reactors, slurry fuels, thoria-based fuel cycles, process heat reactors, and others. However, it is the reprocessing of fuel from the LWRs that is the subject here.

As reactors developed from the original wartime graphite "piles", reprocessing methods have also been changed to suit the new requirements. Production reactors utilized metal fuels, low-atomic-weight metal cladding, quite low burnup in short irradiations, and reasonably short cooling times. Power reactors required, first of all, high-temperature operation, and second, higher fuel burnup to improve economics. For LWRs, this means the fuel is $UO_2$ instead of uranium metal; cladding is a zirconium alloy instead of aluminum or magnesium; and the fuel burnup, which is much higher, is achieved over a period of several years. The different fuel composition requires a different head-end process, since the cladding is no longer easily dissolved. The greatly increased fission product content causes new problems, not just from radiation, but also because the chemical concentration of the fission products is appreciable. As a result, separations methods have had to meet different constraints. Finally, once the wartime pressure ended, there was time to develop processes more appropriate to the requirement, and this led to the demonstration of multistage, countercurrent solvent extraction separations on a large scale.

A listing of most of the significant reprocessing plants is presented in Table 8. The data are condensed from several sources, notably a compilation by Koch,[13] but there

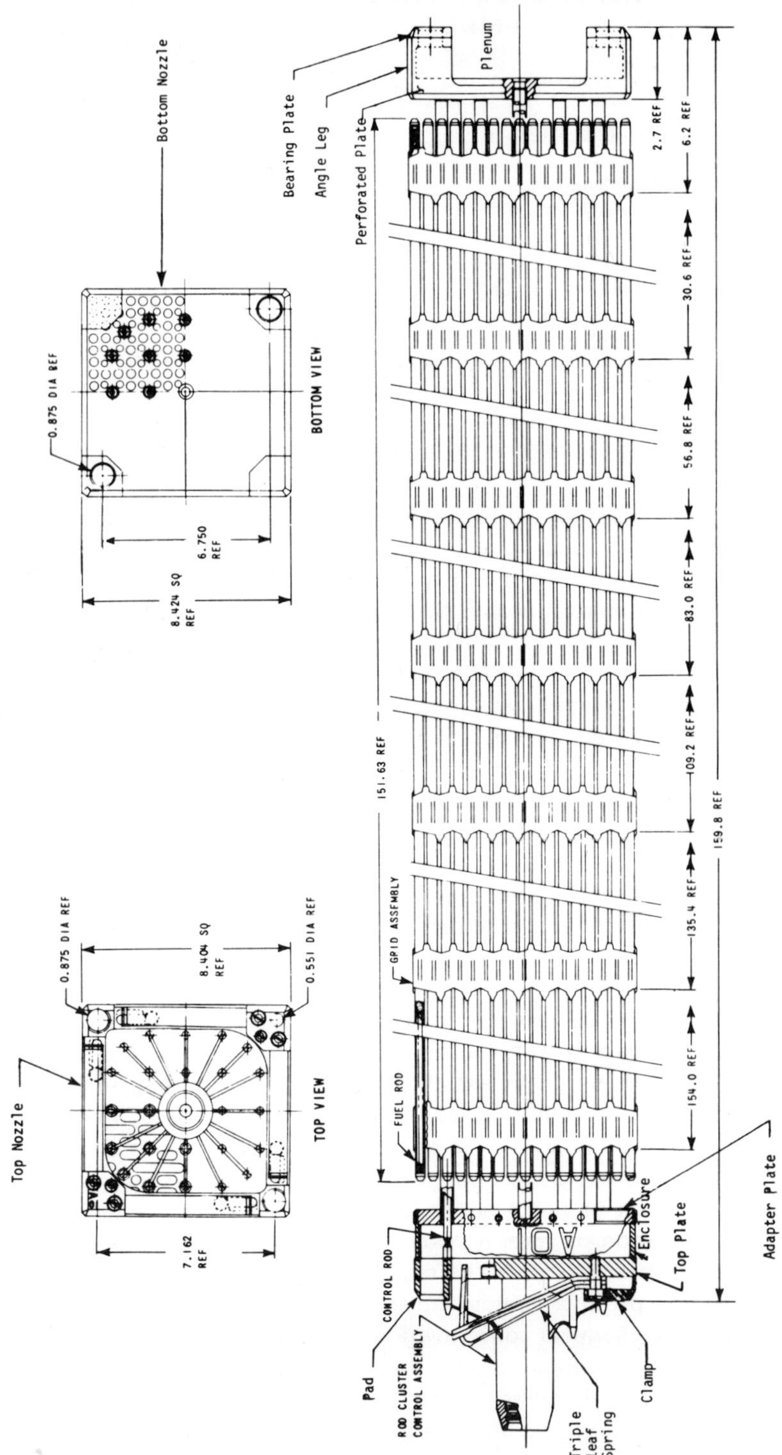


FIGURE 1. PWR fuel assembly.

### Table 2
### MASS OF FISSION-PRODUCT ELEMENTS IN LOW-ENRICHED URANIUM LWR FUEL

| Element | Discharge | 90 D | 160 D | 365 D | 1096 D | 3653 D |
|---|---|---|---|---|---|---|
| H | 5.72E-02 | 5.65E-02 | 5.59E-02 | 5.41E-02 | 4.83E-02 | 3.26E-02 |
| Zn | 7.10E-05 | 7.35E-19 | 9.78E-30 | 1.38E-61 | 0.0 | 0.0 |
| Ga | 3.80E-05 | 3.20E-19 | 4.26E-30 | 6.00E-62 | 0.0 | 0.0 |
| Ge | 6.61E-01 | 6.60E-01 | 6.60E-01 | 6.60E-01 | 6.60E-01 | 6.60E-01 |
| As | 2.03E-01 | 2.00E-01 | 2.00E-01 | 2.00E-01 | 2.00E-01 | 2.00E-01 |
| Se | 5.63E 01 | 5.63E 01 | 5.63E-01 | 5.63E 01 | 5.63E 01 | 5.63E 01 |
| Br | 2.17E 01 | 2.17E 01 | 2.17E 01 | 2.17E 01 | 2.17E 01 | 2.17E 01 |
| Kr | 3.69E 02 | 3.69E 02 | 3.69E 02 | 3.68E 02 | 3.65E 02 | 3.58E 02 |
| Rb | 3.41E 02 | 3.41E 02 | 3.42E 02 | 3.42E 02 | 3.45E 02 | 3.52E 02 |
| Sr | 9.15E 02 | 8.92E 02 | 8.84E 02 | 8.74E 02 | 8.48E 02 | 7.69E 02 |
| Y | 4.71E 02 | 4.63E 02 | 4.60E 02 | 4.57E 02 | 4.56E 02 | 4.56E 02 |
| Zr | 3.54E 03 | 3.53E 03 | 3.52E 03 | 3.52E 03 | 3.55E 03 | 3.63E 03 |
| Nb | 3.90E 01 | 2.44E 01 | 1.33E 01 | 1.70E 00 | 3.82E-03 | 6.72E-03 |
| Mo | 3.22E 03 | 3.28E 03 | 3.30E 03 | 3.33E 03 | 3.33E 03 | 3.33E 03 |
| Tc | 7.76E 02 | 7.80E 02 | 7.80E 02 | 7.80E 02 | 7.80E 02 | 7.80E 02 |
| Ru | 2.40E 03 | 2.33E 03 | 2.31E 03 | 2.26E 03 | 2.20E 03 | 2.18E 03 |
| Rh | 4.27E 02 | 4.62E 02 | 4.69E 02 | 4.72E 02 | 4.72E 02 | 4.72E 02 |
| Pd | 1.20E 03 | 1.23E 03 | 1.24E 03 | 1.28E 03 | 1.35E 03 | 1.37E 03 |
| Ag | 7.78E 01 | 7.73E 01 | 7.72E 01 | 7.69E 01 | 7.67E 01 | 7.66E 01 |
| Cd | 1.06E 02 | 1.07E 02 | 1.07E 02 | 1.07E 02 | 1.07E 02 | 1.07E 02 |
| In | 2.29E 00 | 2.40E 00 | 2.42E 00 | 2.44E 00 | 2.46E 00 | 2.52E 00 |
| Sn | 9.07E 01 | 9.01E 01 | 8.99E 01 | 8.96E 01 | 8.94E 01 | 8.94E 01 |
| Sb | 3.18E 01 | 3.10E 01 | 3.05E 01 | 2.91E 01 | 2.49E 01 | 1.94E 01 |
| Te | 4.79E 02 | 4.73E 02 | 4.73E 02 | 4.74E 02 | 4.78E 02 | 4.84E 02 |
| I | 2.40E 02 | 2.34E 02 | 2.34E 02 | 2.35E 02 | 2.35E 02 | 2.35E 02 |
| Xe | 5.31E 03 | 5.31E 03 | 5.31E 03 | 5.31E 03 | 5.31E 03 | 5.31E 03 |
| Cs | 2.75E 03 | 2.75E 03 | 2.74E 03 | 2.70E 03 | 2.61E 03 | 2.40E 03 |
| Ba | 1.39E 03 | 1.38E 03 | 1.39E 03 | 1.43E 03 | 1.52E 03 | 1.73E 03 |
| La | 1.22E 03 | 1.22E 03 | 1.22E 03 | 1.22E 03 | 1.22E 03 | 1.22E 03 |
| Ce | 2.79E 03 | 2.69E 03 | 2.64E 03 | 2.53E 03 | 2.40E 03 | 2.38E 03 |
| Pr | 1.09E 03 | 1.11E 03 | 1.12E 03 | 1.12E 03 | 1.12E 03 | 1.12E 03 |
| Nd | 3.63E 03 | 3.72E 03 | 3.77E 03 | 3.87E 03 | 4.00E 03 | 4.03E 03 |
| Pm | 1.45E 02 | 1.38E 02 | 1.31E 02 | 1.13E 02 | 6.67E 01 | 1.05E 01 |
| Sm | 7.27E 02 | 7.40E 02 | 7.47E 02 | 7.65E 02 | 8.11E 02 | 8.67E 02 |
| Eu | 1.65E 02 | 1.61E 02 | 1.60E 02 | 1.57E 02 | 1.49E 02 | 1.31E 02 |
| Gd | 7.86E 01 | 8.31E 01 | 8.42E 01 | 8.69E 01 | 9.55E 01 | 1.15E 02 |
| Tb | 2.63E 00 | 2.58E 00 | 2.56E 00 | 2.55E 00 | 2.54E 00 | 2.54E 00 |
| Dy | 1.28E 00 | 1.34E 00 | 1.36E 00 | 1.37E 00 | 1.38E 00 | 1.38E 00 |
| Ho | 1.37E-01 | 1.37E-01 | 1.37E-01 | 1.37E-01 | 1.37E-01 | 1.37E-01 |
| Er | 4.67E-02 | 4.70E-02 | 4.70E-02 | 4.71E-02 | 4.71E-02 | 4.71E-02 |
| Totals | 3.41E 04 | 3.41E 04 | 3.41E 04 | 3.41E 04 | 3.41E 04 | 3.41E 04 |

*Note:* Power = 37.50 MW; burnup = 33000 MWD; flux = 3.25E 14 N $cm^{-2}$ $sec^{-1}$; Element concentration, grams; and bases = one metric ton charge to reactor.

are gaps in the available information, particularly in regard to reprocessing in the U.S.S.R. and mainland China. The type of separation system is also indicated, corresponding to the three extraction cycles of Figure 2. All the plants use tributyl phosphate (TBP) Purex cycles, and some use other purification cycles in addition, including tertiary amine extractions, anion exchange, silica gel, and in one case a precipitation process for neptunium recovery. The concentration of TBP is 30% in all cases except for 20% at Windscale and 25% for the U.S.S.R. Plutonium partitioning is accom-

**Table 3**
**MASS OF HEAVY-ELEMENT NUCLIDES IN LOW-ENRICHED URANIUM LWR FUEL**

| Nuclide | Charge | Discharge | 90 D | 160 D | 365 D | 1096 D | 3653 D |
|---|---|---|---|---|---|---|---|
| He 4 | 0.0 | 2.26E-01 | 2.95E-01 | 3.35E-01 | 4.12E-01 | 5.29E-01 | 8.29E-01 |
| Th230 | 0.0 | 1.08E-03 | 1.21E-03 | 1.30E-03 | 1.58E-03 | 2.58E-03 | 6.18E-03 |
| U233 | 0.0 | 1.44E-03 | 1.47E-03 | 1.50E-03 | 1.58E-03 | 1.89E-03 | 2.89E-03 |
| U234 | 2.90E 02 | 1.79E 02 | 1.80E 02 | 1.80E 02 | 1.81E 02 | 1.83E 02 | 1.90E 02 |
| U235 | 3.20E 04 | 7.94E 03 | 7.94E 03 | 7.94E 03 | 7.94E 03 | 7.94E 03 | 7.94E 03 |
| U236 | 0.0 | 3.96E 03 | 3.96E 03 | 3.96E 03 | 3.96E 03 | 3.96E 03 | 3.96E 03 |
| U238 | 9.68E 05 | 9.44E 05 | 9.44E 05 | 9.44E 05 | 9.44E 05 | 9.44E 05 | 9.44E 05 |
| Np237 | 0.0 | 4.31E 02 | 4.41E 02 | 4.41E 02 | 4.41E 02 | 4.42E 02 | 4.46E 02 |
| Pu238 | 0.0 | 1.26E 02 | 1.31E 02 | 1.33E 02 | 1.35E 02 | 1.36E 02 | 1.28E 02 |
| Pu239 | 0.0 | 4.95E 03 | 5.03E 03 | 5.03E 03 | 5.03E 03 | 5.03E 03 | 5.03E 03 |
| Pu240 | 0.0 | 2.32E 03 | 2.32E 03 | 2.32E 03 | 2.32E 03 | 2.32E 03 | 2.32E 03 |
| Pu241 | 0.0 | 1.22E 03 | 1.21E 03 | 1.20E 03 | 1.17E 03 | 1.06E 03 | 7.61E 02 |
| Pu242 | 0.0 | 4.61E 02 | 4.61E 02 | 4.61E 02 | 4.61E 02 | 4.61E 02 | 4.61E 02 |
| Pu244 | 0.0 | 2.96E-02 | 2.96E-02 | 2.96E-02 | 2.96E-02 | 2.96E-02 | 2.96E-02 |
| Am241 | 0.0 | 3.17E 01 | 4.59E 01 | 5.69E 01 | 8.83E 01 | 1.94E 02 | 4.89E-02 |
| Am242$^m$ | 0.0 | 4.40E-01 | 4.40E-01 | 4.39E-01 | 4.38E-01 | 4.34E-01 | 4.21E-01 |
| Am243 | 0.0 | 8.59E 01 | 8.60E 01 | 8.60E 01 | 8.60E 01 | 8.59E 01 | 8.59E 01 |
| Cm242 | 0.0 | 1.14E 01 | 7.83E 00 | 5.81E 00 | 2.43E 00 | 1.10E-01 | 1.02E-03 |
| Cm243 | 0.0 | 2.75E-01 | 2.74E-01 | 2.73E-01 | 2.70E-01 | 2.58E-01 | 2.22E-01 |
| Cm244 | 0.0 | 1.90E 01 | 1.88E 01 | 1.87E 01 | 1.83E 01 | 1.69E 01 | 1.29E 01 |
| Cm245 | 0.0 | 1.28E-01 | 1.28E-01 | 1.28E-01 | 1.28E-01 | 1.28E-01 | 1.28E-01 |
| Cm246 | 0.0 | 9.13E-02 | 9.13E-02 | 9.13E-02 | 9.13E-02 | 9.13E-02 | 9.12E-02 |
| Subtotal | 1.00E 06 | 9.66E 05 | 9.66E 05 | 9.66E 05 | 9.66E 05 | 9.66E 05 | 9.66E 05 |
| Totals | 1.00E 06 | 9.66E 05 | 9.66E 05 | 9.66E 05 | 9.66E 05 | 9.66E 05 | 9.66E 05 |

*Note:* Power = 37.50 MW; burnup = 33000 MWD; flux = 3.25E 14 N $cm^{-2}$ $sec^{-1}$; nuclide concentration, grams; and bases − one metric ton charge to reactor.

plished by selective stripping of Pu(III) using either ferrous iron or U(IV) to reduce plutonium to Pu(III) in all plants that have operated. Electrolytic reduction has been intensively studied in several countries, and it is proposed for the Allied-General Nuclear Services (AGNS) plant. "Holding" reductants are also normally used, most commonly hydrazine or sulfamic acid. Their purpose is to destroy nitrous acid or nitrite ion which, otherwise, would oxidize both Pu(III) and the reductant, thereby diminishing stripping efficiency, as discussed in Chapter 4.

There are a number of relatively small facilities of the general nature of flexible pilot plants, but not all pilot plants are included. It is apparent that the larger reprocessing plants, in nearly all cases, are installations built originally for weapons production (Marcoule, Hanford, Savannah River, Windscale, and the U.S.S.R.). Outside the U.S. these have evolved into facilities for processing power reactor fuels also. In France, for example, the UP-1 plant at Marcoule processed gas-cooled graphite reactor fuel for military purposes, and a second plant, UP-2, was built at La Hague in 1966 to process gas-cooled power reactor fuel. This latter fuel is being shifted to Marcoule, and a new head-end was put into operation in 1976 at La Hague to permit processing of LWR fuels in UP-2. Two reprocessing campaigns have been completed successfully, one in 1976 and the other in 1977 to 1978, with some 70 t of LWR fuel. An additional plant at La Hague, UP-3, is scheduled for start-up in 1985, with a capacity of 800 t of LWR fuel per year.[14]

## Table 4
## THERMAL POWER OF FISSION-PRODUCT ELEMENTS IN LOW-ENRICHED URANIUM LWR FUEL

| Element | Discharge | 90 D | 160 D | 365 D | 1096 D | 3653 D |
|---|---|---|---|---|---|---|
| H | 1.97E-02 | 1.95E-02 | 1.93E-02 | 1.87E- 02 | 1.67E-02 | 1.12E-02 |
| Zn | 7.53E 01 | 9.38E-16 | 1.25E-26 | 1.76E-58 | 0.0 | 0.0 |
| Ga | 4.63E 02 | 1.89E-14 | 2.51E-25 | 3.54E-57 | 0.0 | 0.0 |
| Ge | 1.70E 03 | 0.0 | 0.0 | 0.0 | 0.0 | 0.0 |
| As | 7.07E 03 | 9.90E-17 | 9.18E-30 | 6.26E-68 | 0.0 | 0.0 |
| Se | 1.47E 04 | 1.02E-04 | 1.02E-04 | 1.02E-04 | 1.02E-04 | 1.02E-04 |
| Br | 4.63E 04 | 3.58E-17 | 1.83E-31 | 0.0 | 0.0 | 0.0 |
| Kr | 5.77E 04 | 1.40E 01 | 1.38E 01 | 1.33E 01 | 1.17E 01 | 7.45E 00 |
| Rb | 1.15E 05 | 2.89E-01 | 2.14E-02 | 1.05E-05 | 1.78E-08 | 1.78E-08 |
| Sr | 1.01E 05 | 9.10E 02 | 4.11E 02 | 1.08E 02 | 8.25E 01 | 6.95E 01 |
| Y | 1.63E 05 | 1.72E 03 | 9.82E 02 | 4.58E 02 | 3.88E 02 | 3.27E 02 |
| Zr | 8.25E 04 | 2.92E 03 | 1.39E 03 | 1.59E 02 | 6.97E-02 | 3.45E-04 |
| Nb | 1.94E 05 | 4.60E 03 | 2.50E 03 | 3.19E 02 | 1.47E-01 | 2.38E-04 |
| Mo | 7.96E 04 | 8.11E-07 | 1.78E-14 | 6.58E-37 | 0.0 | 0.0 |
| Tc | 1.07E 05 | 8.98E-03 | 8.98E-03 | 8.98E-03 | 8.98E-03 | 8.98E-03 |
| Ru | 2.62E 04 | 1.05E 03 | 3.25E 02 | 2.51E 01 | 4.27E 00 | 3.50E-02 |
| Rh | 3.19E 04 | 4.72E 03 | 4.09E 03 | 2.77E 03 | 7.02E 02 | 5.76E 00 |
| Pd | 2.87E 03 | 6.67E-06 | 6.67E-06 | 6.67E-06 | 6.67E-06 | 6.67E-06 |
| Ag | 4.54E 03 | 5.68E 01 | 4.68E 01 | 2.66E 01 | 3.57E 00 | 3.15E-03 |
| Cd | 1.81E 03 | 3.39E 00 | 1.20E 00 | 1.36E-01 | 8.24E-02 | 5.91E-02 |
| In | 7.61E 03 | 6.57E-03 | 2.47E-03 | 1.40E-04 | 5.03E-09 | 2.16E-14 |
| Sn | 1.62E 04 | 2.73E 01 | 1.86E 01 | 6.21E 00 | 1.28E-01 | 9.19E-04 |
| Sb | 6.27E 04 | 4.90E 01 | 4.44E 01 | 3.71E 01 | 2.22E 01 | 3.77E 00 |
| Te | 6.78E 04 | 6.69E 01 | 2.52E 01 | 5.76E 00 | 1.48E 00 | 2.45E-01 |
| I | 1.59E 05 | 1.38E 00 | 3.33E-03 | 1.89E-05 | 1.89E-05 | 1.89E-05 |
| Xe | 7.69E 04 | 1.24E-01 | 1.91E-03 | 1.39E-08 | 6.16E-27 | 0.0 |
| Cs | 1.38E 05 | 1.58E 03 | 1.48E 03 | 1.24E 03 | 6.81E 02 | 1.40E 02 |
| Ba | 8.15E 04 | 4.19E 02 | 3.82E 03 | 3.77E 02 | 3.60E 02 | 3.06E 02 |
| La | 1.40E 05 | 2.22E 02 | 4.99E 00 | 7.48E-05 | 7.00E-13 | 7.00E-13 |
| Ce | 3.36E 04 | 9.48E 02 | 6.12E 02 | 3.31E 02 | 5.55E 01 | 1.09E-01 |
| Pr | 5.04E 04 | 7.49E 03 | 6.29E 03 | 3.81E 03 | 6.42E 02 | 1.26E 00 |
| Nd | 6.65E 03 | 4.46E 00 | 5.31E-02 | 1.29E-07 | 1.22E-27 | 0.0 |
| Pm | 8.37E 03 | 1.37E 02 | 7.31E 01 | 4.01E 01 | 2.31E 01 | 3.63E 00 |
| Sm | 1.44E 03 | 4.53E-02 | 4.52E-02 | 4.50E-02 | 4.43E-02 | 4.21E-02 |
| Eu | 2.32E 03 | 1.29E 02 | 9.79E 01 | 9.21E 01 | 7.79E 01 | 4.35E 01 |
| Gd | 3.49E 01 | 2.88E-02 | 2.35E-02 | 1.30E-02 | 1.59E-03 | 1.01E-06 |
| Tb | 1.74E 01 | 2.47E 00 | 1.26E 00 | 1.77E-01 | 1.60E-04 | 3.63E-15 |
| Dy | 2.58E 00 | 3.96E-10 | 2.49E-16 | 1.67E-34 | 0.0 | 0.0 |
| Ho | 9.51E-01 | 1.79E-05 | 1.79E-05 | 1.79E-05 | 1.79E-05 | 1.78E-05 |
| Totals | 1.89E 06 | 2.71E 04 | 1.88E 04 | 9.83E 03 | 3.06E 03 | 9.08E 02 |

*Note:* Power = 37.50 MW; burnup = 33000 MWD; flux = 3.25E 14 N $cm^{-2}$ $sec^{-1}$; element thermal power, watts; and bases = one metric ton charge to reactor.

In the U.S., in contrast, power reactor fuel reprocessing was determined to be a civilian enterprise in the 1950s, and government-operated military plants were not adapted to this application. Three private plants, NFS, MFRP, and AGNS, have been built, but only the first has operated and only the last has any prospect of ever operating in the future. The entire question of reprocessing is under review in the U.S. at this time.

It is clear that an evolutionary development has been followed, using essentially the same processes and even the same plants (with modifications and additions) to process

## Table 5
## THERMAL POWER OF HEAVY ELEMENTS IN LOW-ENRICHED URANIUM LWR FUEL

| Element | Charge | Discharge | 90 D | 160 D | 356 D | 1096 D | 3653 D |
|---|---|---|---|---|---|---|---|
| Tl | 0.0 | 8.31E-06 | 1.19E-05 | 1.48E-05 | 2.42E-05 | 5.99E-05 | 1.33E-04 |
| Pb | 0.0 | 1.42E-06 | 2.03E-06 | 2.54E-06 | 4.15E-06 | 1.03E-05 | 2.28E-05 |
| Bi | 0.0 | 1.72E-05 | 2.46E-05 | 3.08E-05 | 5.02E-05 | 1.24E-04 | 2.77E-04 |
| Po | 0.0 | 7.41E-05 | 1.06E-04 | 1.33E-04 | 2.16E-04 | 5.35E-04 | 1.19E-03 |
| At | 0.0 | 7.22E-09 | 2.16E-09 | 1.95E-09 | 1.97E-09 | 2.11E-09 | 2.74E-09 |
| Rn | 0.0 | 3.76E-05 | 5.36E-05 | 6.71E-05 | 1.10E-04 | 2.71E-04 | 6.04E-04 |
| Fr | 0.0 | 6.42E-09 | 1.93E-09 | 1.75E-09 | 1.78E-09 | 1.93E-09 | 2.59E-09 |
| Ra | 0.0 | 3.39E-05 | 4.84E-05 | 6.06E-05 | 9.89E-05 | 2.44E-04 | 5.45E-04 |
| Ac | 0.0 | 3.47E-08 | 2.01E-09 | 1.88E-09 | 2.03E-09 | 2.60E-09 | 4.69E-09 |
| Th | 0.0 | 7.52E-04 | 1.72E-04 | 1.84E-04 | 2.20E-04 | 3.60E-04 | 6.49E-04 |
| Pa | 0.0 | 5.12E-03 | 2.04E-03 | 2.04E-03 | 2.04E-03 | 2.04E-03 | 2.05E-03 |
| U | 6.17E-02 | 4.81E 04 | 1.02E-01 | 4.95E-02 | 4.96E-02 | 4.99E-02 | 5.08E-02 |
| Np | 0.0 | 2.88E 04 | 3.15E-02 | 3.15E-02 | 3.15E-02 | 3.15E-02 | 3.16E-02 |
| Pu | 0.0 | 5.41E 02 | 1.04E 02 | 1.05E 02 | 1.06E 02 | 1.06E 02 | 1.01E 02 |
| Am | 0.0 | 7.59E 02 | 5.86E 00 | 7.11E 00 | 1.07E 01 | 2.27E 01 | 5.66E 01 |
| Cm | 0.0 | 1.45E 03 | 1.01E 03 | 7.63E 02 | 3.49E 03 | 6.18E 01 | 3.72E 01 |
| Bk | 0.0 | 3.36E-06 | 3.41E-09 | 2.92E-08 | 1.86E-08 | 3.70E-09 | 1.31E-11 |
| Cf | 0.0 | 2.05E-07 | 1.96E-07 | 1.91E-07 | 1.79E-07 | 1.42E-07 | 7.81E-08 |
| Es | 0.0 | 5.39E-09 | 1.19E-09 | 1.59E-10 | 2.36E-13 | 6.57E-24 | 5.15E-61 |
| Totals | 6.17E-02 | 7.96E 04 | 1.12E 03 | 8.75E 02 | 4.66E 02 | 1.90E 02 | 1.94E 02 |

*Note:* Power = 37.50 MW; burnup = 33000 MWD; flux = 3.25E 14 N $cm^{-2}$ $sec^{-1}$; element thermal power, watts; and bases = one metric ton charge to reactor.

## Table 6
## ACTIVITY OF FISSION-PRODUCT NUCLIDES IN LOW-ENRICHED URANIUM LWR FUEL

| Nuclide | Discharge | 90 D | 160 D | 365 D | 1096 D | 3653 D |
|---|---|---|---|---|---|---|
| H 3 | 5.55E 02 | 5.47E 02 | 5.41E 02 | 5.24E 02 | 4.69E 02 | 3.16E 02 |
| Se 79 | 4.09E-01 | 4.09E-01 | 4.09E-01 | 4.09E-01 | 4.09E-01 | 4.09E-01 |
| Kr 85 | 9.47E 03 | 9.32E 03 | 9.21E 03 | 8.88E 03 | 7.80E 03 | 4.96E 03 |
| Rb 86 | 1.81E 03 | 6.39E 01 | 4.74E 00 | 2.32E-03 | 3.66E-15 | 0.0 |
| Sr 89 | 7.90E 05 | 2.38E 05 | 9.36E 04 | 6.09E 03 | 3.58E-01 | 5.69E-16 |
| Sr 90 | 7.57E 04 | 7.53E 04 | 7.49E 04 | 7.39E 04 | 7.03E 04 | 5.92E 04 |
| Y 90 | 7.91E 04 | 7.53E 04 | 7.49E 04 | 7.39E 04 | 7.04E 04 | 5.92E 04 |
| Y 91 | 1.04E 06 | 3.60E 05 | 1.57E 05 | 1.39E 04 | 2.45E 00 | 1.80E-13 |
| Zr 93 | 2.91E 00 | 2.91E 00 | 2.91E 00 | 2.91E 00 | 2.19E 00 | 2.91E 00 |
| Zr 95 | 1.50E 06 | 5.77E 05 | 2.75E 05 | 3.14E 04 | 1.37E 01 | 2.44E-11 |
| Nb 93M | 2.04E-01 | 2.41E-01 | 2.68E-01 | 3.48E-01 | 6.12E-01 | 1.33E 00 |
| Nb 95 | 1.52E 06 | 9.56E 05 | 5.21E 05 | 6.65E 04 | 3.05E 01 | 5.27E-11 |
| Nb 95M | 1.86E 04 | 7.33E 03 | 3.50E 03 | 3.99E 02 | 1.74E-01 | 3.10E-13 |
| Tc 99 | 1.32E 01 | 1.33E 01 | 1.33E 01 | 1.33E 01 | 1.33E 01 | 1.33E 01 |
| Ru 103 | 1.49E 06 | 3.09E 05 | 9.07E 04 | 2.51E 03 | 6.76E-03 | 2.48E-22 |
| Ru 106 | 5.64E 05 | 4.76E 05 | 4.18E 05 | 2.84E 05 | 7.20E 04 | 5.91E 02 |
| Rh 103M | 1.49E 06 | 3.09E 05 | 9.08E 04 | 2.51E 03 | 6.77E-03 | 2.48E-22 |
| Rh 106 | 6.23E 05 | 4.76E 05 | 4.18E 05 | 2.84E 05 | 7.20E 04 | 5.91E 02 |
| Pd 107 | 1.12E-01 | 1.12E-01 | 1.12E-01 | 1.12E-01 | 1.12E-01 | 1.12E-01 |
| Ag 110 | 1.52E 05 | 4.66E 01 | 3.85E 01 | 2.19E 01 | 2.93E 00 | 2.59E-03 |
| Ag 110M | 4.27E 03 | 3.33E 03 | 2.75E 03 | 1.56E 03 | 2.09E 02 | 1.85E-01 |
| Cd 113M | 5.64E 01 | 5.58E 01 | 5.53E 01 | 5.38E 01 | 4.89E 01 | 3.51E 01 |
| Cd 115M | 2.64E 03 | 6.53e 02 | 2.20E 02 | 9.08E 00 | 1.05E-04 | 5.82E-22 |

# Table 6 (continued)
## ACTIVITY OF FISSION-PRODUCT NUCLIDES IN LOW-ENRICHED URANIUM LWR FUEL

| Nuclide | Discharge | 90 D | 160 D | 365 D | 1096 D | 3653 D |
|---|---|---|---|---|---|---|
| In 114 | 6.53E 00 | 6.89E-01 | 2.59E-01 | 1.47E-02 | 5.28E-07 | 1.51E-22 |
| In 114M | 2.52E 00 | 7.14E-01 | 2.68E-01 | 1.52E-02 | 5.47E-07 | 1.57E-22 |
| Sn 119M | 2.40E 02 | 1.86E 02 | 1.53E 02 | 8.56E 01 | 1.08E 01 | 7.83E-03 |
| Sn 121M | 2.08E-01 | 2.07E-01 | 2.07E-01 | 2.05E-01 | 1.99E-01 | 1.81E-01 |
| Sn 123 | 9.27E 03 | 5.72E 03 | 3.93E 03 | 1.31E 03 | 2.57E 01 | 2.81E-05 |
| Sn 126 | 7.76E-01 | 7.76E-01 | 7.76E-01 | 7.76E-01 | 7.76E-01 | 7.76E-01 |
| Sb 124 | 1.23E 03 | 4.35E 02 | 1.94E 02 | 1.83E 01 | 4.05E-03 | 6.65E-16 |
| Sb 125 | 1.48E 04 | 1.40E 04 | 1.33E 04 | 1.16E 04 | 6.95E 03 | 1.18E 03 |
| Sb 126 | 1.10E 03 | 7.28E 00 | 2.52E-01 | 1.09E-01 | 1.09E-01 | 1.09E-0l |
| Sb 126M | 4.96E 02 | 7.76E-01 | 7.76E-01 | 7.76E-01 | 7.76E-01 | 7.76E-01 |
| Te 123M | 1.43E 01 | 8.51E 00 | 5.68E 00 | 1.73E 00 | 2.51E-02 | 9.46E-09 |
| Te 125M | 3.12E 03 | 3.24E 03 | 3.18E 03 | 2.81E 03 | 1.70E 03 | 2.87E 02 |
| Te 127 | 1.05E 05 | 9.39E 03 | 6.02E 03 | 1.63E 03 | 1.57E 01 | 1.36E-06 |
| Te 127M | 1.64E 04 | 9.59E 03 | 6.14E 03 | 1.67E 03 | 1.60E 01 | 1.39E-06 |
| Te 129 | 2.89E 05 | 7.77E 03 | 1.82E 03 | 2.58E 01 | 6.68E-06 | 6.09E-29 |
| Te 129M | 7.89E 04 | 1.22E 04 | 2.87E 03 | 4.07E 01 | 1.05E-05 | 9.60E-29 |
| I 129 | 3.09E-02 | 3.13E-02 | 3.13E-02 | 3.13E-02 | 3.13E-02 | 3.13E-02 |
| Cs 134 | 1.55E 05 | 1.42E 05 | 1.33E 05 | 1.10E 05 | 5.63E 04 | 5.34E 03 |
| Cs 135 | 3.71E-01 | 3.72E-01 | 3.72E-01 | 3.72E-01 | 3.72E-01 | 3.72E-01 |
| Cs 137 | 1.04E 05 | 1.03E 05 | 1.03E 05 | 1.01E 05 | 9.69E 04 | 8.25E 04 |
| Ba 137M | 9.84E 04 | 9.77E 04 | 9.72E 04 | 9.60E 04 | 9.17E 04 | 7.80E 04 |
| Ba 140 | 1.57E 06 | 1.19E 04 | 2.69E 02 | 4.02E-03 | 2.50E-20 | 0.0 |
| La 140 | 1.63E 06 | 1.37E 04 | 3.10E 02 | 4.63E-03 | 2.88E-20 | 0.0 |
| Ce 141 | 1.50E 06 | 2.21E 05 | 4.97E 04 | 6.31E 02 | 1.09E-04 | 2.41E-28 |
| Ce 144 | 1.21E 06 | 9.72E 05 | 8.19E 05 | 4.97E 05 | 8.37E 04 | 1.65E 02 |
| Pr 143 | 1.32E 06 | 1.49E 04 | 4.17E 02 | 1.19E-02 | 7.39E-19 | 0.0 |
| Pr 144 | 1.22E 06 | 9.72E 05 | 8.19E 05 | 4.97E 05 | 8.37E 04 | 1.65E 02 |
| Pr 144M | 1.45E 04 | 1.17E 04 | 9.83E 03 | 5.96E 03 | 1.00E 03 | 1.97E 00 |
| Pm147 | 1.30E 05 | 1.28E 05 | 1.22E 05 | 1.05E 05 | 6.19E 04 | 9.73E 03 |
| Pm 148 | 2.64E 05 | 4.65E 02 | 1.43E 02 | 4.57E 00 | 2.17E-05 | 5.02E-24 |
| Pm 148M | 3.03E 04 | 6.70E 03 | 2.07E 03 | 6.63E 01 | 3.15E-04 | 7.28E-23 |
| Sm 151 | 3.76E 02 | 3.82E 02 | 3.81E 02 | 3.80E 02 | 3.74E 02 | 3.55E 02 |
| Eu 152 | 6.64E 00 | 6.55E 00 | 6.49E 00 | 6.30E 00 | 5.66E 00 | 3.89E 00 |
| Eu 154 | 1.06E 04 | 1.04E 04 | 1.03E 04 | 9.81E 03 | 8.35E 03 | 4.75E 03 |
| Eu 155 | 6.95E 03 | 6.71E 03 | 6.53E 03 | 6.02E 03 | 4.51E 03 | 1.64E 03 |
| Eu 156 | 1.82E 05 | 3.01E 03 | 1.24E 02 | 1.08E-02 | 3.56E-17 | 0.0 |
| Gd 153 | 5.24E 01 | 4.05E 01 | 3.31E 01 | 1.83E 01 | 2.23E 00 | 1.42E-03 |
| Tb 160 | 9.48E 02 | 4.00E 02 | 2.04E 02 | 2.86E 01 | 2.59E-02 | 5.88E-13 |
| Ho 166M | 2.45E-03 | 2.45E-03 | 2.45E-03 | 2.45E-03 | 2.45E-03 | 2.44E-03 |
| Subtotal | 1.93E 07 | 6.65E 06 | 4.44E 06 | 2.30E 06 | 7.90E 05 | 3.09E 05 |
| Totals | 1.58E 08 | 6.66E 06 | 4.44E 06 | 2.30E 06 | 7.90E 05 | 3.09E 05 |

*Note:* Power = 37.50 MW; burnup = 33000 MWD; flux = 3.25E 14 N $cm^{-2}$ $sec^{-1}$; nuclide radioactivity, curies; and bases = one metric ton charge to reactor.

successively more radioactive fuels, from perhaps $10^3$ MWd/MTU for military reactor fuels to a few thousand MWd per MTU for gas-cooled power reactor fuels, and $30 \times 10^3$ MWd/MTU for LWRs. Although the chemistry of the process is basically the same, the increase in fuel burnup and the attendant increase in radiation intensity change some factors not just in degree, but in kind. For example, fission product concentrations become large enough that these elements are no longer at trace or very

### Table 7
### ACTIVITY OF HEAVY-ELEMENT NUCLIDES IN LOW-ENRICHED URANIUM LWR FUEL

| Nuclide | Charge | Discharge | 90 D | 160 D | 365 D | 1096 D | 653 D |
|---|---|---|---|---|---|---|---|
| Tl 208 | 0.0 | 3.57E-04 | 5.09E-04 | 6.37E-04 | 1.04E-03 | 2.57E-03 | 5.73E-03 |
| Pb 212 | 0.0 | 9.91E-04 | 1.41E-03 | 1.77E-03 | 2.89E-03 | 7.14E-03 | 1.59E-02 |
| Bi 212 | 0.0 | 9.91E-04 | 1.41E-03 | 1.77E-03 | 2.89E-03 | 7.14E-03 | 1.59E-02 |
| Po 212 | 0.0 | 6.34E-04 | 9.05E-04 | 1.13E-03 | 1.85E-03 | 4.57E-03 | 1.02E-02 |
| Po 216 | 0.0 | 9.91E-04 | 1.41E-03 | 1.77E-03 | 2.89E-03 | 7.14E-03 | 1.59E-02 |
| Rn 220 | 0.0 | 9.91E-04 | 1.41E-03 | 1.77E-03 | 2.89E-03 | 7.14E-03 | 1.59E-02 |
| Ra 224 | 0.0 | 9.91E-04 | 1.41E-03 | 1.77E-03 | 2.89E-03 | 7.14E-03 | 1.59E-02 |
| Th 228 | 0.0 | 9.83E-04 | 1.41E-03 | 1.76E-03 | 2.88E-03 | 7.13E-03 | 1.59E-02 |
| Th 231 | 0.0 | 7.30E-01 | 1.70E-02 | 1.70E-02 | 1.70E-02 | 1.70E-02 | 1.70E-02 |
| Th 234 | 0.0 | 3.15E-01 | 3.15E-01 | 3.15E-01 | 3.15E-01 | 3.15E-01 | 3.15E-01 |
| Pa 233 | 0.0 | 2.92E-01 | 3.09E-01 | 3.11E-01 | 3.11E-01 | 3.12E-01 | 3.14E-01 |
| Pa 234M | 0.0 | 3.19E-01 | 3.15E-01 | 3.15E-01 | 3.15E-01 | 3.15E-01 | 3.15E-01 |
| U232 | 0.0 | 5.51E-03 | 6.36E-03 | 6.97E-03 | 8.59E-03 | 1.28E-02 | 1.74E-02 |
| U234 | 1.79E 00 | 1.11E 00 | 1.11E 00 | 1.11E 00 | 1.12E 00 | 1.13E 00 | 1.17E 00 |
| U235 | 6.86E-02 | 1.70E-02 | 1.70E-02 | 1.70E-02 | 1.70E-02 | 1.70E-02 | 1.70E-02 |
| U236 | 0.0 | 2.51E-01 | 2.51E-01 | 2.51E-01 | 2.51E-01 | 2.51E-01 | 2.51E-01 |
| U237 | 0.0 | 8.17E 05 | 8.22E 01 | 2.99E 00 | 2.85E 00 | 2.59E 00 | 1.86E 00 |
| U238 | 3.22E-01 | 3.15E-01 | 3.15E-01 | 3.15E-01 | 3.15E-01 | 3.15E-01 | 3.15E-01 |
| Np 237 | 0.0 | 3.04E-01 | 3.11E-01 | 3.11E-01 | 3.11E-01 | 3.11E-01 | 3.14E-01 |
| Np 239 | 0.0 | 2.00E 07 | 1.65E 01 | 1.65E 01 | 1.65E 01 | 1.65E 01 | 1.65 E 01 |
| Pu 236 | 0.0 | 3.47E-01 | 3.45E-01 | 3.29E-01 | 2.87E-01 | 1.76E-01 | 3.22E-02 |
| Pu 238 | 0.0 | 2.13E 03 | 2.21E 03 | 2.24E 03 | 2.29E 03 | 2.29E 03 | 2.17E 03 |
| Pu 239 | 0.0 | 3.03E 02 | 3.09E 02 | 3.09E 02 | 3.09E 02 | 3.09E 02 | 3.09E 02 |
| Pu 240 | 0.0 | 5.11E 02 | 5.11E 02 | 5.11E 02 | 5.11E 02· | 5.11E 02 | 5.12E 02 |
| Pu 241 | 0.0 | 1.24E 05 | 1.23E 05 | 1.22E 05 | 1.19E 05 | 1.08E 05 | 7.73E 04 |
| Pu 242 | 0.0 | 1.80E 00 | 1.80E 00 | 1.80E 00 | 1.80E 00 | 1.80E 00 | 1.80E 00 |
| Am 241 | 0.0 | 1.09E 02 | 1.57E 02 | 1.95E 02 | 3.03E 02 | 6.63E 02 | 1.68E 03 |
| Am $242^m$ | 0.0 | 4.28E 00 | 4.27E 00 | 4.27E 00 | 4.26E 00 | 4.22E 00 | 4.09E 00 |
| Am 242 | 0.0 | 7.49E 04 | 4.27E 00 | 4.27E 00 | 4.26E 00 | 4.22E 00 | 4.09E 00 |
| Am 243 | 0.0 | 1.65E 01 | 1.65E 01 | 1.65E 01 | 1.65E 01 | 1.65E 01 | 1.65E 01 |
| Cm 242 | 0.0 | 3.78E 04 | 2.59E 04 | 1.93E 04 | 8.05E 03 | 3.63E 02 | 3.37E 00 |
| Cm 243 | 0.0 | 1.27E 01 | 1.26E 01 | 1.26E 01 | 1.24E 01 | 1.19E 01 | 1.02E 01 |
| Cm 244 | 0.0 | 1.54E 03 | 1.52E 03 | 1.51E 03 | 1.48E 03 | 1.37E 03 | 1.05E 03 |
| Cm 245 | 0.0 | 2.25E-02 | 2.25E-02 | 2.25E-02 | 2.25E-02 | 2.25E-02 | 2.25E-02 |
| Cm 246 | 0.0 | 2.82E-02 | 2.82E-02 | 2.82E-02 | 2.82E-02 | 2.82E-02 | 2.81E-02 |
| Subtotal | 2.19E 00 | 2.11E 07 | 1.54E 05 | 1.46E 05 | 1.32E 05 | 1.13E 05 | 8.31E 04 |
| Totals | 2.19E 00 | 4.19E 07 | 1.54E 05 | 1.46E 05 | 1.32E 05 | 1.13E 05 | 8.31E 04 |

*Note:* Power = 37.50 MW; burnup = 33000 MWD; flux = 3.25E 14 N cm $^{-2}$ sec$^{-1}$; nuclide radioactivity, curies; and bases = one metric ton charge to reactor.

dilute levels in process solutions, and interactions involving them can become significant. Similarly, radiation effects are substantial and radiolysis products become important. Thus, application of the old technology to LWR reprocessing is an oversimplification; there are really some different problems that require better solutions.

## B. Feed Preparation

Production reactor fuel in the U.S., consisting of uranium metal cylinders ("slugs") clad with aluminum ("cans"), was entirely dissolved with nitric acid. The fuel required little, if any, preparation; it was simply charged to the dissolver along with the required

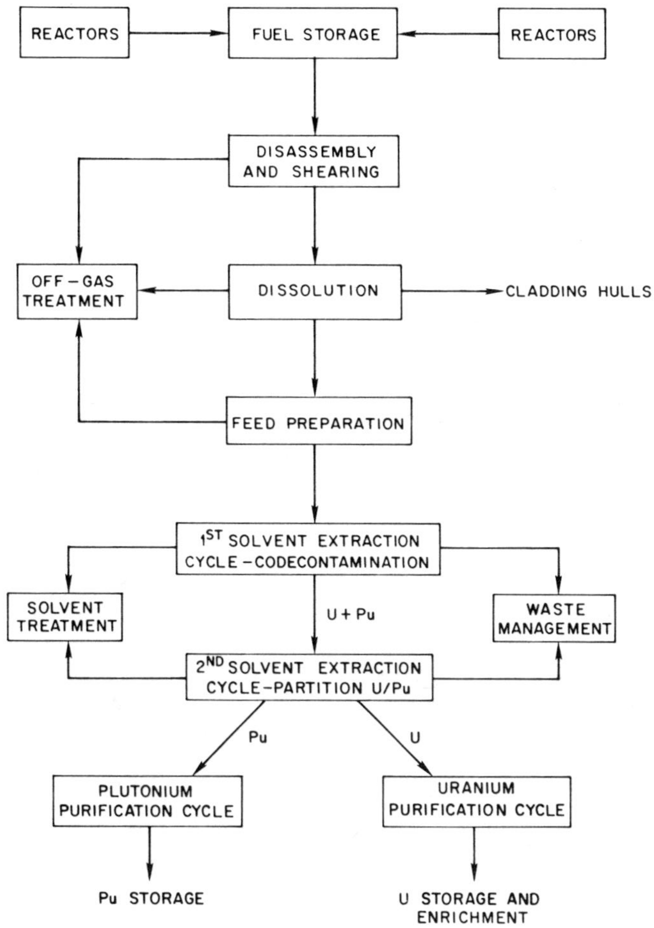


FIGURE 2. Flow sheet for LWR spent fuel reprocessing.

amount of nitric acid. Since nitric acid passivated aluminum it was advantageous to catalyze the dissolution of the cladding by adding a small amount of $Hg(NO_3)_2$. Dissolution of the aluminum can produce hydrogen as well as nitrogen oxides, depending on conditions, and precautions may be necessary to avoid explosive off-gas compositions. Similarly, the magnesium-clad fuel of gas-cooled reactors used in Europe could be dissolved readily or, alternatively, the cladding could be removed mechanically before dissolving the uranium.

Dissolution of uranium metal with excess strong nitric acid yields a mixture of products approximated by Reaction 1; the stoichiometry varies with acidity, giving Reaction 2 for dissolution to final acid concentrations of a few tenths molar.

$$U + 4.5\ HNO_3 \rightarrow UO_2(NO_3)_2 + 1.57\ NO + 0.84\ NO_2 + 0.0005\ N_2O + 0.043\ N_2 + 2.25\ H_2O \qquad (1)$$

$$U + 5.5\ HNO_3 \rightarrow UO_2(NO_3)_2 + 1.25\ NO + 2.25\ NO_2 + 2.75\ H_2O \qquad (2)$$

The addition of oxygen results in oxidation of the NO to $NO_2$ which can be recovered as dilute nitric acid by scrubbing with water. In this case there is little net off-gas, and

**Table 8**
**SUMMARY OF REPROCESSING PLANTS FOR NATURAL AND SLIGHTLY ENRICHED URANIUM FUEL[a]**

| Location[b] | Date of startup | Capacity ($10^3$ kg/year) | Type of contact[c] | Flow scheme[d] (cycle) 1 2 3 | Notes |
|---|---|---|---|---|---|
| Hanford, Wash. | 1956 | ? | Pulse | 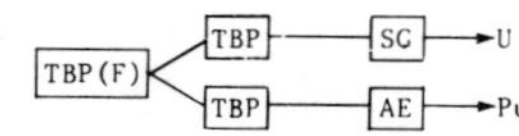 | Initial operation with TBP co-decontamination first cycle |
| Unknown, U.S.S.R. | ? | ? | Mix set | 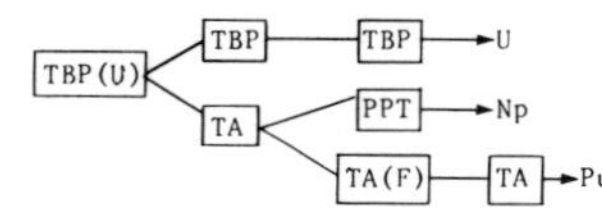 | |
| Windscale II, Seascale, England | 1964 | 2000 metal<br>400 oxide | Mix set | 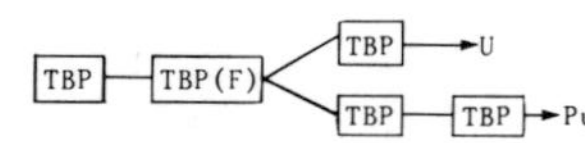 | Used Butex codecontamination first cycle for LWR fuels |
| Marcoule, France | 1960 | 1200 | Mix set<br>pulse | 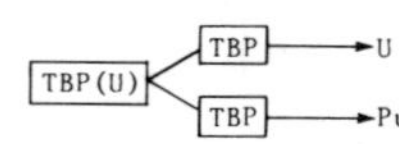 | Other TBP concentrations used also |
| Savannah River, S.C. | 1954 | 4000 | Mix set<br>cent | 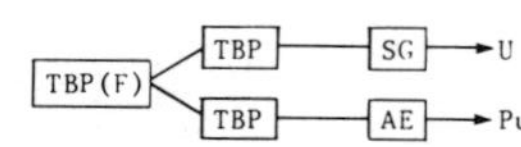 | Centrifugal contactors in first cycle since 1966 |
| Trombay, India | 1965 | 60 | Pulse | 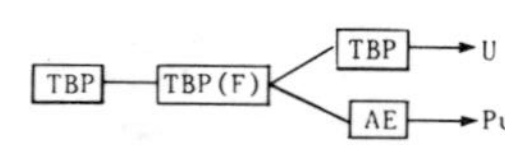 | |
| West Valley, N.Y. (NFS) | 1966 | 360 | Pulse | 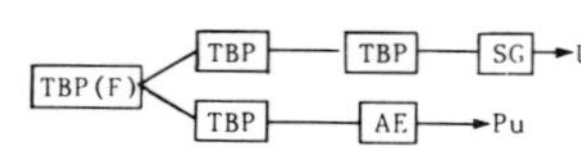 | Shut down indefinitely |
| Mol, Belgium (Eurochemic) | 1966 | 60 | Pulse | 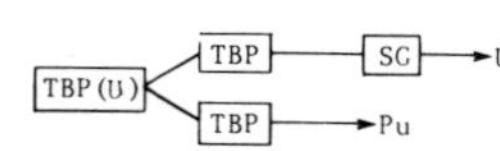 | No operation since 1974 |
| Cap de la Hague, France | 1966 | 800 | Mix set<br>cent | 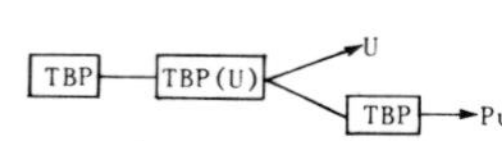 | UP-2 undergoing hot tests |

| | | | | | |
|---|---|---|---|---|---|
| Karlsruhe FRD (WAK) | 1970 | 40 | Mix set | 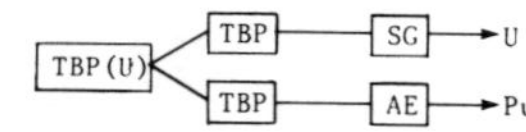 | |
| Tokai Mura, Japan | 1978 | 200 | Mix set | 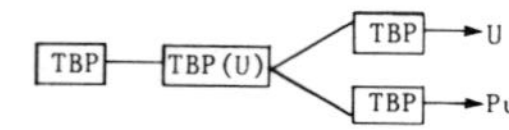 | |
| Morris, Ill. (MFRP) | — | 300 | | 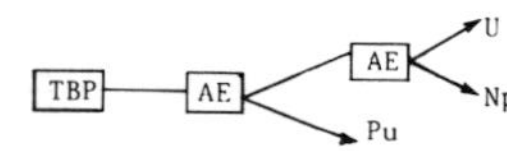 | Operation in present form not planned |
| Barnwell, S.C. (AGNS) | — | 1500 | Pulse cent | 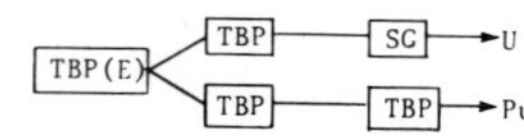 | Operation dependent on regulatory decision |

[a] Early reprocessing of military fuels is omitted (Hanford, Marcoule, Windscale, U.S.S.R.).

[b] Additional plants under construction or planned in India, France, Italy, Japan, UK, and Federal Republic of Germany.

[c] Pulse = pulse column; mix-set = mixer-settlers; cent = centrifugal contactor.

[d] TBP = tributyl phosphate; TA = tertiary amine; AE = anion exchange; SG = silica gel. Reductant for Pu partition in parentheses: (F) = Fe(II); (U) = U(IV); (E) = electrolytic.

the method is referred to as "fumeless" dissolving. In summary, then, the dissolution and feed preparation steps with production reactor fuel were relatively simple and straightforward.

The LWRs, in contrast, use a zirconium alloy cladding that is highly resistant to chemical attack. The fuel assemblies also became quite large and complex, compared to simple metal slugs, and mechanical decladding was not reasonable. It was necessary, therefore, to develop equipment to handle these large fuel subassemblies and break them down physically into smaller parts that could be better dealt with. The cladding had to be dissolved or breached in some way so that the dissolvent could come in contact with the ceramic $UO_2$ fuel. The initial approach was to dissolve the Zircaloy along with the uranium dioxide in a fluoride solution (or, in the case of stainless steel cladding, in dilute aqua regia), by analogy with the complete dissolution of production reactor fuel in nitric acid. For a variety of reasons this chemical method was discarded in favor of mechanical breakup of the subassemblies followed by selective dissolution of the $UO_2$ in nitric acid, the chop-leach method. One important reason for choosing a method that avoided dissolution of the Zircaloy is the steadily increasing incentive to utilize methods that produce less volume of radioactive waste; this has had major effects on reprocessing.

The feed preparation process for Zircaloy-clad $UO_2$ LWR fuels is described in Section III. In general, the goal is to receive the spent fuel, whatever its composition and configuration, and produce a dissolver solution containing all the fertile and fissile material, along with a major fraction of the fission products. This solution should be of a specified composition, regardless of the input fuel, and that composition is established by the separation process. This is, in general, 1 to 1.3 *M* in actinide metal content, 1.5 to 4 *M* in $HNO_3$, and has low but significant concentrations of the accompanying fission products. There is no (or very little) dissolvd cladding, in contrast to the complete dissolution of aluminum-clad production reactor fuel.

Other streams produced during feed preparation, which constitute gas or solid waste streams, must be separately dealt with (see Figure 2). These are the off-gas from fuel handling and dissolution, the leached cladding "hulls", and a solid waste consisting of finely divided insoluble material separated from the dissolver solution. LWR fuels differ from prior fuels with respect to each of these. Because of the higher burnup there is more radioactivity in the off-gas, and there is a much greater amount of insoluble fission products in the dissolver solution. The cladding hulls constitute a waste that did not exist when acid-soluble cladding was used.

Dissolver solutions generally contain small amounts of suspended insoluble materials that may interfere with subsequent separations processes, particularly in solvent extraction, where they tend to collect at the interface between the two phases. With production reactor fuels siliceous solids were commonly present, and removal of these by the usual means, such as filtration or centrifugation, proved difficult. Various methods have been developed for clarifying these solutions, the more successful ones being (1) addition of gelatin to coagulate the solids, or (2) production of $MnO_2$ carrier in the solution by the reaction shown in Equation 3, or both together.

$$3Mn(NO_3)_2 + 2KMnO_4 + 2H_2O \rightarrow 5MnO_2 + 2KNO_3 + 4HNO_3 \qquad (3)$$

The solids from LWR fuel dissolution are different in character, so other means of clarification are required (Section III.C).

Another factor that is particularly significant during fuel dissolution is the possibility of criticality because of the use of enriched uranium. This was not a problem with natural uranium reactors because natural or depleted uranium cannot become critical in ordinary aqueous solutions; however, criticality is possible with the enrichments used in LWR fuels. This problem can be circumvented by appropriate geometric design of the dissolver and subsequent tankage and process equipment (which introduces some degree of awkwardness), or by "poisoning" with neutron absorbers. Heterogeneous neutron absorbers can be used within the tanks, but it has been more common to rely on homogeneous poisoning achieved by dissolving materials with high neutron cross-section in the dissolver solution. The best candidates are boron, cadmium, and gadolinium. There is a chemical problem since it is essential that these poisons not be separated from the fissile material, as by inadvertent precipitation with some other constituent of the solution. In this respect, gadolinium is the preferred poison, although it is also the most expensive.

## C. Chemical Separations Processes

Once the fuel is brought into solution, either with or without simultaneously dissolving the cladding, the chemical separations process can be brought to bear. A number of processes have been studied and demonstrated on the pilot-plant scale, or at least beyond the laboratory scale, including aqueous processes based on precipitation, ion exchange, and solvent extraction, and nonaqueous processes based on volatility of actinide hexafluorides and pyro-chemical methods based on interactions between molten metals, fused salts, and insoluble compounds such as oxides. The first separation process was for the sole purpose of recovering plutonium, and it was based on precipitation. Subsequently, solvent extraction was universally adopted for the primary separation, but individual operations involving ion exchange and precipitation are still frequently included.

Nonaqueous processes have not been adopted, but a large amount of very interesting chemistry has been developed. Although the theoretical basis of some of these processes is attractive, they suffer from engineering problems related primarily to the requirement for remote operation and maintenance of high-temperature systems. The increased emphasis on proliferation resistance in reprocessing has led recently to a reexamination of nonaqueous methods, but they will not be described here because they have not been used and are not under serious consideration for use at this time.

### *1. Precipitation Processes*

The first reprocessing method used on a production basis was the bismuth phosphate precipitation process. It was developed at the Clinton Laboratory (now Oak Ridge National Laboratory) during World War II for separation and recovery of plutonium; the process was used at Hanford, Wash. until 1951. Production reactor fuel differs substantially from power reactor fuel, notably in consisting of aluminum-clad uranium metal "slugs" irradiated to a very low level — only a few percent of that of LWR fuel. The purpose of reprocessing was to recover the plutonium, only, for weapons use, and a separation from radioactive fission products (DF) by a factor of $10^7$ was required for the plutonium product.

It was required that a precipitate be found that would carry plutonium, but at the same time would not precipitate U(VI) and only a small fraction of the fission products should precipitate. It was anticipated that Pu(IV) would form an insoluble phosphate analogous to U(IV), thorium, and zirconium; thus, precipitates of those phosphates were expected to be good carriers for the low-concentration solutions of plutonium. Zirconium was tried first, but the carrying of plutonium was deficient under process conditions (about 1 *M* nitric acid, 0.5 *M* uranyl nitrate, and plutonium on the order of $10^{-4}$ *M*). Other insoluble phosphates were then investigated, including Th(IV), U(IV), Ti(IV), Ce(IV), and Nb(V). For various physical or chemical reasons none was satisfactory.

It is interesting to note that bismuth phosphate was not expected to be a suitable carrier for plutonium. Although bismuth has an insoluble phosphate, its chemistry is substantially different from that of Pu(IV), and its solubility is significant in acid concentrations high enough to prevent uranium phosphate precipitation. However, it was found that digestion at elevated temperatures yielded a fairly complete precipitation of bismuth with nearly quantitative carrying of Pu(IV), and the solid could be readily recovered by filtration or centrifugation. The precipitate could be dissolved with more concentrated $HNO_3$ for further processing.

The processing steps were (1) dissolution of the fuel in nitric acid; (2) adjustment of the valence of plutonium to Pu(IV) with ferrous iron or sodium nitrite; (3) the addition of bismuth and phosphoric acid to precipitate $BiPO_4$ which selectively carried Pu(IV), leaving uranium, aluminum, and most fission products behind; (4) dissolution of the precipitate in nitric acid; (5) oxidation of the Pu(IV) to Pu(VI) with dichromate or bismuthate; and (6) reprecipitation of $BiPO_4$ which again carried most of those impurities precipitated in Step 3 but not the oxidized Pu(VI); and then back to Step 2.[15] The precipitation of $BiPO_4$ with plutonium alternately in the IV and VI oxidation states was repeated for several cycles to attain the desired DF. Uranium phosphate also may precipitate with $BiPO_4$ from nitric acid solutions, but this was prevented by replacement of part of the nitric acid with sulfuric acid, which stablized the uranium in solution by complexation. The initial precipitation carried between 10 and 20% of the fission product activity; the second precipitation, after oxidation of plutonium to Pu(VI), removed most of that activity with very little precipitation of the hexavalent plutonium. It was found useful to add "scavengers" for improving the removal of impurities during the second precipitation of $BiPO_4$ (Step 6 above).

The final part of the process was the concentration and further purification of plutonium by methods such as the well-known $LaF_3$ precipitation, either (1) to remove rare earths by precipitation while plutonium is oxidized to Pu(VI) or (2) to precipitate (and concentrate) plutonium after reduction. After several of these precipitation cycles, DFs of around $10^7$ were obtained with about 95% recovery of the plutonium. Uranium was not recovered, and subsequently work was done to develop precipitation processes for recovery of uranium from the $BiPO_4$ waste. One example is the crystallization of uranyl ammonium phosphate from dilute nitric acid (about 0.1 *M*).[16]

The result of all this is that the first industrial process for recovery of plutonium was based on the precipitation of $BiPO_4$, and that process served successfully for several years until solvent extraction processes were applied. A key feature was the manipulation of the valence of plutonium to change its chemistry in different cycles of the process, and all separations processes to this day utilize such valence changes to produce pure products. Although valence adjustment is not essential (at least in principle) to reprocessing, it permits attainment of the goals of reprocessing — high recovery and high product decontamination at the same time — with much greater ease than would be the case without valence control.

### *2. Ion Exchange Processes*

Ion exchange was considered for fuel reprocessing in the very early days of the Manhattan Project, and in fact it was the alternate to the bismuth phosphate precipitation process (described in the foregoing section) which was developed for use at Hanford. These early processes utilized cation exchange with the organic ion exchange resins newly available at the time. Ion exchange proved invaluable for certain specialized separations and for small-scale work, but it was not adopted for primary fuel reprocessing. The later development of improved resins, and especially of high-capacity anion exchange resins, subsequently provided an alternative process for plutonium recovery and purification. Ion exchange resins have been used for isolation and concentration of plutonium from dilute product streams from solvent extraction processes, and this application has received greatest use on a plant scale.

The early study of ion exchange for plutonium recovery has been summarized by Boyd and Schubert.[17] In 1942 synthetic organic cation exchange resins had just become available commercially, and the strong-acid character of the sulfonic acid exchange groups suggested they might be useful for selectively absorbing higher valence cations such as Pu(IV). Following laboratory studies, a flowsheet was developed in 1943 utilizing deep beds of resin and based on (1) selectively loading Pu(IV) from uranyl nitrate and fission products in nitric acid solution, most of the uranium and fission products passing through the column; (2) selectively eluting the small fraction of uranium which did sorb on the column; (3) selectively eluting certain fission products in an analogous manner; and (4) finally displacing plutonium from the column in a minimum volume. Dilute sulfuric acid was most satisfactory for the uranium elution (Step 2), and dilute oxalic acid removed more than half of the fission product activity, particularly zirconium and niobium (Step 3); both operations are based on complex formation with specific metal ions. The plutonium was finally eluted with either 1.25 $M$ $NaHSO_4$ or 0.4 $M$ oxalic acid. Two such flowsheets were tested on a multi-liter scale before the work was terminated in favor of the bismuth phosphate process. Through ten ion exchange cycles the plutonium recovery was about 97% and the DF averaged 65 for beta-emitters and 85 for gamma-emitters. Because of the low DF and the satisfactory operation of the bismuth phosphate process, ion exchange work on main-line processing was not pursued at that time.

All reprocessing operations to date include separation of a pure plutonium product free of all other metals including uranium. Ion exchange has been applied for the concentration and isolation of the purified plutonium from the dilute product stream from the solvent extraction process.[18-19] Deep-bed cation exchange columns can be loaded with Pu(III) from the dilute nitric acid product stream from solvent extraction; elution with strong nitric acid yields typically a plutonium concentration of 50 to 60 g/$\ell$ in 5 to 6 $M$ $HNO_3$. It is helpful to add a reductant such as sulfamic acid to the eluent to prevent oxidation of Pu(III) to Pu(IV) in the strong nitric acid solution; such oxidation generates nitrogen oxides that form gas bubbles that lead to flow channeling and resultant inefficient elution. Any uranium loaded on the ion exchange column can be eluted with dilute sulfuric acid before the plutonium elution. As in the case of the early cation exchange reprocessing studies discussed above, decontamination is limited (typical beta and gamma DFs are about 3 and 5), and the primary benefit is product concentration.

In contrast to the cation exchange systems, the use of strong base anion exchange offers a substantial product decontamination. The separation is based on the formation of anionic complexes of the actinide metals, e.g., $Pu(IV)(NO_3)_6^{2-}$ in strong nitric acid solutions. A similar neptunium complex also forms. The complex stability is much greater for the tetravalent actinides than for the hexavalent; and it is very small for the trivalent [Pu(III)] and pentavalent [Np(V)]. Some of these anionic complexes have unusually large distribution coefficients in favor of strong base anion exchange resin, quite analogous to their behavior with quaternary amines in solvent extraction. The reactions may be represented by the following equations:

$$Pu^{4+} + n\,NO_3^- \rightleftarrows Pu(NO_3)_n^{(n-4)-} \qquad (4)$$

$$UO_2^{2+} + m\,NO_3^- \rightleftarrows UO_2(NO_3)_m^{(m-2)-} \qquad (5)$$

$$Pu(NO_3)_6^{2-} + 2(resin)NO_3 \rightleftarrows (resin)_2Pu(NO_3)_6 + 2NO_3^- \quad (6)$$

Fission products generally show a very much smaller tendency to form anionic complexes with nitrate, and they are very weakly sorbed by the resin [except for some sorption of Ce(IV), technetium, and certain ruthenium complexes].

The hexanitrate complexes of the tetravalent actinides form in solutions at high nitric acid concentrations, and they increase rapidly as the acid concentration increases between 6 and 10 *M*. At the same time, acid species of the hexanitrate complex are formed at high acidities, as indicated by Reactions 7 and 8:[20]

$$Pu(NO_3)_6^{2-} + H^+ \rightleftarrows HPu(NO_3)_6^- \quad (7)$$

$$HPu(NO_3)_6^- + H^+ \rightleftarrows H_2Pu(NO_3)_6 \quad (8)$$

and these are sorbed much less favorably by the anion exchange resin. There is also competition for resin sites by the nitrate ion. The result is that the distribution coefficient for plutonium increases to a maximum at about 7.7 *M* $HNO_3$ and then decreases at higher concentrations, although less than half the plutonium has been converted to a hexanitrate species at this acidity. Under the same conditions uranium loads much less favorably than plutonium, with a distribution coefficient more than 50 times smaller.[21,22] Formation of the anionic complex is reversed in dilute nitric acid, so the resin can be eluted with 0.35 to 0.75 *M* $HNO_3$.

Unfortunately, the ion exchange process is rather slow, presumably because of the large size of the complex ion. The rate of both loading and elution is increased substantially by increasing the temperature to 40 to 60°C. Product concentrations of 50 to 60 g/ℓ plutonium can be obtained under practical conditions. Plutonium elution can also be accomplished by reduction to Pu(III), but there are serious complications because of gassing, precipitate formation with products of the reduction reaction, introduction of impurities (iron, sulfate), and slow kinetics.

The anion exchange reprocessing concept has been tested with irradiated fuel solutions as well as with partially purified plutonium products. Gamma DFs for plutonium were $2 \times 10^4$ in a single cycle and $4 \times 10^7$ in two cycles; niobium was the limiting activity. This process is also used for the second (or third) cycle of plutonium purification, following Purex solvent extraction.[23] The Midwest Fuel Recovery Plant, in Morris, Ill., included an anion exchange second cycle; this plant was constructed and cold-tested, but not operated with irradiated fuel. Several plants use anion exchange for the purification and concentration of plutonium in the third cycle (Table 8). The same chemical system, anion exchange from strong nitric acid solution, is used for recovery of $^{238}Pu$ and $^{237}Np$ in the special application of producing $^{238}Pu$ for isotopic heat sources.[24]

One cause for concern with anion exchange in nitrate systems is the possibility of a rapid chemical reaction if conditions are not properly controlled. Anion exchange resin loaded with the plutonium nitrate complex contains up to three nitrate ions per exchange site, compared to only one nitrate per exchange site in simple nitrate-form resin. If elevated temperatures are attained within the column, an exothermic reaction can be initiated. The reaction is autocatalytic since it is accelerated by the increased temperature, and the result is an increasingly rapid chemical excursion. Since the actinides generate some heat through radioactive decay, it is necessary to provide for adequate heat removal. Conditions for stable operation have been defined[25] and it is essential that they not be exceeded, as might happen, e.g., during a malfunction or process interruption.

*3. Solvent Extraction Processes*

Although the bismuth phosphate process served adequately for the first several years of fuel reprocessing, it suffered from certain inherent disadvantages, including excessive mechanical manipulations, the necessity of batch operation, the generation of large waste volumes, and the nonrecovery of uranium. There was strong incentive to apply a continuous method, preferably all-liquid, which could alleviate these problems. The study of solvent extraction processes was emphasized in the very early days of the Manhattan Project, and a tremendous variety of such systems has been examined in the years since. Several have been used on a plant scale, but at the present time the Purex process has displaced all others.

Uranium, neptunium, and plutonium form a number of nitrate complexes, depending primarily on the metal valence, the nitrate ion concentration, and the acid concentration. In addition, neutral complexes can be formed which include the metal and nitrate ions along with a variety of organic compounds which are relatively insoluble in water; the important fact is that these complexes may be soluble in an organic liquid. The organic compounds include those containing oxygen, nitrogen, phosphorus, sulfur, or some other element in addition to carbon and hydrogen; examples are ethers, ketones, amines, sulfoxides, and several types of phosphorus compounds. It is interesting that the extraction of uranium from aqueous solutions by diethyl ether had been reported just a hundred years before, in 1842.[26]

These organic "extractants" may be used alone or as a solution in a "diluent". Diluents may be paraffinic or aromatic hydrocarbons chosen primarily for their physical properties (flash point, viscosity, etc.) or, in some cases, halogenated compounds such as carbon tetrachloride; the diluent can strongly affect the properties of the extractant. Most of the early work utilized extractants such as ethers and ketones without a diluent, but, subsequently, stronger extractants were used (amines, phosphates) in solutions containing generally not more than 1 *M* concentration of the extractant. Out of this came a vast amount of chemical information involving thousands of publications covering innumerable combinations of metals and extraction systems. There are comprehensive reviews of this subject, with extensive bibliographies, for those requiring more information.[11,27]

Probably the first solvent extraction process to be applied for plant separations of plutonium was the "Trigly" process at Chalk River, Ontario.[28] Trigly, triglycol dichloride, was used for extraction in the first cycle, and additional plutonium purification cycles utilized methyl isobutylketone and thenoyl trifluoroacetone (TTA). In addition to extraction cycles with these three different extractants, there were two different precipitation-dissolution cycles. Because of its complexity, and the fact that uranium was not recovered, this process was used for only a relatively short time.

Other solvent extraction processes that have seen extensive use are Butex, Redox, Purex, and amine extractions. The Purex process, based on tributylphosphate (TBP) in a diluent, is the standard method used world-wide now, and it is described in Chapter 4. Certain basic properties of countercurrent extraction processes are also described there; this section will be limited to a brief summary of those processes which have been used industrially in the past. The Redox process used undiluted hexone (methylisobutyl ketone); Hexone became a name applied to the process. The Butex process used dibutyl carbitol ($\beta\beta'$ -dibutoxy diethyl ether) for the first separation cycle and for uranium purification, but 20% TBP, similar to Purex, for plutonium purification. Thus, organic compounds containing oxygen, nitrogen, or phosphorus have been used industrially. Laboratory studies have been made and basic chemical data have been accumulated for a large number of possible solvents, but, aside from the processes named above, none of these has shown enough superiority to warrant application. Only the industrial processes will be summarized here.

The Redox process was applied in 1951 at Hanford, Wash., and it continued to be used for some 15 years.[29] It is a continuous countercurrent solvent extraction process, and was the first to purify and recover both plutonium and uranium simultaneously. The name Redox was derived from the use of oxidation and reduction of plutonium to achieve the desired separations, but the same concept had, of course, been used in the original bismuth phosphate process. Hexone extracts both Pu(VI) and U(VI) very strongly, Pu(IV) only moderately, and Pu(III) and most fission products only slightly. The basis of the process is oxidation of the feed solution so both uranium and plutonium are hexavalent, coextraction of both of them into hexone, and selective stripping (partitioning) of plutonium into a reducing solution of high nitrate concentration, to form inextractable Pu(III) while salting the uranyl ion into the organic phase. Uranium is finally stripped with water in a separate contactor. Alternatively, uranium and plutonium can be co-stripped in the hexavalent state with dilute nitric acid containing an oxidant, and then partitioned after the second extraction. Both uranium and plutonium are decontaminated further in separate purification cycles; in the plutonium cycle the plutonium is oxidized to Pu(VI) before extraction and reduced to Pu(III) for stripping to provide additional separation from uranium.

In practice, dissolution was carried out to yield a feed solution about 2 *M* in uranyl nitrate and 0.3 *M* in residual nitric acid. Plutonium was oxidized with excess $Na_2Cr_2O_7$ according to Reaction 9:

$$3Pu^{4+} + Cr_2O_7^{2-} + 2H^+ \rightarrow 3PuO_2^{2+} + 2Cr^{3+} + H_2O \qquad (9)$$

The dichromate concentration was typically 0.1 *M*, and the feed solution was either 0.3 *M* nitric acid (acid flowsheet) or it was adjusted to about 0.2 *M* acid deficient by partial neutralization of aluminum nitrate with NaOH (acid-deficient flowsheet). In both cases additional $Al(NO_3)_3$ was added via the scrub solution, and the high salt concentration [uranyl nitrate in part of the column and $Al(NO_3)_3$ in all of it] assured a high distribution coefficient for uranium and plutonium. Plutonium stripping was essentially identical for both flowsheets, utilizing ferrous sulfamate in 1.3 *M* $Al(NO_3)_3$. The ferrous ion reduces Pu(VI) according to Reaction 10

$$PuO_2^{2+} + 3Fe^{2+} + 4H^+ \rightarrow Pu^{3+} + 3Fe^{3+} + 2H_2O \qquad (10)$$

The purpose of the sulfamate is to destroy any nitrite ion (Reaction 11) which, if present, will autocatalytically oxidize ferrous iron to ferric, and Pu(III) to Pu(IV), preventing reduction of plutonium.

$$NH_2SO_3^- + NO_2^- \rightarrow N_2 + SO_4^{2-} + H_2O \qquad (11)$$

For the most part, the process performed well over a number of years. The overall DF for two cycles averaged about $10^8$ for plutonium and $10^7$ for uranium with the acid-deficient flowsheet; with the acid flowsheet DFs were somewhat lower, especially for uranium. In general, the acid flowsheet gave better recovery of plutonium and, to a lesser extent, of uranium than did the acid-deficient flowsheet, but both were quite good in this respect. The hexone solvent could be readily purified for subsequent reuse by scrubbing with a solution of $Na_2CO_3$, and extended solvent recycle was utilized.

Major disadvantages of the process related to the low flash point and high toxicity of the hexone solvent and the high salt content of the waste. This last results from a property of the solvent: its instability to high concentrations of nitric acid; this required use of a neutral salt (aluminum nitrate) to "salt out" the uranium and plutonium from the aqueous phase, instead of nitric acid as in the Purex process. The aqueous waste stream was about 1.3 *M* in aluminum (in addition to iron, chromium, and sulfate from reagents added for valence control), which resulted in a large volume of high-level waste. Since development of the Redox process, the importance of reducing the permanent waste volume has been increasingly emphasized. This favors use of an extractant that can utilize nitric acid for the salting agent, rather than a neutral salt.

The Butex process was used between 1952 and 1964 at the Windscale Plant in Great Britain.[30] It was somewhat similar to the Redox process in the first cycle, utilizing a polyether rather than a ketone, in a continuous countercurrent process. It had the distinct advantage that uranium and plutonium could be extracted from nitric acid solutions without the addition of a neutral salt to the aqueous phase; this permits the high-level aqueous waste to be concentrated and stored in a much smaller volume. A major disadvantage of the Butex cycle was the rather low DF for ruthenium. The plutonium purification cycles utilized 20% TBP in kerosene to provide better ruthenium decontamination, but the second cycle for uranium purification used Butex again.

Although the chemistry of the two processes is often considered to be similar, it actually differs in several significant respects. As already mentioned, Butex permits operation with nitric acid salting during extraction, rather than with a neutral salt such as the $Al(NO_3)_3$ used in the Redox process. The Butex feed was 3.2 $M\,HNO_3$ and about 300 g/ℓ uranium, and the organic/aqueous phase ratio was adjusted to maintain the uranium concentration below 150 g/ℓ. The relatively low uranium concentration, compared to Redox, is necessary to prevent precipitation of compounds of the type $UO_2(NO_3)_2(H_2O)_3 \cdot C_{12}H_{26}O_3$. The high acid concentration favors formation of the trinitrato nitrosylruthenium complex which is relatively extractable, thereby contributing to the poor ruthenium DF and partially offsetting the advantage of acid salting. The DF for zirconium, in contrast to ruthenium, was better with Butex than Redox, probably at least in part because of the higher acidity.

Butex is a better extractant for Pu(IV) than for Pu(VI) and U(VI), so the oxidation of plutonium to the hexavalent state (required in Redox) was not necessary. Thus, Butex utilized the +3 and +4 states of plutonium, whereas Redox depended primarily on the +3 and +6 states. In both cases, plutonium partitioning was achieved by reduction to the inextractable trivalent state with ferrous sulfamate, but the aqueous salting agent (required to favor uranium extraction into the organic during plutonium stripping) as $Al(NO_3)_3$ in Redox and $NH_4NO_3$ in Butex. The combination of a Butex cycle followed by a TBP cycle gained the advantages of both, and at that time there was much consideration of dual and more complex cycles selected to compensate for each other's weaknesses (the Trigly process, for example). More recently, the trend has been to avoid use of more than one solvent, if possible, because of practical operating problems in keeping the solvents separate.

Extraction with organic amines has received much study, and processes have been operated industrially both in the U.S.S.R. and the French La Hague Plant. There are a large number of amines with widely varying properties, depending on the number of organic groups bonded to the nitrogen atom one to four and the nature of those groups with respect to factors such as branching and aromatic character. Much information and some possible applications were summarized by Coleman.[31,32] Most reprocessing interest has been restricted to tertiary amines such as trilauryl amine, dissolved in hydrocarbon diluents, but primary amines are even stronger extractants. Amines are exceptionally strong extractants for tetravalent actinides such as Pu(IV) from nitric acid solutions, but rather poor for U(VI), so they are better suited for further purification of the plutonium product from some other first-cycle primary separation such as Purex than for the "main-line" processing of both uranium and plutonium. The

French application, for example, was for plutonium purification following a Purex first cycle. An added advantage of the amine cycle for plutonium purification is that the plutonium concentration can be substantially increased over that in the Purex plutonium product, thereby reducing the need for intercycle plutonium concentration processes. Amines are also useful for recovery of plutonium from various waste or recycle streams, and of neptunium from wastes, because of the exceptionally high extractability of the tetravalent state.

Amines are usually used in concentrations of 0.1 to 0.3 *M* in either aromatic (diethylbenzene) or aliphatic (kerosene) diluents, and the nature of the diluent is quite important. Third-phase formation (separation of a relatively concentrated phase containing the metal and amine but little diluent, either liquid or solid, but insoluble in either initial phase) can be a problem. It is common practice to control this by adding a "modifier", such as a few percent of a long-chain alcohol to the solvent. An unfavorable property of amine extraction with respect to plant applications is the relatively slow kinetics of these separations, compared to TBP, ketones, etc. The extraction is somewhat slower with amines, and stripping is considerably slower, so that contact times must be relatively long (for example, stage residence times of several minutes are required). The ion exchange separation with anion exchange resin has the same problem, and it derives from the nature of the large complex ions which are the species involved.

The mechanism for amine extraction is frequently stated to be ion exchange, by analogy with anion resin exchange, as in Equation 12:

$$mR_3NHNO_3 + Pu(IV)\,(NO_3)_{4+m}^{m-} \rightleftarrows (R_3NH)_mPu(IV)\,(NO_3)_{4+m} + mNO_3^- \qquad (12)$$

However, this equilibrium is formally indistinguishable from that for adduct formation between the amine salt and a neutral complex, indicated by Equation 13:

$$mR_3NHNO_3 + Pu(IV)\,(NO_3)_4 \rightleftarrows (R_3NH)_mPu\,(IV)\,(NO_3)_{4+m} \qquad (13)$$

Thus, it is impossible to choose between these mechanisms on the basis of equilibrium measurements since the two equations are thermodynamically equivalent. Both mechanisms, ion exchange and adduct formation, are useful for physical visualization of the extraction process.

Like amines, phosphorous-containing extractants demonstrate a wide variety of properties. The Purex process, which uses TBP, is discussed at length in Chapter 4, so little will be said about these systems here. The common feature is the phosphoryl group, $\equiv P{=}O$, and a number of compounds have been studied, including neutral extractants [such as phosphates $(RO)_3PO$, phosphonates $(RO)_2RPO$, phosphinates $(RO)R_2PO$, and phosphine oxides $R_3PO$, where R represents an organic group] and acidic phosphates (such as alkyl and dialkyl phosphoric acids). Many of these are rather strong extractants. This can be an advantage since favorable distribution coefficients can be obtained with lower concentrations of nitric acid or salting agent in the aqueous phase, and DFs are better in at least some cases. However, it can also be a disadvantage since stripping may be quite difficult, even requiring complexing agents in the aqueous phase to compete with the organic extractant. The combination of properties displayed by TBP is a very favorable compromise, and the large amount of satisfactory experience mitigates against any radical change. There is some interest presently, however, in phosphates similar to TBP except for substitution of larger groups, such as hexyl or 2-ethylhexyl, for the butyl group, a change which should decrease the solubility of water in the extractant.

Although the tendency is to focus on the extractant, the nature of the diluent is also a very important consideration. For the most part, three general classes of diluents are commonly used: namely (1) straight-chain hydrocarbons such as n-dodecane or mixtures of paraffins (kerosene), (2) aromatics such as xylene or diethylbenzene, and (3) halogenated compounds such as $CCl_4$. All have been used in industrial processes. In the last case the organic phase is more dense than the aqueous, so the direction of phase separation is reversed. Several properties of the aqueous-organic two-phase system, such as distribution coefficients and third-phase formation, are dependent on the nature of the diluent as well as the extractant. The diluent also may be degraded and enter into reaction with solvent degradation products; and there is evidence that high-molecular-weight degradation products, derived partially from the diluent as well as the extractant, such as butyllaurylphosphate, accumulate in the system because they are not removed by the processes commonly used to remove simpler degradation products (mono or dibutyl phosphate).[33]

The chemical bond is clearly limited in its ability to withstand radiation damage; some compounds are more resistant than others, but none is immune. The important factors, then, are not only resistance to radiation damage but, even more, the degree to which the products of the damage interfere with fuel reprocessing objectives. In the second instance, phosphorus-containing compounds (e.g., TBP) are not apparently superior; their degradation products are deleterious since they are extremely strong extractants and some of them are not readily removed during solvent regeneration processes. On the other hand, the oxygen-containing compounds (ketones, ethers) form degradation products that appear to be relatively innocuous and easily removed. The amines include such a variety of compounds that generalization may not be advisable, but some are not degraded to the extent that TBP is.

In reviewing the history of the development of solvent extraction, an unmistakable fact is that a TBP-based process (Purex) has monopolized the fuel reprocessing industry. Purex plants have been constructed in the U.S., U.S.S.R., Japan, France, England, India, Belgium, and West Germany. The Purex process was really developed for recovering plutonium from low-burnup fuel from military production reactors, and it was later adapted for higher-burnup power reactor fuels. For good reasons, Purex displaced Redox and Butex, just as they had earlier displaced the bismuth phosphate process. It would now require some really extraordinary circumstance to cause the displacement of Purex.

It is conceivable that events may develop in the future that will change the criteria on which processes are selected. Examples of technical reasons could be reduced limits for effluents from reprocessing plants or, if sufficiently strong incentive develops to process fuel after higher burnup and shorter decay time, the substantially higher radiation level and correspondingly increased radiation damage effects. There may also be major changes in the economic or political ground rules regarding reprocessing, based on factors such as economics, resource utilization, or the growth rate of nuclear power.

Thus, it is useful to keep in mind the advantages and disadvantages of all these separations systems, even though one of them so completely dominates the industry now, because changes in requirements may occur that could shift the choice to a different system — perhaps one less advantageous under ideal conditions, but less susceptible to interference or degradation under practical conditions that occur industrially. The tremendous range of solvent properties that has already been delineated provides an excellent base for selection of extraction systems. However, it must be remembered that the Purex process is so firmly established that some really definitive new factor would be required to allow a challenge to Purex.

## III. FEED PREPARATION CHEMISTRY

The two basic types of LWRs, pressurized water and boiling water reactors, utilize the same materials in the fuel subassemblies, but the physical configurations are somewhat different (Section I). The fuel is normally $UO_2$ containing about 3% $^{235}U$, in the form of cylindrical pellets enclosed inside cladding tubes. Mixed oxide fuel (MOX), containing 3 to 4% $PuO_2$ for fissile material, with the remainder natural or depleted $UO_2$, has been considered for eventual use (plutonium recycle) and small quantities have been irradiated for test purposes. In all cases the cladding is a zirconium alloy, such as Zircaloy-2 or Zircaloy-4. There are also other materials associated with the subassemblies, such as end fittings, grid spacers, and perhaps shrouds and poison elements.

The individual fuel rods are about 4 m long in large LWRs (1000 to 1300 MWe). In PWRs the rods are about 1 cm in diameter, and in BWRs they are larger, about 1.3 cm. In PWRs subassemblies are generally larger and contain more fuel rods, such as a 15 × 15 square array, compared to 7 × 7 for a BWR. The cladding thickness is somewhat under 1 mm — less for the PWR than the BWR. The starting material for reprocessing, then, is a rather long assembly of Zircaloy tubes filled with ceramic oxide, weighing a few hundred kilograms in all. Both ends consist of sturdy metal fixtures that do not contain fuel. These, and other structural components of the subassembly, can interfere with reprocessing operations; therefore, there is incentive to remove the end-fittings and other hardware prior to reprocessing.

### A. Fuel Pretreatment

Fuel pretreatment involves the steps required to prepare the fuel for dissolution. A necessary step is to expose the oxide fuel so that it can be attacked by nitric acid, since the fuel is completely enclosed by cladding which is highly resistant to nitric acid. Other operations have been considered, but not applied to date, for purposes of removing certain radioactive materials which might be removed more easily prior to dissolution, or to modify the fuel in some way to benefit subsequent processing operations.

#### *1. Fuel Degradation*

The essential step is to expose the fuel material (primarily $UO_2$, but containing also about 3% fission products and a smaller fraction of plutonium) so that it can be dissolved in nitric acid. Several options are available, including complete dissolution of cladding plus fuel in hydrofluoric and nitric acids, selective dissolution of cladding followed by fuel dissolution, or mechanical break-up of the subassembly to expose the fuel which can then be selectively dissolved. This last can be done either directly or after a chemical treatment to make mechanical break-up easier (clad embrittlement). Although all these options may be technically feasible, the universal approach is direct mechanical "chopping" with a large hydraulic shear.

A number of methods have been investigated for dissolving fuels containing zirconium and certain alloys under consideration for power reactor fuels.[34] Since zirconium forms a volatile tetrachloride, a chlorination or hydrochlorination reaction could be used to remove cladding from oxide fuel. Both liquid-phase and gas-phase processes were developed under the name Zircex.[35] In the gas-phase process anhydrous HCl gas was used at 300 to 600°C to volatilize zirconium as $ZrCl_4$ either from the cladding, from a metal phase of a cermet, or from an alloy (U-Zr) fuel. Since $UO_2$ is not chlorinated by hydrogen chloride, subsequent dissolution of the oxide yields a dissolver solution nearly free of chloride. In the liquid-phase reaction a low-melting composition such as $AlCl_3 \cdot NH_4Cl$ is used during hydrochlorination to dissolve the $ZrCl_4$ product and remove the heat of reaction. The reaction is carried out at about 400°C, at which temperature $ZrCl_4$ volatilizes; subsequently the $AlCl_3 \cdot NH_4Cl$ salt is recovered by distillation, leaving the less volatile fuel material behind. There was substantial interest in these processes in the late 1950s, especially for metal alloy or cermet fuels. However, the engineering and chemical problems associated with the Zircex process, the evolution and adoption of Zircaloy-clad oxide fuels, and the demonstration of other methods suitable for such fuels resulted in the abandonment of this approach.

The only chemical approach that has been used on a plant scale is the Zirflex process, which is based on dissolution of zirconium in a solution of $NH_4NO_3$ and $NH_4F$.[36] This process has been used at the Idaho Chemical Processing Plant. Ammonium fluoride can be used for dissolution, but large amounts of hydrogen and ammonia gas are evolved; Equation 14 is an approximation of the stoichiometry.

$$Zr(s) + 6NH_4F \rightarrow ZrF_6^{2-} + 2NH_4^+ + 4NH_3(g) + 2H_2(g) \tag{14}$$

With addition of $NH_4NO_3$, hydrogen production is suppressed, and more ammonia is generated; Equation 15 approximates the stoichiometry.

$$Zr(s) + 6NH_4F + 0.5\ NH_4NO_3 \rightarrow ZrF_6^{2-} + 2NH_4^+ + 5NH_3(g) + 1.5\ H_2O \tag{15}$$

In both cases the product is ammonium fluozirconate, with each zirconium atom essentially complexing six fluorides. The Zircaloy dissolves readily, along with a small fraction of the fuel (about 1%). The residual oxide fuel is then dissolved separately in nitric acid. This approach is not viewed with favor for power reactor fuels because of problems introduced by the large amount of fluoride (corrosion during waste concentration, solidification, and storage is particularly serious), the large amount of ammonia released into the off-gas, and the loss of fuel to the decladding solution. Clearly, the large volume of solution containing zirconium, fluoride, and ammonium nitrate poses a more challenging waste disposal problem than does the Zircaloy metal cladding itself. Thus, processes that avoid clad dissolution are favored.

A variety of chemical methods can be applied to weaken and embrittle the cladding so that it may be more easily broken up. These include reactions with gases (such as hydrogen to form brittle zirconium hydride, or with air at 800 to 1000°C which causes the cladding to expand away from the contained fuel, become brittle, and finally break up into insoluble granular material) and reactions with liquid metals (such as zinc at high temperature which causes embrittlement by entering into compound formation with zirconium or by dissolution and penetration of grain boundaries). The primary disadvantage of such processes is the difficulty of remote processing at high temperatures, but this could be handled if necessary.

The fact of the matter is that simple mechanical shearing has proved to be quite satisfactory. Large hydraulic shears are used to cut repeatedly through the whole subassembly at once. The subassembly is advanced a few centimeters between successive cuts, so clad pieces of fuel rods are produced in the range of 2 to 10 cm long. This method has been used in the U.S. (NFS plant), France (La Hague), and elsewhere, and is included in new plant designs in these and other countries. An alternative is to break down the subassembly into individual rods, and shear these singly or a few at a time. This permits removal of the extraneous structural components of the subassembly, but at the expense of the difficult remote operations of disassembly.

A primary concern with shearing is the possibility that the cladding may be deformed in such a way that the ends of some pieces will be pinched shut, thereby sealing in the contained fuel and interfering with its subsequent dissolution. In practice, complete closing of the ends has not been a problem, although some degree of such a "pillowing" effect is commonly observed. A secondary concern is that parts of the subassembly other than the fuel rods (notably grid spacers) may jam the shear or, since they could yield pieces relatively large in two dimensions, that they might interfere with subsequent process operations involving solids transport. Again, this has not been a serious problem and, with proper attention to design, it need not be one.

The present and anticipated practice, then, is to use mechanical shearing to break up the fuel assembly into pieces of individual fuel rods a few centimeters long, either with or without prior removal of some of the fuel subassembly components that do not contain fuel. This is the first part of what is called the "chop-leach" method. The Zirflex process is a demonstrated technology that could be applied to LWR fuels, also. Other chemical means are technically feasible, although they generally exhibit engineering problems; none has progressed beyond laboratory-scale study.

*2. Tritium Removal*

There has been substantial interest in the U.S., but much less to date in other countries, in the removal of a major fraction of the tritium prior to fuel dissolution. The incentive is to reduce tritium release to the environment. If not so removed, tritium contained in the oxide fuel enters the dissolver solution as tritiated water (HTO) at the time of fuel dissolution, and thereby becomes part of a high-volume stream. In all operations to date, no attempt has been made to control tritium and it has been carried into all process streams, thereby increasing the volume even more. There are proposals to maintain the tritium in the aqueous stream of the first cycle of reprocessing, and to recover and concentrate it from that stream by an isotopic separation process.[37] However, if tritium is to be recovered and concentrated for storage for decay, there is incentive to do so before the fuel dissolution step which dilutes it with a large volume of water.

Various small fractions of several radioactive gases are released when the fuel cladding is ruptured. These include krypton, iodine, $^{14}C$, and $^{3}H$. In any case, the off-gas must be treated appropriately to control release of these isotopes, as discussed in the next chapter, so there might be advantage to routing more (or all) of them to the off-gas. In practice, tritium is the only isotope that has been quantitatively removed into the off-gas.

The quantities of some volatile radioisotopes found in the fuel are subject to uncertainty, partly because of their formation mechanisms and partly because of the possibility of reaction with the zirconium cladding. Only krypton behavior is really straightforward, since it is formed as a fission product and it is not chemically reactive; it is released quantitatively during fuel dissolution. Iodine is also a fission product; it is predominantly released during fuel dissolution, as will be discussed subsequently. Tritium results both from fission and from neutron capture by lithium impurity. The lithium content is not generally determined for reactor fuel or cladding, and it is certainly low. However, 1 ppm lithium would contribute an amount of tritium comparable to that from fission. It is well known that zirconium absorbs tritium (hydrogen) at reactor temperatures. It is observed that the cladding contains about as much tritium as the fuel,[38] and less tritium is found in the fuel than should be produced, based on physics calculations (about 300 Ci/MTU).[39] Accordingly, it is likely that a substantial fraction of the tritium diffuses into the cladding.

Carbon-14 is produced primarily by an (n,p) reaction on $^{14}N$, and nitrogen is an impurity in fuel in amounts ranging from a few to hundreds of parts per million, depending on fabrication methods. A secondary production route is the $(n,\alpha)$ reaction on $^{17}O$. In a number of measurements with different LWR fuels, usually of unknown initial nitrogen content, the amount of $^{14}C$ found (about 0.36 Ci/MTU) generally corresponded to that expected if the fuel contained initially about 10 ppm nitrogen (within a factor of two).[39] As in the case of tritium, nitrogen could react with the Zircaloy cladding, with the result that most of the $^{14}C$ would be produced in the cladding and not in the fuel.

Thermodynamic considerations and very limited data suggest that $H_2$, $H_2O$, and $O_2$ would react rapidly with zirconium in the temperature range of 300 to 400°C (which is the operating temperature of the cladding) and that $N_2$ would react more slowly but still rapidly compared to the fuel exposure time.[40] The transport of these materials through the fuel and to the cladding is a more complex problem because of the temperature gradient in the fuel. Depending on operating conditions, somewhat over one half of the fuel is at a temperature above 1000°C, but that fuel nearest the cladding is cooler. Outgassing of $UO_2$ is rapid at temperatures above 1000°C, but the transport rate would decrease as the gases approached the cladding. In any case, the "gettering" action of zirconium might decrease substantially the amounts of these volatile materials in the fuel and, as a result, decrease the amount released during reprocessing.

The voloxidation process has been developed specifically to remove tritium from the fuel prior to dissolution.[41,42] The chemical basis of the process is oxidation of the $UO_2$ in the sheared fuel to form $U_3O_8$ by reaction with air or oxygen at 450 to 600°C for a period of a few hours. During this reaction tritium is released from the fuel to the extent of 99% or more, and smaller fractions of other volatile radioisotopes are also released, notably up to 6% of the krypton, 50% of the $^{14}C$, but only about 1% of the iodine. The volatile isotopes may be selectively trapped or condensed from the off-gas to yield a low-volume waste suitable for storage or further treatment. At the same time, because of the phase change, the $UO_2$ is transformed from relatively large fragments and pellets in the cladding to a $U_3O_8$ powder, and nearly all (>99.9%) of the powder may be mechanically separated from the cladding, which is unaffected. An interesting observation is that oxygen consumption and the release of tritium and krypton are not simple functions of time, but the rate curves often show a double peak. The minimum between the two peaks occurs when the fuel has reached the $U_4O_9$ composition. Thus, the process appears to be enhanced by the phase changes occurring in the solid, first from $UO_2$ to $U_4O_9$, and then to $U_3O_8$; the change in form of the fuel (powdering) occurs only at the end, when $U_3O_8$ is formed.

There has been recent study of the removal of tritium by nonoxidative heat treatments which avoid the phase change and powdering of $UO_2$ fuels. This approach is also appropriate for fuel compositions that do not oxidize or undergo phase changes, such as $ThO_2$-based fuels or $UO_2$ with high $PuO_2$ content. Hot cell tests have demonstrated tritium removal in a few hours at temperatures of 900 to 1000°C. The rate of removal is much more rapid from fragmented fuel than from clad sections containing dense fuel, as would be expected for a diffusion-controlled process. The appearance of the fuel is not changed, in contrast to voloxidation which always yields a powder. At these high temperatures, oxidation must be avoided to prevent release into the off-gas of certain fission products that form volatile oxides, notably ruthenium and cesium. It may be noted that there is the possibility of fractionation of certain fission products such as these, should that be desired, by controlling the oxygen activity (or the oxygen/metal ratio of the fuel). Also, such heat treatments, including voloxidation, increase the fraction of noble metal fission products which is insoluble in nitric acid, discussed below.

### B. Aqueous Dissolution

After the integrity of the cladding has been destroyed in a head-end process, most of the exposed oxide fuel material is readily dissolved in nitric acid. Some fission products are gaseous or volatile, and a fraction of certain fission products is insoluble, but the largest part dissolves in nitric acid. In practice, dissolution has been conducted in a batch or semi-batch mode, but there is substantial interest in continuous dissolution for large plants projected for the future.

The stoichiometry of the dissolution reaction varies with the oxidation state of the fuel, the nitric acid concentration, and some other variables, but for $UO_2$ it can be approximated by Equation 16.

$$2UO_2 + 6HNO_3 \rightarrow 2UO_2(NO_3)_2 + NO_2 + NO + 3H_2O \tag{16}$$

Typically, the acid is initially about 7 *M* $HNO_3$ (possibly containing a soluble neutron poison such as gadolinium nitrate) and the dissolver product is about 3 *M* in $HNO_3$ and 1.3 *M* in heavy metal (uranium plus plutonium) nitrates. The aqueous composition during dissolution depends on the type of dissolver used, approximating that of the final product for a batch-continuous dissolver, decreasing from 7 *M* $HNO_3$ initially to the product composition for a simple batch dissolver, and being different in different parts of a countercurrent continuous dissolver. The stoichiometry varies somewhat for these different cases.

Dissolution occurs readily with urania fuels, and it is normally carried out at a temperature in the range of 90°C to boiling, although appreciable dissolution rates are obtained at lower temperatures, especially with finely divided fuel. The time required for dissolution is typically one to a very few hours. Frequently, dissolution is followed by a "soak" at boiling or near boiling to assure completion of the reaction; this can affect the chemical state of several elements. It is also common practice to bubble air through the solution, or to distill away some water, or both, in order to favor transfer of iodine out of the dissolver solution and into the off-gas. The air oxidizes NO to $NO_2$; $NO_2$, in equilibrium with $N_2O_4$, hydrolyzes to form nitric and nitrous acids, according to Reactions 17, 18, and 19:

$$2NO(g) + O_2(g) \rightarrow 2NO_2(g) \tag{17}$$

$$2NO_2(g) \rightleftarrows N_2O_4(g) \tag{18}$$

$$N_2O_4(g) + H_2O(l) \rightarrow HNO_3(l) + HNO_2(l) \tag{19}$$

Nitrous acid decomposes according to Reaction 20:

$$3HNO_2(l) \rightleftarrows HNO_3(l) + 2NO(g) + H_2O(l) \tag{20}$$

The result is that there will be an appreciable concentration of nitrous acid in the dissolver solution. The chemical environment, then, is a moderately high nitric acid concentration (3 to 5 *M*) along with some nitrite ion, some NO, $N_2O_4$, and probably air in the gas phase, and an elevated temperature, possibly boiling.

Following dissolution, the solution is exposed to a variable not ordinarily encountered in chemical processes: an extremely intense radiation field. This causes radiolysis of water and nitric acid with the obvious result of continuous production of nitrogen oxides leading to nitrous acid in the aqueous phase. There is a much greater production of nitrogen oxides during the dissolution reaction itself, but these may be removed by a post-dissolution boil-off or air sparge. In contrast, radiolysis provides a steady source which will regenerate and maintain the nitrous acid concentration. Nitrous acid and nitrite ion react with several important species, particularly neptunium, plutonium, and ruthenium, and some of these reaction systems are quite complex.

*1. Actinides*

Uranium is quantitatively oxidized to the hexavalent state during dissolution, and it is stable throughout as the uranyl ($UO_2^{2+}$) ion. Plutonium exists as a mixture of Pu(IV) and Pu(VI) under dissolution conditions, but the post-dissolution heat treatment favors oxidation to Pu(VI) (high temperature, moderately strong nitric acid, low nitrous acid). On the other hand, nitrous acid reduces plutonium to Pu(IV), so if nitrous acid is not removed following dissolution or if enough time is allowed for radiolytically produced nitrous acid to be generated, the tetravalent state will be favored. It is observed that, given sufficient time (one or a few days, depending on radiation level), plutonium is spontaneously reduced to Pu(IV) by radiolysis products in normal LWR dissolver solution. With fuels of lower radiation level, feed adjustment has normally included a step to convert plutonium to Pu(IV), and that step will probably continue to be used for LWR fuels also, as a guarantee that all plutonium is tetravalent.

Neptunium presents a more complex situation since it can exist in three valence states simultaneously (IV, V, and VI), depending on the acidity and oxidation potential of the solution. The tetravalent state is stable only if it is complexed, in equilibrium with other valences, or in the presence of reducing agents (hydroxylamine, hydrazine, $Fe^{2+}$). In nitric acid concentrations below 2 *M* it is slowly oxidized to Np(V) in the presence of nitrous acid. Pentavalent neptunium is the stable form in lower acid concentrations, <2 *M*; at higher acid concentrations it disproportionates to Np(IV) and Np(VI). The presence of nitrous acid concentrations comparable to the neptunium concentration catalyzes the oxidation of Np(V) to NP(VI) according to Equation 21,

$$NpO_2^+ + 1/2\ NO_3^- + 3/2\ H^+ \rightleftarrows NpO_2^{2+} + 1/2\ HNO_2 + 1/2\ H_2O \tag{21}$$

the reaction being favored by increasing $HNO_3$ concentration. At higher concentrations of nitrous acid, however (a few hundredths molar), the reverse reaction dominates because of the decrease in oxidation potential. The hexavalent state is the stable form in high concentrations of nitric acid (>5 *M*), but it is reduced rapidly to Np(V) by nitrous acid, even in high nitric acid concentrations.[43]

During fuel dissolution the presence of large amounts of nitrogen oxides, which form nitrous acid, favors the existence of neptunium in the pentavalent state, even if the nitric acid concentration is high enough to stabilize Np(VI). Following dissolution, if the solution is maintained at or near boiling under conditions that remove nitrous acid, the neptunium will oxidize to the (VI) state, according to Equation 21. Subsequently, the solution is cooled and maintained at a lower temperature for a number of hours during which accountability measurements, feed adjustments, and other operations occur. During this time, formation of nitrogen oxides and nitrous acid by radiolysis will again convert some of the Np(VI) to Np(V) which, in turn, disproportionates, yielding some Np(IV). The result is that all three valence states may be present, the relative amounts changing with time and depending on actual conditions. If a feed adjustment step for plutonium valence control is used (commonly addition of nitrogen oxides or nitrite ion), the neptunium will be converted largely to Np(V).

Following dissolution of MOX fuels, there may also be some insoluble material containing $PuO_2$. Pure plutonia is insoluble in nitric acid, and plutonium-rich mixed oxide phases are soluble only with difficulty. Dissolution can be achieved by adding a small amount of fluoride as a catalyst, but this may cause problems elsewhere, especially from corrosion in acid recycle evaporators and waste treatment. A preferred approach is to fabricate fuel in such a way that the plutonium exists as a true solid solution with $UO_2$, rather than as $PuO_2$ particles dispersed in $UO_2$. With nonhomogeneous fuel containing about 4% $PuO_2$, as much as 10% of the plutonium has been found insoluble in boiling 10 $M$ $HNO_3$, whereas with homogeneous fuel prepared by a coprecipitation method dissolution was essentially complete.[44]

If plutonium-rich insoluble solids are encountered, there will be little choice but to recover the solids and recycle them to a secondary dissolution. Recent studies have suggested two possible dissolution reagents in addition to $HNO_3$–HF, mentioned before. These are either $HNO_3$ with Ce(IV)[45] or HI solutions.[46] The reactions are not well understood, but they apparently involve complexation and oxidation/reduction.

*2. Fission Products*

Most fission product elements dissolve along with the fuel and behave chemically in a straightforward manner. Only those that demonstrate nontypical behavior or cause problems will be discussed in any detail. These are the gaseous or volatile fission products, the noble metals which are partially insoluble in nitric acid, and those that present chemical difficulties, notably ruthenium.

The noble gases, krypton and xenon, are released quantitatively during dissolution, and they must be dealt with in the off-gas. Only the isotope $^{85}Kr$ is a long-lived gamma-emitter, and methods for recovering it from the off-gas are discussed in Chapter 5. Nearly all the tritium in the fuel is converted to water during the dissolution, but a small fraction, typically a few tenths of 1%, is released as the elemental gas, HT. Most of the $^{14}C$ is evolved from the dissolver solution as $CO_2$, but a small fraction (the order of a percent) may remain in the solution, possibly as organic compounds. Iodine has two radioisotopes that are significant, mass 131 which has a half-life of 8 days and is a major biological hazard in short-cooled fuel, and mass 129 with a half-life of 17 million years, which is a potential long-term hazard. Nearly all the iodine is evolved from the dissolver solution, most of it carried out by the nitrogen oxides produced by the dissolution reaction, and the rest by either an air sparge, steam stripping, or boiling. Typically, less than 1% of the iodine is left in the dissolver solution.[13,39]

A small fraction of the fuel, about 0.2%, does not dissolve in nitric acid. This insoluble material has received considerable attention in several countries, and there is general agreement that it consists of small particles containing primarily molybdenum and ruthenium, with smaller amounts of technetium, rhodium, and palladium. The composition of the dissolver residue appears to be somewhat variable, depending on the characteristics of the fuel, the irradiation history, and the particular dissolution procedure. Representative compositions are (in wt%) Mo, 10 to 30; Tc, 2 to 5; Ru, 20 to 45; Rh, 2 to 6; and Pd, 4 to 10. Typically, there is very little fission product zirconium, but there may be some natural zirconium from cladding, either fines produced by the shear or fragments of oxide film. There are, of course, variable amounts of other common impurities including silica, structural metal oxides, and others.

The solids account for only a part of the total amount of these fission products produced in the reactor, typically 20 to 30% of the ruthenium and less of the others; the remainder dissolves with the other fission products. However, voloxidation causes a much greater fraction of the noble metals to be insoluble. Such noble-metal phases are observed in highly irradiated fuel, and in fuels with higher temperatures and higher burnup the particles may become relatively large.[47] Under LWR operating conditions, however, the particle size ranges generally from a few microns downwards, and there may be no realistic lower limit. Particles of such noble metals, if sufficiently large, are insoluble during fuel dissolution.

Noble metals are insoluble in the $UO_2$ fuel, and their ability to form discrete particles within the fuel is limited by the rate with which they can diffuse together. Thus, the quantity of such particles in irradiated fuel will increase with time, temperature, and burnup or concentration; it would be quite small for the low-burnup fuels encountered prior to LWRs. Conversely, it is indeed greater in high-burnup, high-temperature fast reactor fuel. Thus, this is essentially a new problem, not encountered in prior reprocessing.

The insoluble residue is of importance because it is extremely radioactive and, therefore, a heat source. The radioactivity results almost entirely from $^{106}Ru$ and, in the case of short-cooled fuel, $^{103}Ru$. Fortunately, elements other than the noble metals are not associated with this material in any significant amounts; uranium, plutonium, and transplutonium isotopes, in particular, are found only to the extent of a hundredth or a few thousandths of 1% of the total present.[48] Thus, it does not present a process loss or criticality problem.

It should be pointed out that several phases other than $UO_2$ have been observed in irradiated oxide fuel, in addition to these noble-metal inclusions.[49] There are separate phases that contain uranium, for example, alkaline earth zirconates; these may require somewhat longer dissolution times than one would expect for pure $UO_2$. However, nitric acid apparently dissolves all these phases except the noble metals, given sufficient time. Thus, the solid-state reactions that occur during irradiation are effectively nullified during fuel dissolution, with the single exception of the noble metal inclusions.

The fate of the fission products during dissolution can be summarized as follows: noble gases, iodine, and $^{14}C$ are normally evolved into the off-gas; tritium, if not removed in a head-end process, enters the dissolver solution as water; and a fraction of the noble metals, including ruthenium in particular, remains insoluble as a very small amount of solid residue.

### C. Feed Clarification

Following dissolution of $UO_2$ fuel with nitric acid the dissolver solution is opaque and black because of the insoluble residue of noble metals (and possibly some $PuO_2$). Although the amount of solids is small, about 2 kg/t of fuel (0.6 $g/\ell$ of solution), the particle size is also very small, so clarification poses some problems. The amount of residue increases with burnup, more rapidly than linearly,[50] and it is probably also dependent on the temperature-time history of the fuel as well. Additional solids may result from cladding fines or from fragments of the oxide film from the Zircaloy cladding, small particles being generated during the shearing operation. There may also be a small amount of solids rich in $PuO_2$ in the case of plutonium recycle fuels, as discussed in the previous section.

An additional possible source of solids in process solutions results from interactions of fission products (and actinides) with each other. Although there are several possible compounds between the many fission product elements which are highly insoluble, the only one that has been found in practice contains predominantly zirconium and molybdenum. This is a crystalline solid that has the composition $ZrMo_2O_8 \cdot nH_2O$; plutonium, which forms a similar molybdate, will coprecipitate with it.[51] Formation of this compound is inhibited by (1) higher acid concentration, (2) lower temperature, and (3) lower concentrations of zirconium and molybdenum (lower burnup). Further, it will not form in fresh dissolver solution under the same conditions that result in its formation in solutions a few days old.

Under normal process conditions of acidity, temperature, and aging time, this solid is not encountered, but it would be formed in the event of a process shutdown, recycle, or other unusual operation that results in heating of aged process solutions, especially at lower acidity. At temperatures below 60°C and acidities above 3 *M*, solutions are stable for weeks; at the boiling point an aged solution may precipitate in minutes, but fresh dissolver solution does not precipitate during the dissolution which includes temperatures near boiling for as long as a few hours. Conditions during waste evaporation or storage are favorable for precipitate formation. The solids are crystalline and relatively easy to remove from solution, but they tend to adhere to surfaces so their removal from process equipment may be more of a problem. Unlike the other solids which are present initially in dissolver solutions, these solids only appear later in the operation, if at all.

Clarification of the dissolver solution is required because solids tend to collect at the aqueous phase-organic phase interface during solvent extraction, thereby interfering with that operation. For practical purposes, the solids most difficult to remove are the insoluble residue consisting of noble-metal fission products. Because of the very small sizes of some of the particles clarification is not easy. Filtration has been used with mixed success, both in the laboatory and larger-scale work. A photograph of some dissolver residue solids (Figure 3) collected on a 0.1-$\mu$m Millipore® filter indicates the impermeability of the filter cake; the 0.1-$\mu$m filter pores are clearly visible, but neither porosity nor individual particles are visible in the filter cake.[52] Centrifugation has been used to clarify the solutions, generally with reasonable success. The use of a surfactant has also been reported to promote solution clarification markedly.[53] All these observations are in accord with the concept that these solids include a size range extending down to extremely small particles and that particle aggregation enhances clarification.

Finally, it must be remembered that these solids are a substantial heat source, so that proper account must be taken of this in the design of the clarification equipment and the means to recover the solids and store them for decay.

### D. Feed Adjustment

Among the operations performed following dissolution are (1) accountability measurement, (2) clarification, and (3) feed adjustment, not necessarily in that order. At least some of these operations are carried out on a batch basis, and this, along with other holdup volumes for operational purposes, results in a holdup time between dissolution and solvent extraction of at least several hours, and possibly a day or two. As discussed before, the solutions may be modified spontaneously during this time because of radiation effects, particularly from formation of nitrogen oxides and nitrous acid. In addition, depending on the specific design, there may be efficient contact of the solutions with air for solution mixing or air-lift transfer, and there may be operations at elevated temperature. All these affect the chemical state of certain elements, especially plutonium, neptunium, and ruthenium.

The primary purpose of feed adjustment is to modify the dissolver solution composition to that desired for solvent extraction. This, again, is somewhat variable, depending on the particular process flowsheet. Typical compositions have been about 1 $M$ total uranium plus plutonium and 2 $M$ $HNO_3$, which can be achieved by diluting the dissolver solution with water or nitric acid. In the event of recycle of recovered uranium or plutonium that does not meet the required purity, it may be necessary to increase the concentration; therefore, a capability for evaporation may be included. Recent studies indicate that a higher acidity tends to diminish some of the operational problems in the solvent extraction system, and a concentration of 3 $M$ $HNO_3$, or even somewhat higher, has been recommended.[54]

The other requirement of feed adjustment, aside from gross concentration control, is valence adjustment. Attention has focused almost entirely on plutonium in the past, but neptunium is receiving more attention because its valence is affected by the means applied for plutonium valence control. For best extractability, it is desired that all plutonium be tetravalent. Hexavalent plutonium is extractable, but somewhat less so. As discussed above, the plutonium valence is spontaneously changed to +4 because of radiolytic products, but this may require too much time. Historically, reprocessing was done with fuels of lower radiation level, so the radiolysis effects were much less significant. Accordingly, a valence adjustment was normally made as a routine matter. The method most used over the years has been the reduction of plutonium wih ferrous sulfamate, producing Pu(IV) and, since excess reductant must be used, Pu(III). Since Pu(III) is inextractable, it was necessary to oxidize it back to Pu(IV), which was commonly done by adding $NaNO_2$.

More recently there has been increasing incentive to use only reagents that do not contribute to the permanent solid waste disposal problem. Since ferrous sulfamate introduces large amounts of iron and sulfate, and $NaNO_2$ is converted to $NaNO_3$, these additives were substantial (frequently primary) contributors to the waste. Accordingly, other chemicals were sought for this purpose. The usual reductants, such as hydroxylamine nitrate, are rather ineffective at the high acidity of the dissolver solution. The favored approach at this time is to sparge the solution with nitrogen oxides, which form nitrous acid in the solution (Reaction 19 and the reverse of Reaction 20). This, in effect, simply enhances the spontaneous reaction from the radiolysis products. The intricacies of valence control are discussed in some detail in Chapter 4.

The final result is the preparation of a clarified solvent extraction feed solution that contains nitric acid at a predetermined concentration probably between 1.5 and 4 *M*, uranium as uranyl ion at a concentration near 1 *M*, plutonium as Pu(IV) at a concentration near 0.01 *M* (or a few times larger for plutonium recycle fuel), neptunium as a mixture of (IV), (V), and (VI) but predominantly Np(V) at a concentration near 0.0004 *M*, a variety of fission products at concentrations ranging from about 0.01 *M* (Nd, Zr, Mo) downward to 0.001 *M* (Rb, Cs, Ba, Sr, several rare earths, Tc, Te, Ru, Rh, and Pd), with others at still lower concentrations. Most of the fission products exist as the species that would be expected in moderately strong nitric acid solution, but a few demonstrate chemical properties that are less simple. Zirconium and niobium (the latter present at only very low concentrations, about $10^{-5}$ *M*) both tend to form hydrolyzed species. Ruthenium exists as several complexes of the general form

$$\{Ru(NO)(NO_3^-)_w(NO_2^-)_x(H_2O)_y(OH^-)_z\}^{[3-(w+x+z)]+}$$

with $w + x + y + z = 5$, the relative amounts depending on the acid, nitrate, and nitrite concentrations, and some of them reaching equilibrium only slowly.[55] Ruthenium chemistry is described in more detail in the next chapter.

## IV. PRODUCT CONVERSION

Product conversion operations are required to couple the reprocessing plant products to the next step of the fuel cycle. There are two products, one a pure uranyl nitrate solution in dilute nitric acid and the other, plutonium. Until recently the plutonium product was presumed to be highly purified $Pu(NO_3)_4$ in nitric acid, but concern relating to proliferation has led to examination of fuel cycles in which the plutonium is never purified; it would always be mixed with excess uranium, radioactive materials, or some substance that would interfere with its diversion out of the proper fuel cycle. Thus, although prior considerations have assumed a pure plutonium nitrate product, alternative products might be a mixture containing about 20% plutonium and 80% uranium, possibly with other materials.

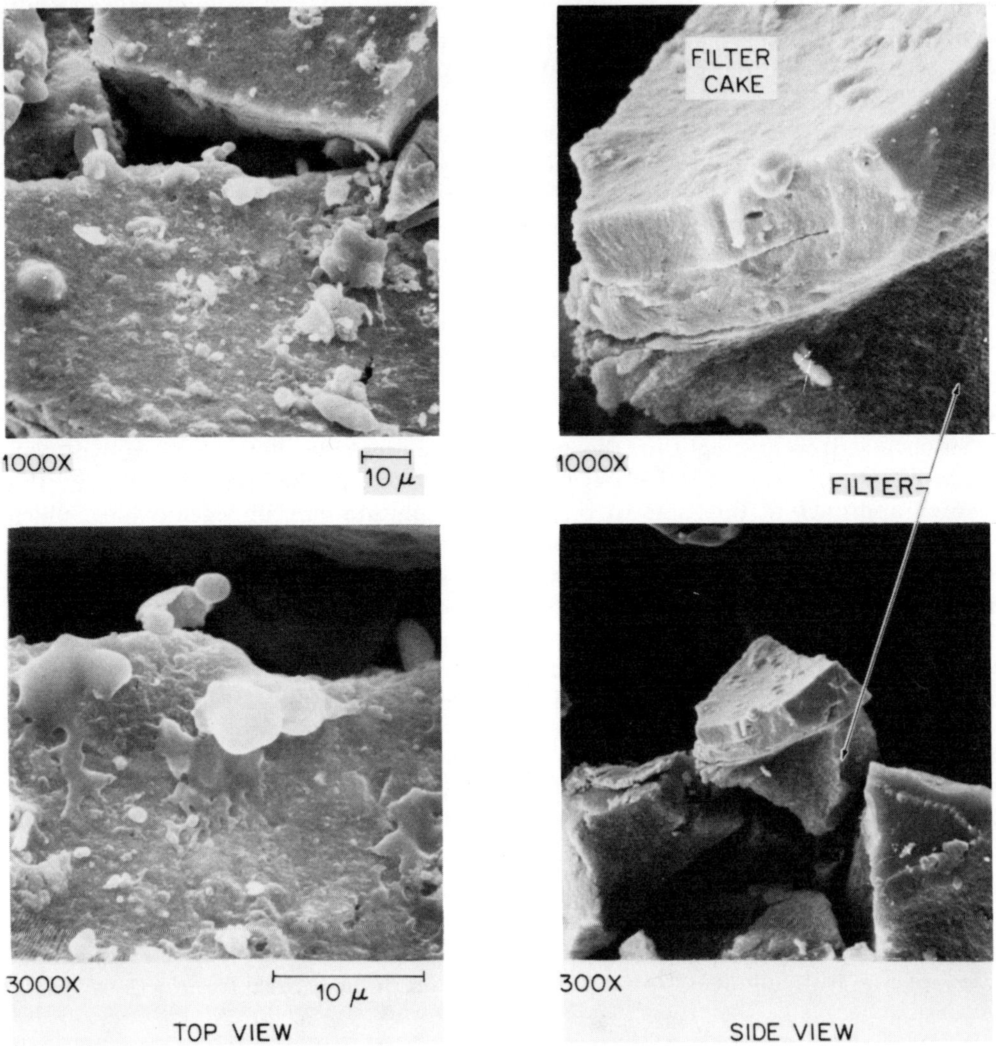


FIGURE 3. Filter cake of dissolver residue solids.

The next step for these products is enrichment in the case of uranium and MOX fuel fabrication in the case of plutonium. Enrichment requires conversion of the uranyl nitrate solution to $UF_6$, and the procedure is essentially identical to that for conversion from ore to $UF_6$. MOX fuel fabrication requires production of an oxide containing about 4% $PuO_2$ in $UO_2$, with suitable ceramic properties such as density, homogeneity, particle size, and geometry. Prior experience centers on the production of $UO_2$ fuels, and the incorporation of plutonium into recycle fuel introduces some important changes. In fact, the recycle of uranium and plutonium from LWR fuel back to LWR fuel has never been carried out on a significant scale, mainly because so little fuel has been reprocessed compared to the amount of uranium mined and fresh fuel produced.

The conversion to $UF_6$ is a routine industrial process, described in Chapter 2. The sequential steps are (1) concentration of the uranyl nitrate by evaporation, (2) denitration to $UO_3$, (3) reduction of $UO_3$ to $UO_2$ with hydrogen, (4) hydrofluorination of $UO_2$ to $UF_4$ with HF, (5) fluorination of $UF_4$ to $UF_6$ with fluorine gas, (6) freezing and resubliming the $UF_6$ in appropriate cold traps, and (7) packaging the $UF_6$ in transport cylinders. There are alternative ways of carrying out some of the steps, and typical plants for conversion of uranium from ore to $UF_6$ have been described.[56,57]

Preparation of a plutonium product is not a developed industrial process. In early reprocessing work plutonium was the product in demand for weapons. However, plutonium from LWR fuel is not well suited to weapons production, and its ultimate use is open to several possibilities, including recycle to other LWRs, use as fuel in a different type of reactor such as a fast breeder (where its fuel value is much greater), or even conversion to a stable form for long-term management as a hazardous waste. For purposes of the fuel cycle, conversion to a MOX fuel for use in either an LWR or an FBR is the goal, and that will be considered here.

MOX fuel is normally made by preparing $UO_2$ and $PuO_2$ powders separately, mechanically blending the two in the required ratio, pressing the blended powders into pellets which are then sintered to give suitable properties, grinding to specified dimensions if necessary, and finally loading the pellets in cladding tubes. Certain alternatives are under consideration, two primary ones being (1) mixing the uranium and plutonium solutions before making the oxide (coprecipitated oxide) and (2) utilizing compact beds of fragments or microspheres of the oxide instead of pellets.

Mechanically blended fuel utilizes $UO_2$ from sources identical to those for pure $UO_2$ fuel, discussed in Chapter 2. The $PuO_2$ is usually prepared by oxalate precipitation of the plutonium nitrate product from reprocessing; ignition of the oxalate to a temperature in the range 550 to 900°C gives a suitable oxide. The particle size and surface area of both the $UO_2$ and $PuO_2$ are very important variables that affect the sinterability of the oxide powders and the properties of the final fuel, such as homogeneity, density, and porosity. Other methods for preparing $PuO_2$ can be used, including peroxide or hydroxide precipitation and thermal denitration. However, working out a specific process that consistently yields good fuel pellets is a long and rather empirical operation that cannot be predicted on the basis of simple chemical considerations.

Coprecipitated fuel would be expected to have a clear advantage in regard to homogeneity as long as the uranium and plutonium are really coprecipitated. However, this is not usually the case, since the uranium is hexavalent and the plutonium is usually tetravalent. Although they are precipitated rapidly and at the same time, they do not form mixed crystals. Precipitation with ammonia, for example yields separate crystals of ammonium diuranate and $Pu(OH)_4$. Oxalate is not suitable for coprecipitated fuel since uranyl oxalate is rather soluble; thus, the usual process is ammonia precipitation or thermal denitration. One process utilizing this method is "Coprecal", which consists of precipitation with excess ammonia followed by evaporation of the slurry (all of the water and supernate as well as the precipitate).[58] If only the precipitate is recovered and converted to oxide there is a possibility that the Pu/U ratio may deviate from that of the initial solution. Both mechanically blended and coprecipitated mixed oxide fuels have been prepared on a substantial scale, while Coprecal is still under development.

An alternate group of processes that has received considerable study and has been used to prepare modest amounts of test fuel is termed "gel precipitation".[59] In these processes uranium, plutonium, or a suitable mixture, as a nitrate solution, is rapidly converted to spherical droplets of gel by increasing the pH to form ammonium diuranate and plutonium hydroxide. Chemically, this is the same as the coprecipitation process described above. This may be done by internal gelation (analogous to homogeneous precipitation) by adding reagents such as hexamethylenetetramine (to produce ammonia) and urea (to complex uranium) and heating, or by external gelation by diffusing ammonia into droplets of the solution containing a polymer (gel support) to increase viscosity. By injecting the solution into a heated organic liquid in the first case, or into ammonia gas in the second, it is possible to produce very regular microspheres of controlled sizes up to more than a millimeter in diameter. The ammonium nitrate product of the neutralization reaction and other soluble constituents can be washed out of the microspheres by contacting them with dilute $NH_4OH$ solution. The spheres can then be dried and calcined in an appropriate atmosphere to yield the mixed oxide product, and sintering temperatures are generally lower than with mechanically blended mixed oxide. An alternative process is to press calcined microspheres into a pellet form prior to sintering, thereby yielding a pellet fuel of excellent homogeneity.

An earlier version of the gel microsphere process, termed "sol-gel", utilized a concentrated sol produced by ammonia precipitation of the hydroxide followed by extended washing of the precipitate until it peptized. The sol was then injected into an organic alcohol which has a limited solubility for water, such as 2-ethyl-1-hexanol; the alcohol extracts water from the sphere as it falls through a column or is suspended in an upflowing stream, yielding dense gel microspheres. The gel sphere is then dried and sintered to give the product.[60] There is a problem of valence control with uranium and plutonium, but separate sols made from Pu(IV) and U(IV) can be mixed prior to gel formation. This approach is more appropriate for tetravalent elements (thorium and plutonium) than for hexavalent uranium.

The microspheres can be poured into cladding tubes which are vibrated to settle the spheres into a dense bed ("Sphere-Pac"). By use of three sizes of spheres, nominally 800, 200, and 40 $\mu$m in diameter, selected so that the smaller occupy the interstitial volume between the larger, bulk densities of about 87% can be achieved. These densities are adequate for most reactor fuel applications. The advantage of this process is that it is much better suited to remote operation, compared to a pellet process, because it eliminates several steps and most of the mechanical manipulations, and it avoids production of fine powders required for pellets. The materials are in the form of solutions or of microspheres which can be transported by fluidization with liquids or gases, so conventional solids handling is not necessary. Since fabrication of fuels containing recycle plutonium will almost certainly require remote operations (and possibly remote maintenance of the facility), these handling advantages are significant. Yields of useful products can be very high also, in the range of 95 to 100%, thereby minimizing recycle. Both pellet fuel and Sphere-Pac fuel have given good results in irradiation tests, but nearly all LWR fuel used commercially has been pellet fuel that did not contain plutonium.

Since the $UO_2$ pellet is the standard commercial LWR fuel at this time, there is a substantial momentum to continue with it. Just as the Purex process is established and would be displaced only for some outstanding reason, so a fuel form other than pellets will not be easily accepted. There are potentially valid reasons for a change to Sphere-Pac fuel in the future, primarily related to the complications introduced into pellet manufacture as radiation levels increase in recycle fuel, but also because the process may yield a better fuel at a lower cost. Thus, the evolution of MOX fabrication processes will likely depend on the fuel recycle scheme that finally emerges, and particularly on the radiation level of the fuel being recycled. Gel fuel fabrication is under study in England, Holland, Germany, Switzerland, Italy, Japan, Taiwan, Czechoslovakia, the U.S., and the U.S.S.R.

Reactor fuel is normally tailored to meet the requirements of reactor irradiation, with little consideration of problems that might be introduced into other parts of the fuel cycle. For example, highly insoluble fuels are preferred in case of leakage of water into the cladding, but the same fuel must be readily dissolved for reprocessing. MOX fuels present a particular problem in regard to solubility. Plutonium oxide is essentially insoluble in nitric acid. Thus, a fuel containing discrete $PuO_2$ particles may present a dissolution problem, although it is perfectly satisfactory for reactor irradiation. Mechanically blended fuel, containing small enough $PuO_2$ particles and sintered adequately, approaches a homogeneous state, but plutonium-rich regions may still be difficult to dissolve. Experimentally, MOX dissolves satisfactorily in nitric acid if the plutonium content (on a microscopic scale) is less than about 40%, but dissolution is increasingly more difficult with localized plutonium concentrations greater than this. Homogenization is promoted by using small-sized $PuO_2$ particles and by high temperatures during sintering and reactor operation. Temperatures above 1400°C are required for significant homogenization, and most of the fuel in LWR reactors is well below this temperature; thus, irradiation may improve solubility properties to only a limited extent. Experimentally, it has been reported that a few percent of the plutonium in mechanically blended mixed oxide LWR fuel may be insoluble in boiling 10 *M* $HNO_3$, but other similar fuels may have satisfactory dissolution characteristics.[54] The dissolution problem is discussed further in Section III.B.

Another potential problem that is affected by fuel fabrication methods derives from the nitrogen content of the fuel. Nitrogen is the predominant source of $^{14}C$ produced during irradiation, and $^{14}C$ is a long-term biological hazard that is receiving increased attention. The nitrogen content of reactor fuel can vary greatly, depending on the production methods used.[61] A major source of nitrogen is ammonia which is used in some processes for precipitation of ammonium diuranate in the conversion of uranyl nitrate to $UO_3$, or in the Coprecal process mentioned above, and also as cracked ammonia for reduction of the oxide to $UO_2$. In any case, processes that do not unnecessarily introduce nitrogen into the final fuel are advantageous with respect to this important environmental factor.

## LITERATURE SOURCES

Most of the better information on the chemistry of fuel reprocessing is not in the recent literature except by reference or in review articles. The reader is referred to the abundance of excellent publications issued during the decade starting in 1955. This resulted from declassification of information accumulated in the military programs of several countries, stimulated largely by the first Geneva Conference, and the ongoing work generated by an intense interest in nuclear power. Subsequently, the literature diminished because of the withholding of proprietary information for commercial and competitive purposes and the decision in the U.S. that reprocessing was a developed technology that no longer required government support. Although methods and techniques have improved since the bulk of the work was done, particularly in instrumentation, any shortcoming is more than compensated for by the capabilities of the earlier investigators.

It is the practice today to do literature surveys by computer, but, unfortunately, computer bases generally do not extend back to this prolific period. Some general sources that summarize much of the information and give more specific references are listed here. For broad-based, general interest, the *Proceedings of the International Conferences on Peaceful Uses of Atomic Energy, Geneva, Switzerland* are useful, especially appropriate volumes of the first two conferences held in 1955[3] and 1958,[4] and each published by the United Nations. The specific subject of reprocessing is included in the four volumes of *Progress in Nuclear Energy, Series 3, Process Chemistry,* published by Pergamon Press in 1956, 1958, 1961, and 1970.[5-8] Other good summaries are included in the *Reactor Handbook*[9] and the *Plutonium Handbook.*[10] A recent review is presented in *Gmelin Handbuch der Anorganischen Chemie,* Volume 21.[11] For more specific topics the proceedings of various conferences are of interest, one outstanding example being the International Solvent Extraction Conferences held triannually, most recently in Toronto in 1977. Finally, there are several review articles that are referred to in the text.

Additional sources of information that should present a reasonably comprehensive review of the reprocessing state-of-the-art, as well as alternatives thereto, are being prepared in two major studies that are scheduled for completion in 1979-1980, namely NASAP, the *Nonproliferation Alternative Systems Assessment Program,* prepared in the U.S., and INFCE, the *International Nuclear Fuel Cycle Evaluation,* which includes participation by some 40 countries. A recent publication of this type is the *Barnwell Nuclear Fuels Plant Applicability Study*[12] published by the U.S. Department of Energy in 1978, particularly Appendix C (DOE/ET-0040/3), which describes a number of fuel cycle variations that might be utilized in the Barnwell Nuclear Fuels Plant.

## REFERENCES

1. Barnwell Nuclear Fuels Plant Applicability Study, Vol. 2, Rep. DOE/ET-0040/2, U.S. Department of Energy, Washington, D.C., 1978, Sections 5.2 to 5.4.
2. Barnwell Nuclear Fuels Plant Applicability Study, Vol. 3, Rep. DOE/ET-0040/3, U.S. Department of Energy, Washington, D.C., 1978, Appendix B.
3. Proc. Int. Conf. Peaceful Uses Atomic Energy, Geneva,1955, Vols. 7 to 9, United Nations, New York, 1956.
4. Proc. 2nd United Nations Int. Conf. Peaceful Uses Atomic Energy, Geneva, 1958, Vols. 4 and 17, United Nations, New York, 1958.
5. *Prog. Nucl. Energy Ser. 3,* Vol. 1, 1956.
6. *Prog. Nucl. Energy Ser. 3,* Vol. 2, 1958.
7. *Prog. Nucl. Energy Ser. 3,* Vol. 3, 1961.
8. *Prog. Nucl. Energy Ser. 3,* Vol. 4, 1970.
9. **Stoller, S. M. and Richards, R. B., Eds.,** *Reactor Handbook,* Vol. 2, 2nd ed., Interscience, New York, 1961.
10. **Wick, O. J., Ed.,** *Plutonium Handbook: A Guide to the Technology,* Vol. 2, Gordon & Breach, New York, 1967.
11. **Koch, G.,** *Transurane, Gmelin Handbuch der Anorganischen Chemie,* Vol. 21, 8th ed., (Supplement), Springer-Verlag, Berlin, 1975.
12. Barnwell Nuclear Fuels Plant Applicability Study, Vol. 3, Rep. DOE/ET-0040/3, U.S. Department of Energy, Washington, D.C., 1978, Appendix C.
13. **Koch, G.,** The chemical reprocessing of irradiated nuclear fuels, *Chem. Ztg.,* 101(2), 64, 1977; ORNL-tr-4379, Oak Ridge National Laboratory, Oak Ridge, Tenn., 1977.
14. *Commis. Energ. At. Fr. Rapp. Annu.,* 1977.
15. **Thompson, S. G. and Seaborg, G. T.,** The first use of bismuth phosphate for separating plutonium from uranium and fission products, *Prog. Nucl. Energy Ser. 3,* 1, 163, 1956.
16. **Lang, D. M.,** Uranium purification by uranyl ammonium phosphate precipitation, *Prog. Nucl. Energy Ser. 3,* 2, 387, 1958.
17. **Boyd, G. E. and Schubert, J.,** The first use of organic and inorganic ion exchangers for separating plutonium from uranium and the fission products, *Prog. Nucl. Energy Ser. 3,* 4, 319, 1970.
18. **Bruce, F. R.,** The concentration and purification of uranium and plutonium by ion exchange, *Prog. Nucl. Energy Ser. 3,* 2, 363, 1958.
19. **Tober, F. W.,** Concentration and purification of uranium, plutonium, and neptunium by ion exchange in nuclearly safe equipment, Paper P/520, in Proc. 2nd United Nations Int. Conf. Peaceful Uses Atomic Energy, Geneva, 1958, Vol. 17, United Nations, New York, 1958, 574.
20. **Ryan, J. L.,** Species involved in the anion-exchange absorption of quadrivalent actinide nitrates, *J. Phys. Chem.,* 64, 1375, 1960.
21. **Wheelwright, E. J.,** Recovery and purification of plutonium by anion exchange, *Ind. Eng. Chem.,* 51, 60, 1959.

22. **Wheelwright, E. J. and Ryan, J. L.**, Application of anion exchange to the reprocessing of plutonium, Paper P/1915, in Proc. 2nd United Nations Int. Conf. Peaceful Uses Atomic Energy, Geneva, 1958, Vol. 17, United Nations, New York, 1958, 137.
23. **Cleveland, J. M.**, *The Chemistry of Plutonium*, Gordon & Breach, New York, 1970, 498.
24. **Groh, H. J. and Schlea, C. S.**, The recovery of neptunium-237 and plutonium-238, *Prog. Nucl. Energy, Ser. 3*, 4, 507, 1970.
25. **Miles, F. W.**, Ion-exchange-resin system failures in processing actinides, *Nucl. Saf.*, 9, 394, 1968.
26. **Peligot, E.**, Recherches sur l'uranium, *Ann. Chim. Phys.*, Ser. 3, 5, 5, 1842.
27. **Peppard, D. F.**, Liquid-liquid extraction of the actinides, in *Annual Reviews of Nuclear Science*, Segre, E., Ed., Annual Reviews, Palo Alto, 1971, 365.
28. **Campbell, W. M.**, Canadian fuel reprocessing, *Nucleonics*, 14, 92, 1956.
29. **Lawroski, S. and Levenson, M.**, Redox process — a solvent-extraction reprocessing method for irradiated uranium, *Prog. Nucl. Energy Ser. 3*, 2, 258, 1958.
30. **Howells, G. R., Hughes, T. G., and Saddington, K.**, The chemical processing of irradiated fuels from thermal reactors, in Proc. 2nd United Nations Int. Conf. Peaceful Uses Atomic Energy, Geneva, 1958, Vol. 17, United Nations, New York, 1958, 3; *Prog. Nucl. Energy Ser. 3*, 3, 151, 1961.
31. **Coleman, C. F.**, Amine extraction in reprocessing, *Prog. Nucl. Energy Ser. 3*, 4, 233, 1970.
32. **Coleman, C. F.**, Amine extraction in reprocessing, *At. Energy Rev.*, 2(2), 3, 1964.
33. **Becker, R., Baumgartner, F., and Stieglitz, L.**, Zur Aufklarung von komplexbildenden Radiolyseprodukten des Purex-systems (20% TBP-Dodekan-$HNO_3$), Rep. KFK-2304, Kernforschungszentrum Karlsruhe, Germany, 1976.
34. **Blanco, R. E.**, Preparation of power reactor fuels for processing by solvent extraction, *Prog. Nucl. Energy Ser. 3*, 2, 223, 1958.
35. **Savolainen, J. E. and Blanco, R. E.**, Preparation of power reactor fuels for aqueous reprocessing, *Chem. Eng. Prog.*, 53, 78, 1957.
36. **Swanson, J. L.**, The selective dissolution of zirconium or zircaloy cladding by the Zirflex process, Paper P/2429, in Proc. 2nd United Nations Int. Conf. Peaceful Uses Atomic Energy, Geneva, 1958, Vol. 17, United Nations, New York, 1958, 154.
37. **Koch, G., Ochsenfeld, W., Schmieder, H., and Weinlander, W.**, Chemical aspects in the design of a flowsheet for a large-scale LWR reprocessing plant, *Kerntechnik*, 18, 253, 1976.
38. **Webster, D. S., Jonke, A. A., Levitz, N. M., Miller, W. E., Steindler, M. J., and Vogel, R. C.**, Chemical Engineering Division Fuel Cycle Technology Quarterly Report, July-September, 1972, Rep. ANL-7972, Argonne National Laboratory, Argonne, Ill., 1972, 58.
39. **Campbell, D. O., Buxton, S. R., and Pattison, W. L.**, in LWR Fuel Reprocessing and Recycle Quarterly Report for Period April 1 to June 30, 1977, Vondra, B. L., Program Manager, Rep. ORNL/TM-5987, Oak Ridge National Laboratory, Oak Ridge, Tenn., 1977, Sect. 2.2.2.
40. **Beyer, C. E. and Hann, C. R.**, Release of Indigenous Gases from LWR Fuel and the Reaction Kinetics with Zircaloy Cladding, Rep. BNWL-1956, Battelle Northwest Laboratory, Richland, Wash., 1977.
41. **Vaughen, V. C. A.**, in Alternate Fuel Cycle Technologies Program Quarterly Report for Period July 1 to September 30, 1977, Vondra, B. L., Program Manager, Rep. ORNL/TM-6076, Oak Ridge National Laboratory, Oak Ridge, Tenn., 1977, Sect. 2.3.4.
42. **Stacy, R. G. and Goode, J. H.**, Voloxidation and dissolution of irradiated LWR fuel, in Proc. American Nuclear Society Topical Meeting, Back-End of the LWR Fuel Cycle, CONF-780304, Savannah, Ga., March 19 to 23, 1978, X-11.
43. **Bathellier, A.**, Transuranium elements in the reprocessing of nuclear fuel, *Bull. Inf. Sci. Tech. Commis. Energ. At. (Fr.)*, 217, 7, 1976.
44. **Nicholson, E. L., Jr.**, The $PuO_2$ Dissolution Problem for LWR Plutonium Recycle and LMFBR Fuels: Fabrication and Reprocessing Problems and Their Resolution, Rep. ORNL/TM-5903, Oak Ridge National Laboratory, Oak Ridge, Tenn., July 1977.
45. **Wilson, A. S.**, Method of Dissolving Plutonium Dioxide in Nitric Acid Using Cerium Ions, U.S. Patent 3,005,682, October 24, 1961; Horner, D. E., Crouse, D. J., and Mailen, J. C., Cerium-Promoted Dissolution of $PuO_2$-$UO_2$ in Nitric Acid, Rep. ORNL/TM-4716, Oak Ridge National Laboratory, Oak Ridge, Tenn., 1977.
46. **Tallent, O. K., Mailen, J. C., and Williams, K. D.**, Study of the dissolution of refractory $PuO_2$ in aqueous hydriodic acid solutions, in Proc. American Nuclear Society Topical Meeting on the Plutonium Fuel Cycle, Bal Harbour, Florida, May 2 to 4, 1977, 6, 4-1.
47. **Bramman, J. I. and Powell, H. J.**, Distribution of fission products in irradiated oxide fuel pins, *J. Br. Nucl. Energy Soc.*, 14(1), 63, 1975.
48. **Campbell, D. O., Buxton, S. R., Beatty, R. L., and Pattison, W. L.**, in LWR Fuel Reprocessing and Recycle Quarterly Report for Period July 1 to September 30, 1976, Vondra, B. L., Program Manager, Rep. ORNL/TM-5660, Oak Ridge National Laboratory, Oak Ridge, Tenn., 1976, Sect. 2.2.3.

49. **Bazin, J., Jouan, J., and Vignesoult, H.,** Behavior and physico-chemical state of fission products in fuel elements for pressurized water reactors, *Bull. Inf. Sci. Tech. Commis. Energ. At. (Fr.),* 196, 55, 1974.
50. **Boudry, J. C. and Miguel, P.,** Adaptation of the Purex process to the reprocessing of fast reactor fuels, in *Proc. Int. Solvent Ext. Conf.,* Vol. 2, Society of Chemical Industry, London, 1974, 1551.
51. **Campbell, D. O. and Buxton, S. R.,** in LWR Fuel Reprocessing and Recycle Quarterly Report for Period April 1 to June 30, 1976, Vondra, B. L., Program Manager, Rep. ORNL/TM-5547, Oak Ridge National Laboratory, Oak Ridge, Tenn., 1976, Sect. 2.2.5.
52. **Campbell, D. O., Buston, S. R., and Pattison, W. L.,** in LWR Fuel Reprocessing and Recycle Quarterly Report for Period April 1 to June 30, 1977, Vondra, B. L., Program Manager, Rep. ORNL/TM-5987, Oak Ridge National Laboratory, Oak Ridge, Tenn., 1977, Sect. 2.2.4.
53. **Plodinec, M. J.,** Clarification of LWR dissolver solutions, in Proc. American Nuclear Society Topical Meeting, Back-End of the LWR Fuel Cycle, CONF-780304, Savannah, Ga., March 19 to 23, 1978, III-27.
54. **Ochsenfeld, W., Baumgartner, F., Bauder, U., Blayl, H. J., Ertel, D., and Koch, G.,** Experience with the reprocessing of LWR, Pu recycle, and LMFBR fuel in the MILLI facility, in *Proc. Int. Solvent Ext. Conf.,* Rep. KFK-2558, CIM Spec. Vol. 21, Canadian Institute of Mining and Metallurgy, Montreal, 1979.
55. **Rudstam, G.,** Studies on nitrosylruthenium complexes in nitric acid using repeated extractions, *Acta Chem. Scand.,* 13, 1481, 1959.
56. **Ruch, W. C., Peterson, D. A., Gaskill, E. A., and Tepp, H. G.,** Production of pure uranium hexafluoride from ore concentrates, *Chem. Eng. Prog. Symp. Ser.,* 56(15), 35, 1960.
57. **Powell, C. A.,** Current manufacturing processes used in the United States for mass production of $UF_6$ from purified $UO_3$, in Proc. 2nd United Nations Intern. Conf. Peaceful Uses Atomic Energy, Geneva, 1958, Vol. 4, United Nations, New York, 1958, 165.
58. **Shoat, R. J., Graf, W. A., Jr., and Miller, G. P.,** Criteria and concepts for coprecal converstion in an LWR reprocessing refabrication complex, in Proc. American Nuclear Society Topical Meeting, Back-End of the LWR Fuel Cycle, CONF-780304, Savannah, Ga., March 19 to 23, 1978, VII-15.
59. **Beatty, R. L., Norman, R. E., and Notz, K. J.,** Gel-Sphere-Pac Fuel for Thermal Reactors — Assessment of Fabrication Technology and Irradiation Performance, Rep. ORNL-5469, Oak Ridge National Laboratory, Oak Ridge, Tenn., 1979.
60. **Ferguson, D. E.,** Sol-gel technology in the nuclear reactor fuel cycle, *Prog. Nucl. Energy Ser. 3,* 4, 37, 1970.
61. **Davis, W., Jr.,** Carbon-14 production in nuclear reactors, in *Management of Low Level Radioactive Waste,* Vol. 1, Carter, M. W., Moghissi, A. A., and Kahn, B., Eds., Pergamon Press, New York, 1979.

Chapter 4

# PUREX SOLVENT EXTRACTION CHEMISTRY

**Walter D. Bond**

## TABLE OF CONTENTS

# I. INTRODUCTION

The chemistry utilized in Purex solvent extraction makes it a near ideal process for use in remote operations where high radiation levels are present. The selectivity of tributylphosphate (TBP) affords sharp separations in multistage, countercurrent, liquid-liquid extraction. In addition, TBP is one of the more resistant extractants to degradation by radiation and by chemical attack. The principal reagents used in the process (TBP, nitric acid, and water) can be sufficiently purified by relatively simple methods to permit a high degree of recycle. The versatility of the process has been demonstrated in reprocessing plants throughout the world.[1,2] Plants having capacities of up to 14 MT/day of irradiated uranium have been operated.[1,3]

# II. LIQUID-LIQUID EXTRACTION EQUILIBRIA

## A. Neutral Complex Formation with Nitrate Salts

The separations accomplished by the Purex process are based on differences in the abilities of nitrate salts of cations to form neutral complexes with TBP. These neutral complexes are lipophilic and are soluble in the immiscible TBP phase where they are essentially unionized.[4,5] Typical neutral complex species formed with nitrate salts[6] are shown in Table 1. Nitrate salts of cations in which the metallic elements have valences of +4 or +6 are more strongly complexed by TBP than are those salts containing metallic elements with +1, +2, +3, and +5 valences. Metal nitrate salts in which the metal valence is +1, +2, or +3 are virtually inextractible under Purex conditions. Actinide elements in the +4 and +6 valence form stronger complexes than do any of the other elements in the periodic table; this feature provides the essential basis for Purex separations. Under Purex process conditions, the only fission products which show any appreciable tendency to form complexes with TBP are zirconium, ruthenium, niobium, and technetium. However, the complexes with +4 and +6 actinides are considerably more stable than those of these fission products and good separations are obtained.

## B. Extraction Equilibria Parameters

The extraction of cations from aqueous solutions of nitrate salts into an immiscible TBP phase is depicted by the following equation:

$$M^{z+}_{(aq)} + z\,NO^-_{3\,(aq)} + n\,TBP_{(org)} \rightleftarrows M(NO_3)_z \cdot (TBP)_{n(org)} \quad (1)$$

**Table 1**
**TYPICAL NEUTRAL COMPLEXES FORMED IN THE EXTRACTION OF NITRATE SALTS FROM AQUEOUS SOLUTIONS**

$H_2O \cdot TBP$
$HNO_3 \cdot TBP$
$LiNO_3 \cdot 2\,TBP$
$Ca(NO_3)_2 \cdot 3\,TBP$
$(Y, Ce, Pm, Eu, Tb, Tm, Lu, Am)(NO_3)_3 \cdot 3\,TBP$
$(Zr, Th, Np, Pu) \cdot 2\,TBP$
$(U, Np, Pu)O_2 \cdot 2\,TBP$

Modified from McKay, H. A. C. and Healy, T. V., *Prog. Nucl. Energy Ser. 3*, 2, 546, 1958. With permission.

where M = the cation ($H^+$, $UO_2^{2+}$, $Pu^{4+}$, $Pu(NO_3)^{3+}$, $Pu(NO_3)_2^{2+}$, etc.) and z = charge of the cation. Subscripts (aq) and (org) indicate the aqueous and organic phases, respectively. The equilibrium constant, K, for the extraction is given in Equation 2:

$$K = \frac{\left[M(NO_3)_z\,(TBP)_{n(org)}\right]}{\left[M^{Z+}_{(aq)}\right]\left[NO^-_{3(aq)}\right]^z\left[TBP_{(org)}\right]^n} \cdot F \qquad (2)$$

The brackets denote equilibrium concentration of the species and F denotes the factor which converts the ratio of concentration terms to a ratio of activity terms. Mass action parameters of Equation 1 may be manipulated to influence the degree of extractability and to maximize the separation factors among the various cations. Values of $[TBP_{org}]$ which is commonly called "free" TBP, can be varied from its maximum value of 3.7 *M* in the pure extractant to any desired lower concentration by using an inert diluent such as normal dodecane (NDD). Physical properties of TBP and its diluent mixtures have been reviewed by several authors.[6-8] In most chemical studies, concentrations of about 1 *M* (30 vol %) and less are preferred because of the ease of phase separations and of recovery of U and Pu from the extractants. Values of $[NO^-_{3(aq)}]$ that are most useful for achieving the desired separations in the Purex process are in the range of $10^{-2}$ to about 3 *M*.

Equilibrium constants defined by Equation 2 have not been determined for the various TBP complexes over a wide range of concentrations because of the difficult task of determining the concentrations of all of the species which participate in the equilibrium and their activity coefficients. This difficulty stems primarily from the usual inability to directly measure the equilibrium species depicted in Equation 2. In the TBP extraction of cations from aqueous nitric acid solutions, several complex species of the cation may coexist in the aqueous phase. Species present may include nitrato, nitro, hydroxy, and nitro-nitrato complexes depending on the cation and the nitric acid concentration. Several different TBP complexes may also exist in the organic phase. However, with the exception of Zr and Ru complexes, under Purex process conditions it is more usual to have only a single complex formed with other extractable elements of interest such as U and Pu. Evaluating $[TBP_{(org)}]$ is further complicated by corrections for the extraction of $HNO_3$ (Figure 1) with the formation of the complexes $(HNO_3)_2 \cdot TBP$ and $HNO_3 \cdot TBP$. Under most conditions in Purex the $(HNO_3)_2 \cdot TBP$ can be neglected as can the extraction of water. Several authors[9-15] have pointed out the cor-

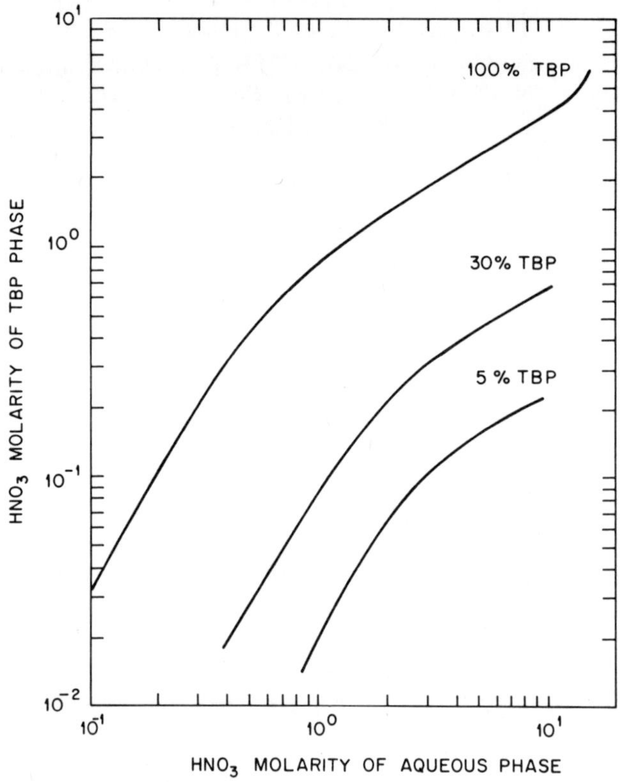


FIGURE 1. Equilibrium distribution of $HNO_3$ between nitric acid and TBP phases at 25°C.[11,14]

rection factors that are necessary for calculating equilibrium concentrations of the species depicted in Equation 2 from the total concentrations present.

Because of the difficulties in measuring the equilibrium concentrations of the various species, chemists have found it more practical to estimate the strengths of the various neutral complexes formed between cations and TBP from measurement of distribution ratios. The distribution ratio, D, for a given cation is defined as the ratio of its total concentration in the TBP phase, $C_{M(org)}$ to that in the aqueous phase, $C_{M(aq)}$, according to the following equation:

$$D = \frac{C_{M\ (org)}}{C_{M(aq)}} \tag{3}$$

This definition of D will be used throughout all subsequent discussions. Since the salting agent, $HNO_3$, is extracted by TBP in addition to metal cations, care must be taken to maintain $[TBP_{(org)}]$ and $[NO^-_{3(aq)}]$ constant in experiments designed to obtain values of D that are used for comparisons of the extractabilities of cations or in determinations of n in Equation 1. In such experiments, aqueous phases containing tracer levels of the metal cation and a fixed concentration of $HNO_3$ are contacted with the TBP phase. Since tracer levels of the test cation are employed, the fraction of nitric acid extracted is approximately constant.

In the separations chemistry dealing with macro amounts of actinides and fission products, a considerable number of D values have been measured in relation to

$[TBP_{(org)}]$ and to nitric acid and metal cation concentrations of the aqueous phases.[16] Such studies permit the evaluation of the effects of metal cation loading of the extractant and the effects of $HNO_3$ on extractant loading. Various computer programs[17-21] utilize these measured values of D to compute empirical values of K which, in turn, are used to calculate the distribution behavior of uranium, plutonium, and nitric acid in multistage, countercurrent extraction systems.

## III. EXTRACTION CHEMISTRY OF ACTINIDES

### A. Effect of Actinide Valence

The distribution behavior of actinides in their various valence forms is reasonably well understood. There are complexities due to valence control in the separation of uranium, plutonium, and neptunium from fission products and from each other, and the redox reactions employed are discussed in Section VI. Only the tetra- and hexavalent actinides show significant extractability under Purex process conditions. The pentavalent actinides are virtually inextractable and trivalent actinides have low distribution ratios. Distribution ratios for pentavalent actinides are sparse in the literature because of the difficulty in maintaining all of a given element in the +5 state.[22] Usachev and Chaikhorskii[23] have determined that the D value of Np(V) is of the order of $10^{-3}$ in 0.2 to 2 *M* $HNO_3$. Trivalent actinides have D values of a magnitude of a few hundreths under Purex conditions.[24]

Plutonium is the most extractable +4 actinide whereas it is the least extractable +6 actinide. The extractability of the +6 actinides decreases with increasing atomic number (Table 2). Trivalent actinides are not appreciably extracted by TBP except at very high concentrations of nitric acid or of inextractable metal salts such as sodium nitrate.[26]

### B. Effect of Nitric Acid Concentration

The effects of $HNO_3$ on the extractability of actinides[15,27-30] (Figures 2 and 3) are in qualitative agreement with equilibrium considerations. The formation of nitrate complexes of the actinides in the aqueous phase and of the TBP complexes with nitric acid in the extractant phase introduces complications that prevent quantitative interpreta-common ion effects of $[NO_3^-]$. On further increases of $HNO_3$ (>2 *M*) the formation of nitrate complexes in the aqueous phase becomes noticeable and the $HNO_3$ concentration in the extractant phase increases to the extent that the free TBP concentration decreases appreciably. Both effects cause the distribution ratio to pass through a maximum and then fall again with increasing $HNO_3$ concentration. Extraction behavior of $Pu^{3+}$ is in good accord with the above results and with that for trivalent lanthanides.[31,32] For 19% TBP[26] the D value for $Pu^{3+}$ increases from an initial value of about $10^{-3}$ at 0.2 *M* $HNO_3$ and reaches a maximum value of about 0.02 at about 2 to 3 *M* $HNO_3$.

### C. Effect of TBP Concentration

The complexes formed in the extraction of actinides have been shown to be $MO_2(NO_3)_2 \cdot 2TBP$,[24] $MO_2(NO_3) \cdot TBP$,[23] $M(NO_3)_4 \cdot 2TBP$,[24] and $M(NO_3)_3 \cdot 3TBP$[24] for the +6, +5, +4, and + 3 actinides, respectively. The ratio of TBP to metal nitrate salts is determined from the variation of measured D values as a function of TBP concentration (Figure 4). When log D is plotted vs. TBP concentration, the slope of the line gives the number of TBP molecules per actinide molecule in the complex. A typical plot for $Pu^{4+}$ extraction[30] is given in Figure 4.

**Table 2**
**VARIATION OF DISTRIBUTION RATIOS OF TETRA- AND HEXAVALENT ACTINIDES: 20°C, 1 *M* TBP IN XYLENE, 3 *M* $HNO_3$[25]**

| | Distribution ratio, D | | | |
|---|---|---|---|---|
| Valence | Th | U | Np | Pu |
| +4 | 2.13 | 1.59 | 3.63 | 22.0 |
| +6 | — | 23.6 | 27.1 | 5.25 |

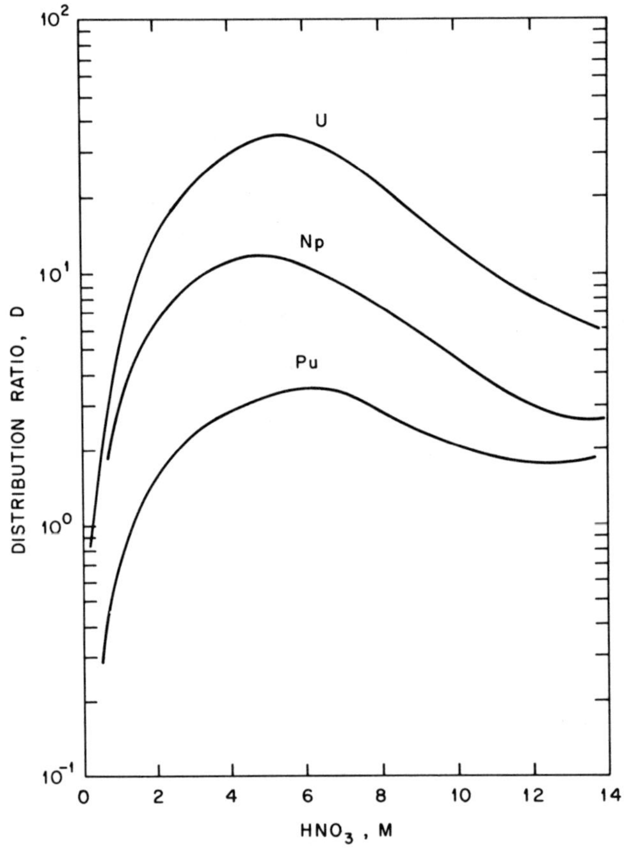


FIGURE 2. Extraction of tracer concentrations of hexavalent actinides by 19% TBP as a function of the nitric acid concentration of the equilibrium aqueous phase at 25°C (kerosine diluent).[27]

Evidence exists that complexes of the type $[(UO_2(NO_3)_2]_a$ $[HNO_3]_b$ $[TBP]_c$ are formed when all of the extractant capacity of 0.2 *M* TBP in NDD is utilized.[33] Infrared absorbance measurements obtained after successive dilutions of the loaded extractant phase indicated this complex was dissociating into the usual uranium and nitric acid complexes. However, as the authors have stated, more work needs to be done to confirm the existence of such mixed complexes.

Measurements of the effect of actinide loading on the distribution of actinides show

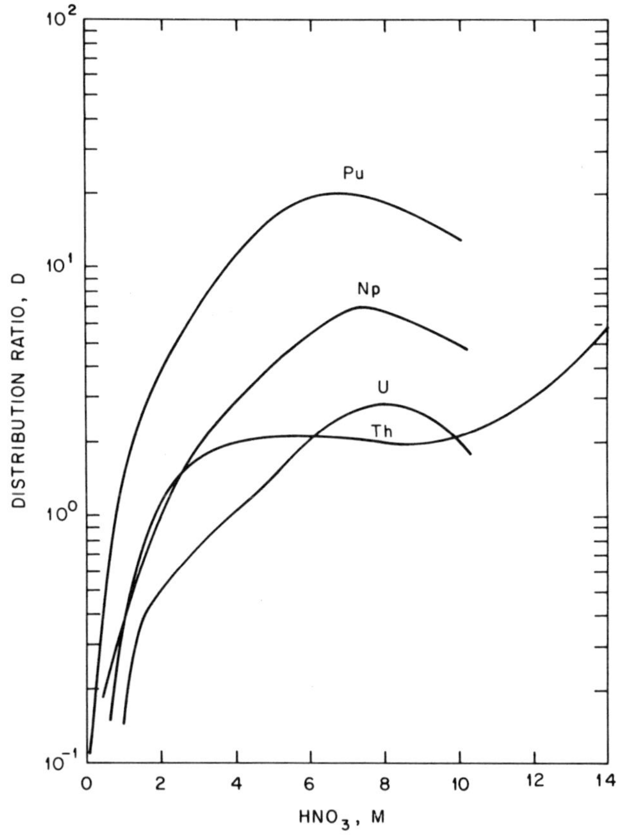


FIGURE 3. Extraction of tracer concentrations of tetravalent actinides by 19% TBP as a function of the nitric acid concentration of the equilibrium aqueous phase at 25°C (kerosine diluent).[27-29]

that limiting values of the actinide concentration in the organic extractant phase are approached as the actinide concentration of the equilibrium aqueous phase is increased. Different limiting values are approached depending on the equilibrium aqueous phase acidity[8,34] (Figure 5). The limiting value that is approached by each of the curves shown in Figure 5 corresponds theoretically to complete complexing of TBP by the formation of both the $UO_2(NO_3)_2 \cdot 2$ TBP solvate and the nitric acid solvates. In the case of hexavalent uranium, the maximum load in the extractant (about 0.5 *M* in 1 *M* TBP) can be nearly approximated by using high concentrations of inextractable metal nitrate salting agents such as $Al(NO_3)_3$. Often the uranium content of the extractant phase is expressed as a percentage of the theoretical amount necessary for the complete complexation of TBP as $UO_2(NO_3)_2 \cdot 2\ H_2O$.

The TBP complexes of tetravalent actinides have a limited solubility in the extractant phase, and process conditions must be controlled to prevent the presence of two organic phases in equilibrium with the aqueous phase.[35] When certain concentrations of Pu are exceeded, the extractant phase separates into two organic phases. The two phases consist of a heavier one that principally contains $Pu(NO_3)_4 \cdot 2TBP$ and TBP but little dodecane, and a lighter one that contains chiefly the diluent and little of the TBP or plutonium complex. The three-phase boundaries in the system $Pu(NO_3)_4$-TBP-$HNO_3$-kerosine[15,36] are shown in Figure 6. The boundaries of the three-phase system

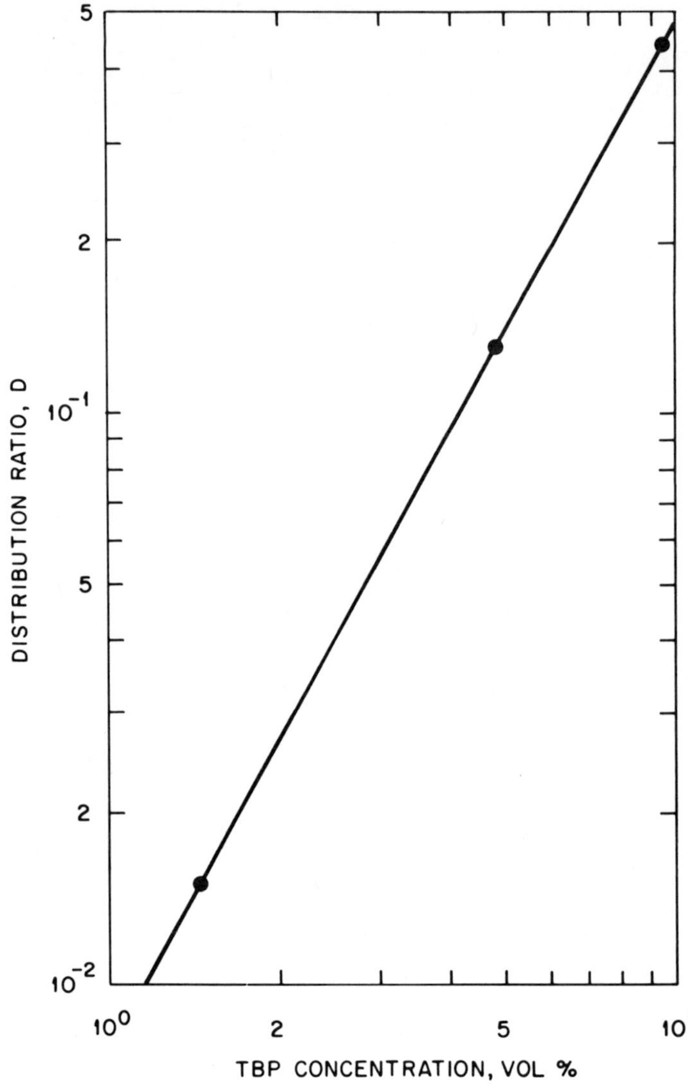


FIGURE 4. Dependence of the distribution ratio of Pu(IV) in trace concentrations on the TBP concentration at 25°C (initial $HNO_3$ concentration = 1 *M*.)[30]

are dependent on the plutonium concentration of the organic phase, the nitric acid concentration, and the temperature. Three phases exist for the area above the isotherms and two phases exist below. Boundary conditions also depend on the diluent used and the TBP concentration employed. Aliphatic diluents like dodecane have a higher tendency toward three-phase formation than do aromatic diluents such as diethylbenzene. Thorium[35] and hexavalent plutonium[36] have been shown to more readily form three phases than does tetravalent plutonium. It is also known that $HNO_3$[35] forms three phases at aqueous concentration of 17 to 20 *M*, but this is of no concern since nitric acid concentrations of this magnitude are not employed in Purex. It is likely that hexavalent uranium would also show three-phase behavior at high nitric acid concentrations and high loading of the extractant, but three phases have not been reported with the $HNO_3$ concentrations and the uranyl nitrate loadings employed in the Purex

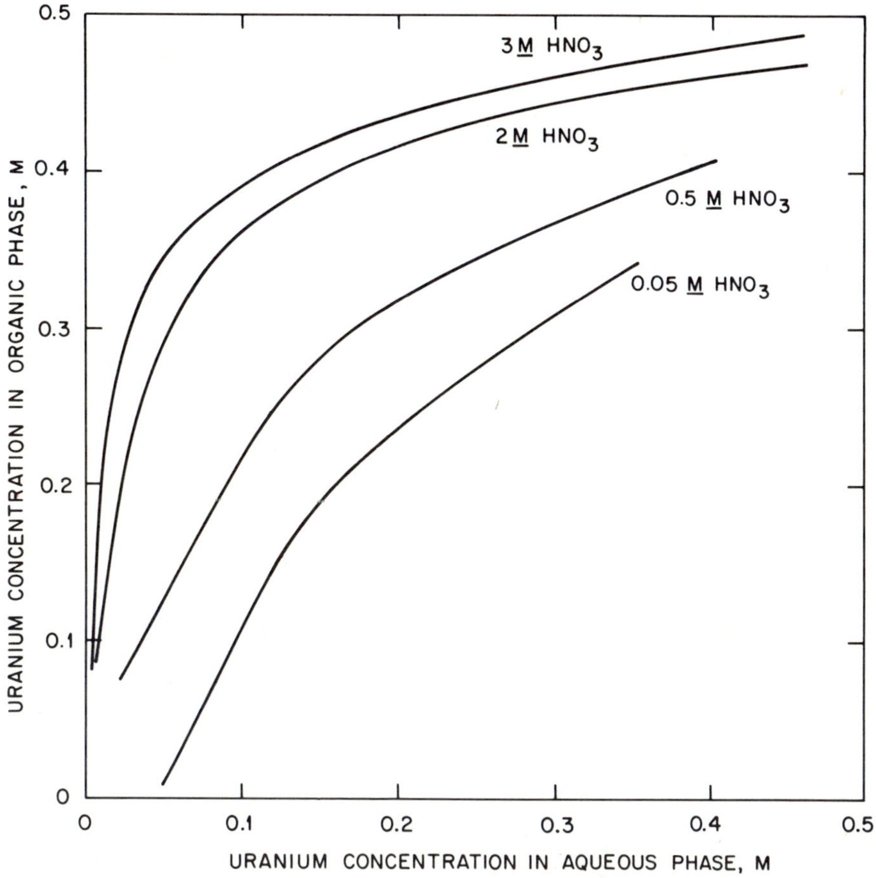


FIGURE 5. Effect of the equilibrium aqueous phase uranyl nitrate concentration on the uranium loading of the organic phase at several equilibrium $HNO_3$ concentrations (30% TBP-kerosine; 25°C).[34]

process. To date, three-phase systems have been avoided in Purex separations because of the difficulty of operating extraction equipment with the coexistence of three liquid phases.

The fact that the distribution ratio decreases with increasing loading of the extractant (i.e., decreasing "free TBP") is utilized to advantage in the recovery and purification of the most extractable ions, $UO_2^{2+}$ and $Pu^{4+}$. This effect is of considerable importance in extractive purification processes because impurities such as zirconium which are less extractable are removed from the extractant more effectively as the load of the most extractable elements is increased. Mathematical statements of solvent load effects based on Equation 1 (Section II) are difficult to derive because of the many parameters that must be considered and thus can be accomplished with only partial success in practice with the help of semiempirical and empirical relationships.

### D. Effect of Temperature

The enthalpy change for the extraction of the +4 and +6 actinide complexes is relatively small (Table 3). The enthalpy change is exothermic for the +6 actinides whereas, with the exception of thorium, it is endothermic for the +4 actinides. Because the enthalpy changes are rather small, only modest increases in the extractability of

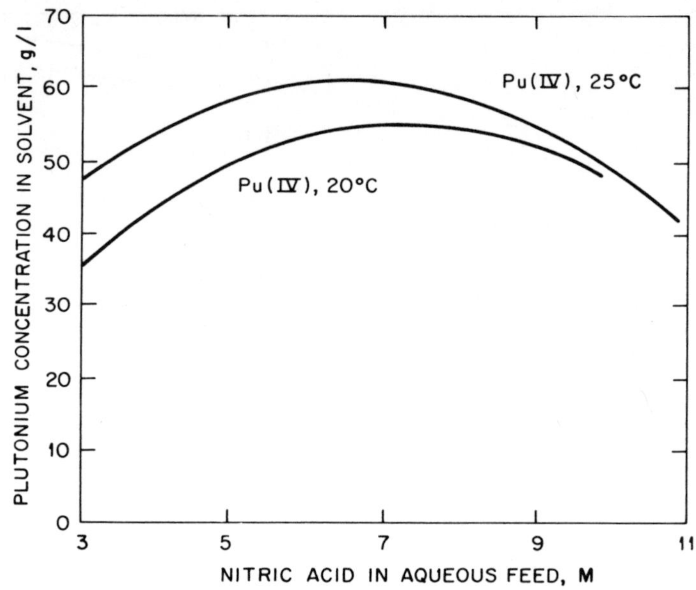


FIGURE 6. Three phase boundaries for the system $Pu(NO_3)_4$-20 vol % TBP-nitric acid as a function of the aqueous phase nitric acid concentration (TBP diluent is kerosine). (Adapted from Donaldson, D. M., Hartley, K., Lees, P., and Parkinson, N., *Trans. Metall. Soc. AIME*, 227, 191, 1963. With permission.)

**Table 3**
**ENTHALPY CHANGES FOR THE EXTRACTION OF ACTINIDES FROM 3.0 *M* NITRIC ACID WITH 1 *M* TBP IN XYLENE[a,25]**

Temperature range 20 to 60°C
Tracer level actinides

| Actinide ion | Enthalpy change, kcal $mol^{-1}$ |
|---|---|
| Th(IV) | −2.57 |
| U(IV) | 1.81 |
| Np(IV) | 2.51 |
| Pu(IV) | 1.44 |
| U(VI) | −2.85 |
| Np(VI) | −4.81 |
| Pu(VI) | −3.11 |

[a] Enthalpies determined from plots of log D versus the reciprocal of absolute temperature.

Pu(IV) or modest decreases in the extractability of U(VI) are attained on increasing the extraction temperature. The magnitude of the enthalpy change depends on the nitric acid concentration of the aqueous phase, the degree of saturation of the organic phase with extractable metal ions and nitric acid, the diluent employed, and the temperature range. Although xylene rather than paraffinic hydrocarbons was used as the diluent for the enthalpy change measurement of Table 3, these data were cited[25] because they were the only data available in which comparisons of these actinides were made under identical conditions. There is some disagreement in the literature[25] as to the numerical values of the enthalpy changes for the extraction in 30% TBP of U(VI) and Pu(IV) by TBP diluted with hydrocarbons. Enthalpy changes reported for Pu(IV) extraction are of the order of about 1 to 2 kcal $mol^{-1}$ except for low acidity conditions where $Pu^{4+}$ hydrolysis make the results suspect. Enthalpy values for U extraction range from about −6.3 to −6.0 kcal $mol^{-1}$. The enthalpy changes for all of the individual chemical reactions are needed to resolve differences among various investigations. The enthalpy changes associated with the dehydration of the actinide ion, the formation of nitrate complexes of the actinide ion in the aqueous phase, and the formation of the TBP complex with the extractant phase are all needed to understand the overall enthalpy change in a meaningful way. Recently, Patil and co-workers[10] have determined the thermodynamic value for the net enthalpy of U(VI) extraction into TBP-dodecane (2.5 to 30% TBP) to be −5.0 ± 0.5 kcal $mol^{-1}$. The value was obtained by extrapolating enthalpy values that were determined for the extraction of uranyl ion from 0.1, 0.25, 0.5, and 4.0 *M* $HNO_3$ to zero concentration of $HNO_3$. The value obtained by this method is in remarkable agreement with the −5.1 ± 0.2 kcal $mol^{-1}$ value[37] determined from changes in the equilibrium constant (Equations 1 and 2) with temperature.

The effect of temperature on the extraction of actinides is diminished by increasing loading of the extractant phase[8,38] (Figure 7). The enthalpy change for the extraction of actinides is relatively insensitive to the nitric acid concentration of the aqueous phase.[38] Since it is the overall enthalpy that is reflected by extraction equilibria, nitric acid would not be expected to affect the actinide enthalpies. The distribution ratio of nitric acid remains approximately constant as the temperature is increased.[39]

A study[40] of the variation of Pu(IV) distribution ratios over the temperature range of 5 to 80°C illustrates the need for a better understanding of the role of shifts in the relative percentages of aqueous ionic species as the extraction temperature is increased. The observed maximum in the value of D is reached at about 60°C, after which there is a uniform decrease. Extractions were carried out using an aliphatic diluent. Maxima were observed at nitric acid concentrations of 1.7 and 3.0 *M* in the initial aqueous phase, but not at 0.5 *M*. The authors interpret the results on the basis of shifts in the relative amounts of the aqueous nitrato complexes of Pu(IV) as the temperature is increased. However, to verify this interpretation, it must be ascertained that significant hydrolysis of Pu(IV) did not occur and that the enthalpy changes for the shifts among the aqueous nitrato complexes are not endothermic.

### E. Extraction Kinetics

Equilibrium is attained rather quickly for both the forward and reverse transfer of actinides between the aqueous nitric acid phase and the TBP phase.[41,42] Rates of transfer of U and Pu from the aqueous to organic phase and for the reverse transfer are shown in Figure 8. Although uranium is more rapidly extracted but more slowly stripped than plutonium, both rates are too rapid to permit practical separations of them based on the kinetics. The kinetic data suggest a transport mechanism in which a "boundary surface complex" is formed between metal cations, nitrate ions, and TBP. The concentration of the "boundary surface complex" can be described by a Langmuir-type adsorption isotherm, and the transfer rate is governed by the desorption rates into the aqueous or organic phases. Complete transfer of U and Pu occurs in about 5 to 10 sec in centrifugal contactors[43,44] where very high interfacial areas are obtained.

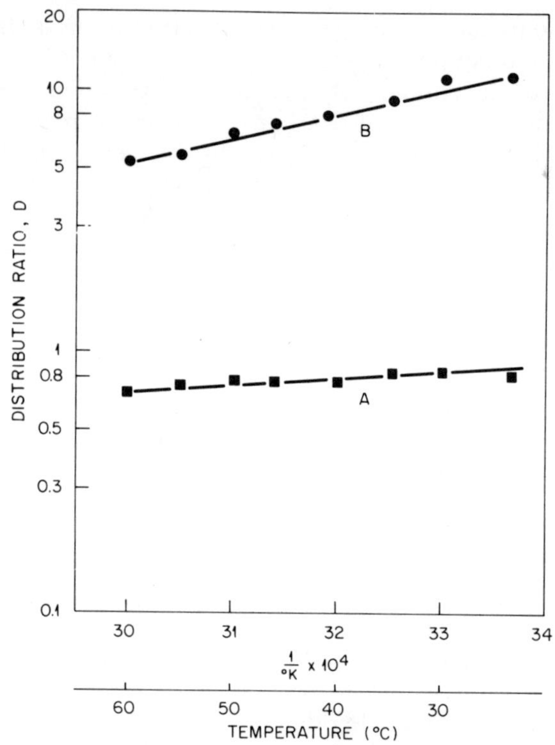


FIGURE 7. Variation of the uranium distribution ratio with temperature at different uranium loadings of 15% TBP-normal dodecane (initial aqueous phase before extraction: curve A—3 *M* $HNO_3$, 94 g/ℓ U; curve B—3 *M* $HNO_3$, 12 g/ℓ U; Volume ratio of organic to aqueous of 1). (Adapted from Horner, D. E., Formation of Third Phases and the Effect of Temperature on the Distribution of Plutonium and Uranium in Extractions by Tri-n-Butyl Phosphate, Rep. ORNL-4724, Oak Ridge National Laboratory, Oak Ridge, Tenn., 1971.)

## IV. EXTRACTION CHEMISTRY OF FISSION PRODUCTS

The only fission products which are moderately extractable under Purex process conditions are zirconium, ruthenium, niobium, and technetium. Ruthenium exhibts the most complex extraction behavior of any fission product element. Ruthenium and zirconium have been the most troublesome fission products in the operation of Purex processing plants. High loadings of TBP with actinides and back scrubbing of extracts with $HNO_3$ are utilized to suppress fission product extraction.

### A. Zirconium

Zirconium extraction is complicated because of the numerous complex species that can exist in aqueous solutions. The aqueous complex that is most extractable by TBP is the neutral tetranitrato complex which is in equilibrium with lower cationic nitrato complexes of the general formula $[Zr\ (H_2O)_X\ (NO_3)_{4-X}]^{(4-X)+}$, neutral mixed complexes of the type $Zr(OH)_m(NO_3)_{4-m}$, and the anionic complexes $Zr(NO_3)_5^-$ and $Zr(NO_3)_6^{2-}$. Direct identification of these complexes in aqueous solution and measurements of all their formation constants have not been accomplished. Rather, these complexes have been inferred from the compositions of salts in equilibrium with aqueous phases of varying nitric acid concentration or from the ion exchange or solvent extraction behavior of aqueous nitric acid solutions containing soluble zirconium.

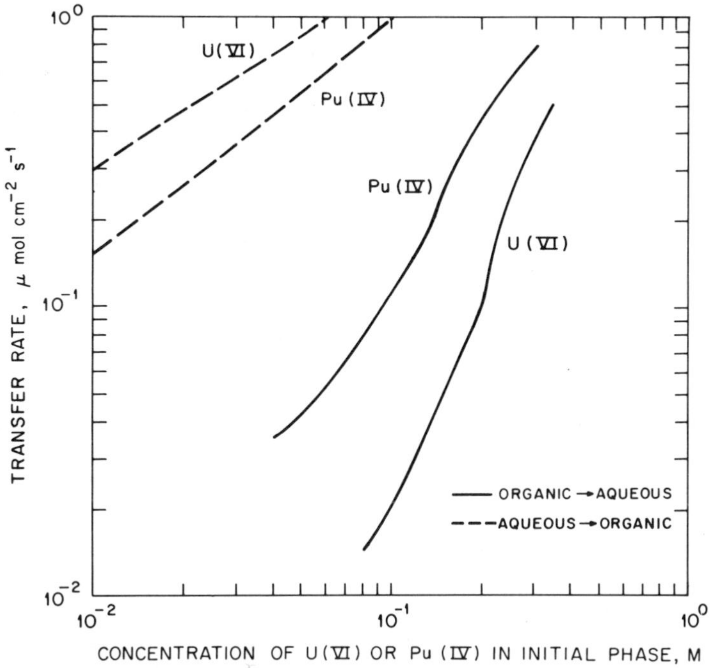


FIGURE 8. Initial transfer rates of Pu(IV) and U(VI) between an aqueous 3 *M* $HNO_3$ phase and a 20% TBP-normal dodecane phase. (Adapted with permission from Baumgartner, F. and Finsterwalder, L., *J. Phys. Chem.*, 74, 108, 1970. Copyright 1970, American Chemical Society.)

In addition to the above-mentioned nitrato complexes, there can exist in aqueous solutions of zirconium soluble polymeric species which are virtually inextractable by TBP and which are not sorbed by ion exchange resins.[45,46] When zirconium is present at concentrations of about $10^{-2}$ *M* or greater, very stable polynuclear complexes are formed which are not in rapid equilibrium with monomeric forms. The relative percentage of the soluble zirconium in this form depends on the zirconium nitrate salts used to prepare solutions and on the dissolution procedure employed.[47] High acid concentrations and elevated temperatures are required to convert these inextractable forms to monomeric extractable forms. For example, in 7 *M* $HNO_3$ about 10 hr at 90°C was required to achieve 90% conversion of an aqueous soluble inextractable species to an extractable species.[46] The conversion proceeded with first order kinetics and an activation energy of about 25 kcal $mol^{-1}$.

The zirconium present in fuel dissolver solutions appears to exist primarily in nonpolymeric nitrato complexes. Under Purex process conditions, zirconium is extracted mainly as the complex $Zr(NO_3)_4 \cdot 2TBP$ which is considered the normal species of solvated zirconium. However, another equilibrium exists between the normal extracted species and another solvate described by Orth and co-workers[48] as $[Zr]_s$. The slow rate of exchange between the normal species and $[Zr]_s$ determines the efficiency of transfer of zirconium from the extractant to the aqueous phase.

The inextractable species of zirconium are not of major concern in Purex process applications since it is desired that the fission products be rejected to an aqueous raffinate. After aqueous feed adjustment and clarification, nitric acid concentrations and temperatures used in extraction are maintained at conditions to prevent precipitation of polymeric species during bulk separation of the actinides from fission products.

There is, however, a tendency to form the polymeric species in the low acid conditions used in back-extraction of the actinides.

The distribution ratio for the extraction of the extractable species has been measured by many workers. The effect of nitric acid concentration on the distribution is typified by the results of Alcock and co-workers.[49] Since distribution measurements are based on measurement of the total zirconium concentration of the aqueous and extractant phases, care was taken to isolate the extractable complexes by first extracting them from mixtures of extractable and inextractable complexes using TBP and then back-extracting them into an aqueous phase. Using this procedure, stock solutions of extractable species in about 8 $M$ $HNO_3$ are usually prepared for distribution ratio measurements. The relative amounts of the extractable nitrato complexes and the inextractable polymeric complexes remain essentially unchanged when solutions are diluted with nitric acid and extracted immediately. However, even using this procedure, small increases in D values are observed on back-extraction of the extractable species.[49] Nevertheless, such experiments show the approximate effect of $HNO_3$ concentration on the distribution ratio for the extractable nitrato complexes (Figure 9). In contrast to the extraction of actinides, values of D for zirconium extraction do not show maxima at a certain $HNO_3$ concentration but continuously increase as the $HNO_3$ concentration is increased. It is seen that the D value of Zr is only of the order of 0.1 at 2 to 3 $M$ $HNO_3$ and satisfactory separation from the more highly extracted actinides can be accomplished.

Distribution ratios of zirconium have been shown to depend only on the total nitrate concentrations by varying the proportions of inextractable $NaNO_3$ and $HNO_3$ in aqueous phases of constant total nitrate ion concentration. Other elements show increases in D values as the concentrations of inextractable salting agents are increased relative to nitric acid because more free TBP is available. The unusual extraction behavior of zirconium with respect to either nitric acid or inextractable salting agents has not been satisfactorily explained. Evidently it is related to differences in the degree of nitrate complexing for the two systems. However, it is difficult to comprehend how this effect could always exactly compensate for changes in the free TBP content in experiments where the total nitrate concentration is constant and the molar ratio of $HNO_3$ to $NaNO_3$ is varied.

Results of early work indicated only one solvate, $Zr(NO_3)_4 \cdot 2TBP$, was formed.[49] More recent work[50-52] indicates that in addition to the disolvate a monosolvate, $Zr(NO_3)_4 \cdot TBP$, is formed, and the monosolvate is the dominant complex at TBP concentrations of less than about 0.3 $M$. By means of thermodynamic analysis of experimental data, Solovkin[52] in his later work has indicated the formation of the following compounds in extractant phases in equilibrium with aqueous phases when nitrate concentrations are <6$M$:

$$Zr(OH)_2(NO_3)_2 \cdot 2\,TBP,\ Zr(OH)(NO_3)_3 \cdot 2\,TBP,\ \text{and}\ Zr(NO_3)_4 \cdot 2\,TBP$$

Calculated values of distribution ratios were in remarkable agreement with published values, notwithstanding the uncertainties in the assumptions involved with equations used to calculate activity coefficients. The $Zr(NO_3)_4 \cdot 2TBP$ complex is, of course, dominant at high acidities. It is interesting to note that $Zr(NO_3)_4 \cdot TBP$ does not appear in the thermodynamic analysis. Higher solvates may exist at aqueous nitrate concentrations of >6$M$.

Zirconium extraction shows unusual behavior with respect to extraction temperatures. Shevchenko and Fedorov[40] have shown that in 3 $M$ $HNO_3$, D values decrease by approximately fourfold when the extraction temperature is increased from 5 to

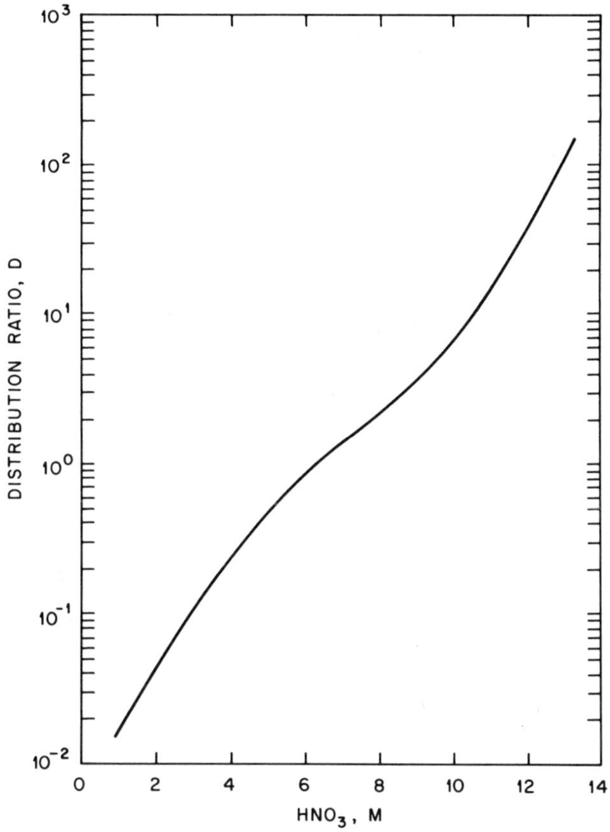


FIGURE 9. Extraction of zirconium by 19% TBP as a function of the nitric acid concentration of the equilibrium aqueous phase at 20 to 23°C (kerosine diluent, $10^{-11}$ to $10^{-12}$ $M$ $^{95}Zr$.)[49]

about 30°C. After attainment of a minimum value at about 30° C, values of D increase uniformly with increasing temperature giving a fivefold increase in D when 80°C is reached. The authors surmise that the temperature behavior may be explained by two competing effects: (1) shifts in the equilibrium of the aqueous nitrato complexes in the direction of lower nitrato complexes and hydrolytic species with increasing temperature and (2) increasing extractability of $Zr(NO_3)_4 \cdot 2TBP$ with increasing temperature. The first effect is presumed dominant in the 5 to 30°C range, and the second effect is dominant for the 30 to 80°C range. However, the authors conclude that further work is required to confirm this interpretation.

## B. Ruthenium

The extraction behavior of ruthenium from nitric acid solutions is different from that of actinides or zirconium. The extractable aqueous species, which consist of the higher nitrato complexes and the higher mixed ligand complexes containing both nitro and nitrato groups, are not in rapid equilibrium with their lower complex forms containing these ligands. Therefore, the common ion effect of the aqueous ligand cannot shift the equilibrium in the direction of the higher complexes as their extraction proceeds during an acceptable contacting time for the extractant and aqueous phases. Similarly, slow equilibrium between the stable TBP solvate and more labile TBP solvates of the extracted ruthenium complexes prevents effective back-extraction of ruthenium into aqueous phases.

During the dissolution of irradiated, light water reactor (LWR) fuel in nitric acid some of the ruthenium (70 to 80 %, see Chapter 3) is dissolved and aqueous complexes are formed between the very stable ruthenium nitrosyl ion, $RuNO^{3+}$, and nitrato and nitro ligands which have the following general formulas:

$$RuNO(NO_3)_x(H_2O)_{5-x} \qquad x \lesssim 4$$

$$RuNO(NO)(NO_2)_y(H_2O)_{5-y} \qquad y \lesssim 2$$

$$RuNO(NO_3)_x(NO_2)_y(H_2O)_{5-x-y} \qquad x + y \lesssim 3$$

Identification of these complexes has not been carried out using actual LWR dissolver solutions but their presence has been demonstrated by maintaining ruthenium in solutions that simulate actual process conditions. The nitro complexes have the highest stability but are slowly converted to the nitrato species by mass action using excess nitrate ion.[53] Isolation and identification of these aqueous complexes and the determination of their distribution ratios were pioneered by Fletcher and co-workers[54-59] at Harwell, and since that time additional studies have been reported by workers at other laboratories.

The complexes of nitrosyl ruthenium have an octahedral structure. Positions relative to the ruthenium atom are conventionally depicted by the structure notation in Figure 10. The NO group is always located at Position 1 and the position that is trans to the NO group (Position 6) is very labile and is believed to always be occupied by a water molecule when the complexes are present in aqueous phases. Using paper chromatography, ion exchange, extraction chromatography, and solvent extraction, the aqueous complexes listed in Table 4 have been isolated and identified in solutions that are closely similar to Purex process solution.[54-62] Of the complexes, the tri- and tetranitrato complexes show the greatest extractability into TBP. The tri complexes containing both nitro and nitrato groups are of moderate extractability and the di and mono complexes are of low extractability. Since the nitrato complexes show the highest extractability, more detailed studies have been made of these complexes. It has been estimated that of the complexes present in actual plant solutions (2 to 3 $M$ $HNO_3$ and $10^{-3}$ to $10^{-2}$ $M$ $HNO_2$), up to 70 mol % may contain nitro groups.[58] Decontamination of actinides from ruthenium in processing plants seems to increase in inverse proportion to the fraction of ruthenium present as the higher nitrato complexes.[59]

Studies of the extraction mechanism of the higher nitrato complexes show the stable species in the extractant phase is a solvate of the neutral trinitrato complex which is in slow equilibrium ($t½ \approx 10$ min) with two different labile solvates, one containing the trinitrato complex and another containing the tetranitro complex.[59] Likely structures of these complexes are given in Figure 11. The slow step in transforming the labile solvates is the movement of a TBP molecule with displacement of a $H_2O$ molecule from the trinitrato solvate and a nitrato group from the tetranitrato solvate. The nonlabile, trinitrato is responsible for the ineffectiveness of back-extracting ruthenium into aqueous $HNO_3$ scrubbing solutions at acceptable contacting times of the immiscible phases. Fletcher [59] has indicated that it may be possible to replace nitrato groups in the nonlabile, trinitrato solvate with aquo groups during back-washing with dilute $HNO_3$ while still maintaining the Ru-TBP bond. This conversion might explain the low decontamination factors observed for ruthenium in additional purification cycles. However, confirmation of the existence of nonlabile solvates containing <3 nitrato groups has not been reported.

The equilibrium distribution of the various aqueous nitrato complexes is shown in

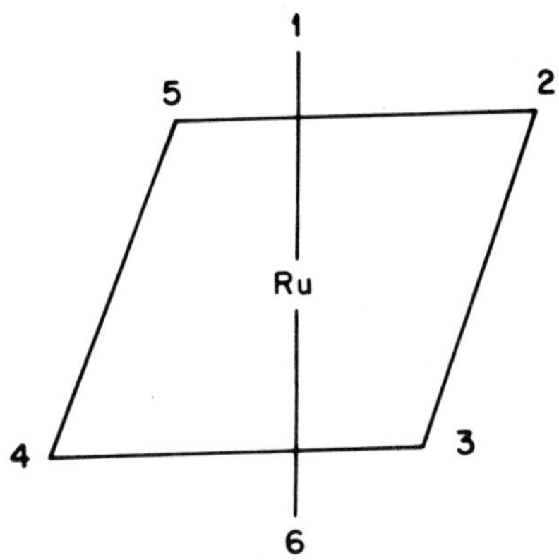


FIGURE 10. Octahedral structure representation of nitrosyl ruthenium complexes. (Position 1 is occupied by the NO group and is stable; position 6 is labile; positions 2, 3, 4, and 5 are stable with respect to complexing ligands such as $NO_3^-$ and $NO_2^-$.)

## Table 4
## TYPES OF COMPLEXES FORMED IN SIMULATED PROCESS STREAMS WITH NITROSYL RUTHENIUM ION[54,61]

| Type | Number | Charge[a] | Common designation | Tenative structure designation[b] |
|---|---|---|---|---|
| nitrato | 1 | +2 | mononitrato | 2-nitrato |
| | 2(cis) | +1 | cis-dinitrato | 2,3-dinitrato |
| | 2 (trans) | +1 | trans-dinitrato | 2,4-dinitrato |
| | 3 | 0 | trinitrato | 2,3,4-trinitrato |
| | 4 | −1 | tetranitrato | 2,3,4,5-tetranitrato |
| nitro | 1 | +2 | mononitro | 2-nitro |
| | 2 (cis) | +1 | dinitro | 2,3-dinitro |
| | 2 (trans) | +1 | dinitro | 2,4-dinitro |
| mixed | 2 | +1 | none | 2-nitro-3-nitrato |
| | 2 | +1 | none | 2-nitro-4-nitrato |
| | 3 | 0 | none | 2,3-dinitro-4-nitrato |
| | 3 | 0 | none | 1,4-dinitro-3-nitrato |
| | 3 | 0 | none | 2-nitro, 3,4-dinitrato |

[a] Assumes trans-position relative to NO is occupied by a water molecule.

[b] Joon's[61] nomenclature for the nitro and mixed nitro-nitrato complexes was changed to correspond with that of Fletcher[54] and to that of Figure 10.

Figure 12. Because of the slow rates of addition of nitrato ligands to lower nitrato complexes, increasing the $HNO_3$ concentration does not increase the distribution ratios. Instead, the ratios decrease because of the extraction of $HNO_3$ with its attendant reduction of the free TBP concentration. Rudstam[62] did observe maxima in D values at about 0.5 *M,* although not very pronounced, for the extraction of the aqueous trinitrato and dinitrato complexes (Figure 13). The D values illustrated in Figure 13 for the mixture of the cis- and trans-dinitrato complexes are meant to indicate the upper

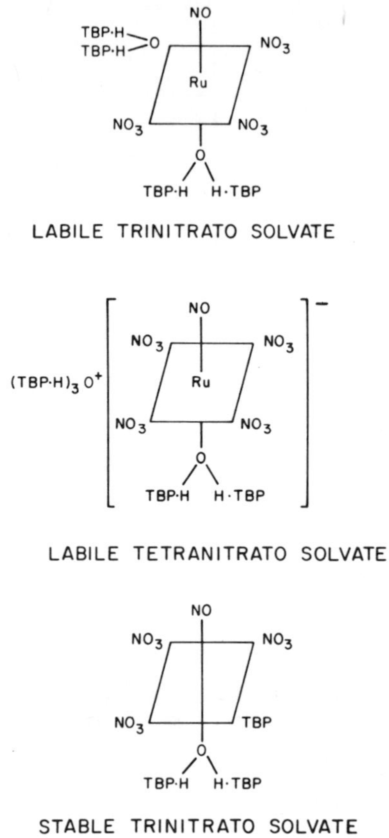


FIGURE 11. Proposed structures for the solvates formed between TBP and the aqueous tri- and tetranitrato complexes of nitrosyl ruthenium.[55,59]

limit for the extraction, since small quantities of the higher complexes are always present due to incomplete separations of the species. Values of the distribution ratios given for the various complexes are for a 1-min contacting of the phases. All values used throughout subsequent discussions are for a 1-min contact time unless otherwise stated. Distribution values based on back-extraction of ruthenium from extractant phases containing only the stable trinitrato solvate[59] are shown in Figure 14. It is seen that once the stable solvate is formed, it is not effectively back-extracted into $HNO_3$ regardless of the $HNO_3$ concentration using a 1-min contact time.

Studies[54,62] of the rates of interconversion of the various aqueous nitrato complexes show that the rates are slow (Table 5) and that the dinitrato omplex and lower complexes would not be expected to show appreciable extractability. The reactions are first order with respect to the concentration of the complexes. The rate constants for nitration are dependent on the nitric acid concentration, whereas the generally slower aquation reactions are independent, an exception being the tetranitrato conversion. The rates of interconversion are increased by increasing temperature. Rudstam's data[62] indicate that the activation energies are of the order of 15 to 20 kcal $mol^{-1}$. Therefore, temperature of extraction should affect distribution ratio values. The temperature effect on the distribution ratios for the extraction of the trinitrato complexes is shown in Figure 15. It is seen that significant decreases in the extraction coefficients are ob-

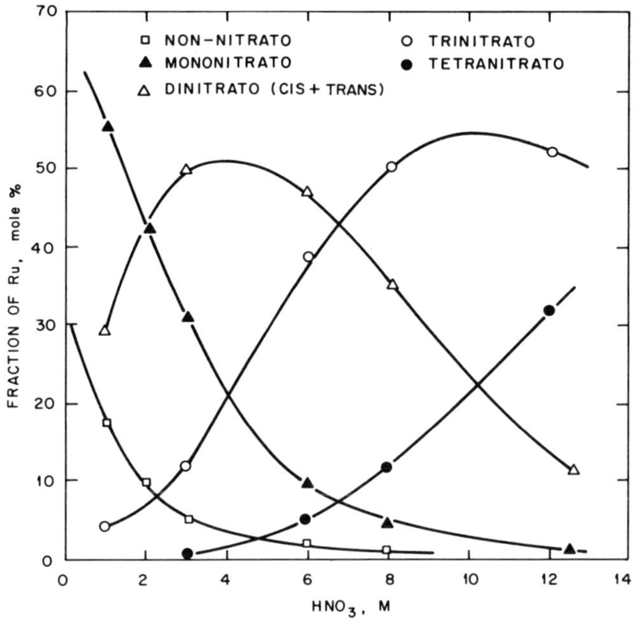


FIGURE 12. Variation in the relative quantities of the different nitrato-nitrosyl ruthenium complexes in nitric acid at 20°C (total ruthenium = $10^{-3}$ *M*.) (Adapted from Scargill, D., Lyon, C. E., Large, N. R., and Fletcher, J. M., *J. Inorg. Nucl. Chem.*, 27, 161, 1965. With permission.)

tained by increasing the temperature. This effect is no doubt produced by increasing the rates of interconversions among the species present in both phases and perhaps also by decreasing the molecular attraction between TBP and the TBP-nitrato solvates in the extractant phase.

Kinetic studies of the extraction of ruthenium from equilibrium mixtures of the nitrato complexes show increasing extraction with increasing mixing time of the phases. In 3 $M$ $HNO_3$, Fletcher et al.[56] obtained the following results at 25° C using 1 $M$ TBP:

| Time (min) | 0.5 | 10 | 30 | 60 | 180 | 660 |
|---|---|---|---|---|---|---|
| $D_{Ru}$ | 0.074 | 0.093 | 0.119 | 0.146 | 0.167 | 0.167 |

In shorter contact times, the extraction of Ru is reduced even further. The rate-limiting step for interfacial transfer is not diffusion control but slow rates of conversion of aqueous and organic phase species. Studies[63] at short contact times indicate that the extractable species initially present in the aqueous phase are extracted in $< 1$ sec and that slow conversion of the species occurs thereafter.

Joon[61] has made studies of the nitro, mixed nitro, and nitrato complexes previously shown in Table 4. His results indicate the same general trend for the interconversion rates and the extractabilities of these complexes as that exhibited by the series of nitrato complexes. The notable difference, of course, is the higher extractability of the nitrato species shown in Table 4. Only one mononitro and two dinitro complexes have been identified in nitric acid solutions by the chromatographic techniques available. It is interesting that the presence of nitrous acid in the extractant phase does not facilitate conversion by displacement of nitrato groups from those species which contain nitro ligands.

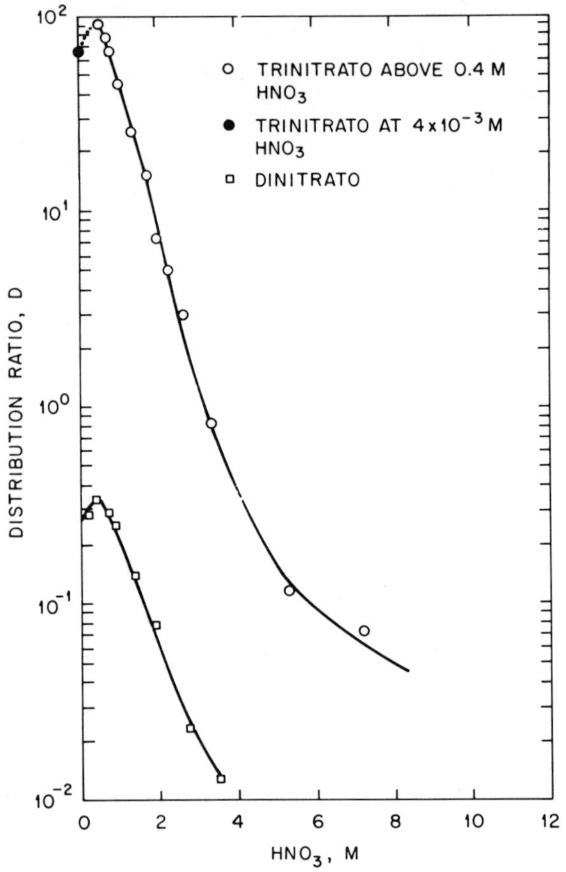


FIGURE 13. Extraction of the dinitrato and trinitrato complexes of nitrosyl ruthenium as a function of the nitric acid concentration of the equilibrium aqueous phase at 20°C (organic phase: 30% TBP in nysolvin 100/200 — a paraffinic kerosine.)[16]

Because of the perplexing behavior[8,58,64] of ruthenium extraction, various chemical treatments of the aqueous feed solutions have been tried for the purpose of reducing its extractability. Chemical treatments with ozone, nitrous gases, hydrogen peroxide, and hydrazine have all been tried but without much success. Ozone and other powerful oxidizing agents such as Ce (IV) oxidize ruthenium to the inextractable IV state, but unfortunately the reaction cannot be stopped there. The volatile tetroxide is also formed which distributes ruthenium activity throughout the processing plant in an undesirable manner. Nitrous gases convert nitrato complexes to the less extractable nitro species but the process is slow (about 3 hr at 60°C) and does not go to completion. Further decontaminations from ruthenium are limited to a factor of about 2. Hydrogen peroxide decomposes too rapidly in process solutions to be of value. Hydrazine and other reducing agents are unsuitable in the presence of plutonium because it is also undesirably reduced, but are of some value in the purification of uranium from ruthenium.

Although variables can be adjusted in multiple cycles of countercurrent extraction to attain a high degree of decontamination of actinides from ruthenium, fission product ruthenium still presents a challenge to separations chemists. As pointed out by

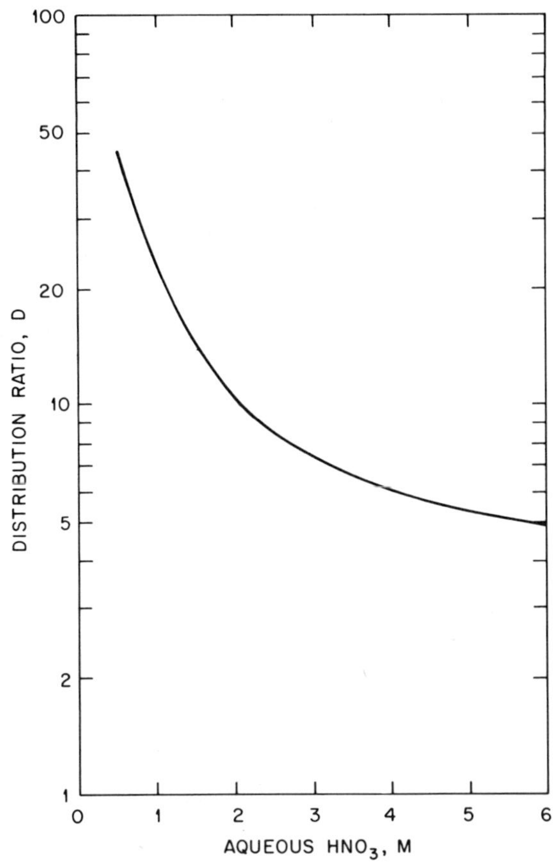


FIGURE 14. Distribution ratios for the back-extraction of ruthenium from an initial 20% TBP-kerosine phase containing only the stable solvate at 25°C.[59]

**Table 5**
**RATES OF NITRATION AND AQUATION REACTIONS OF NITRATO-NITROSYL RUTHENIUM COMPLEXES AT 25°C**

| Reaction | Rate constant (× $10^3$ $min^{-1}$) | | | |
|---|---|---|---|---|
| | Nitration | | Aquation | |
| $HNO_3$ | 6.1 *M* | 8.1 *M* | 6.1 *M* | 8.1 *M* |
| mono- ⇌ cis-di- | 1.65 | 3.0 | 0.70 | 0.72 |
| mono- ⇌ trans-di- | 8.1 | 13.5 | 4.3 | 4.5 |
| cis-di- ⇌ tri- | 17.9 | 32.5 | 11.1 | 11.7 |
| Trans-di- ⇌ tri- | 74.0 | 144 | 36.5[a] | 37.4 |
| tri- ⇌ tetra- | 17.0 | 68 | 102 | 270 |

[a] Rudstam[62] found 25 × $10^{-3}$ $min^{-1}$ for this rate constant at 20°C in 0.95 *M* $HNO_3$.

From Scargill, D., Lyon, C. E., Large, N. R., and Fletcher, J. M., *J. Inorg. Nucl. Chem.*, 27, 161, 1965. With permission.

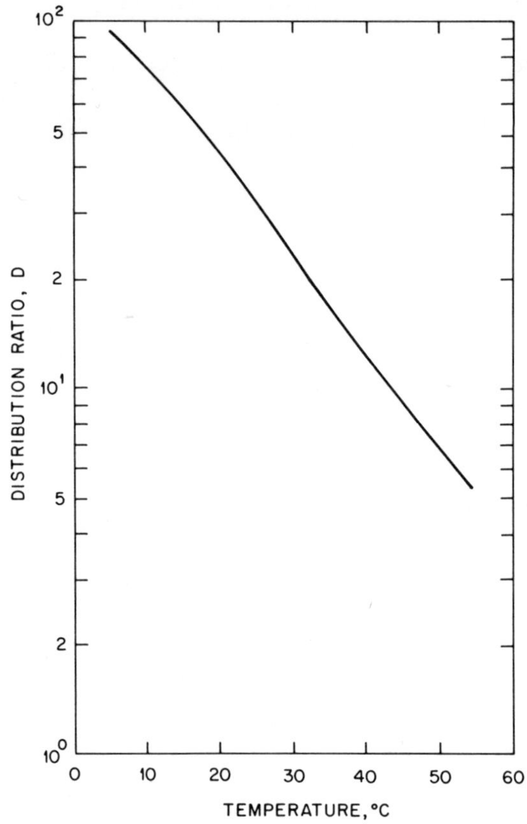


FIGURE 15. Effect of temperature on the distribution ratio of the trinitrato nitrosyl ruthenium complex between 30% TBP — kerosine and 1 $M\,HNO_3$.[59]

McKay,[64] first we need better understanding of its behavior, and then hopefully it will be in the realm of chemical behavior to make it all go where it is wanted — in the high-activity fission product wastes.

### C. Niobium

Niobium shows very low extractability into pure TBP-diluent phases.[65,66] However, its aqueous solution and extraction behavior are not well understood. Interaction of polymeric or colloidal niobium species with the species of the various other elements present in process solutions has received limited study. Degradation products formed in the extractant phase by radiation and hydrolysis are the primary reason for the slight extractability of niobium in the Purex process.

At concentrations of niobium encountered in dissolver solutions ($\sim 10^{-5}$ *M*), niobium is present in inextractable polymeric forms of Nb(V). Only a small fraction is present in chemical forms which show any extractability into TBP and their extractability is very low ($D \sim 5 \times 10^{-4}$ with 2 $M\,HNO_3$/20% TBP). Hardy and Scargill[65] reported that reproducible extraction could only be attained with niobium concentrations that were $< 10^{-7}$ *M*. At $10^{-5}$ *M* concentrations, larger irreproducible amounts of polymeric species are attained even after heat treatment in strong nitric acid.

Hardy and Scargill[65] carried out their work in glassware that had been waterproofed by coating it with hydrophobic silicone compounds. Orth[48] also found that niobium had virtually no extractability provided no silicate compounds were added and glassware was thoroughly waterproofed. However, addition of silicate compounds or failure to adequately coat glassware caused perceptible increases in niobium extraction. It was presumed that the niobium was extracted as a silicate complex or was adsorbed on colloidal silica which is somehow transferred to the organic phase. Both niobium and zirconium are known to absorb on silica gel columns; this method is used to achieve further purification of uranyl nitrate solution.[8] Niobium can be removed from nitric acid solutions containing zirconium by its adsorption on unfired Vycor® glass.[67]

### D. Technetium

The levels of decontamination of actinide products from reprocessing plants with respect to technetium have not been determined. This is because its radioactivity in the product is negligible in comparison to Zr, Nb, and Ru. It may, however, be important as a chemical impurity in the conversion of uranium products to $UF_6$ and in the subsequent separation of uranium isotopes.

Technetium is present in nitric acid solutions as the pertechnetate ion ($TcO_4^-$). Siddall's[68] results indicate the extraction mechanism from pure nitric acid solution is the formation of a trisolvate according to Equation 4:

$$H^+_{(aq)} + TcO^-_{4(aq)} + 3\ TBP_{(org)} \rightleftarrows HTcO_4 \cdot 3\ TBP_{(org)} \tag{4}$$

However, when uranyl nitrate is present in the aqueous phase, Siddall noted the principal mode of extraction appears to be described by Equation 5:

$$UO^+_{2(aq)} + NO^-_{3(aq)} + TcO^-_{4(aq)} + 2TBP_{(org)} \rightleftarrows UO_2NO_3TcO_4 \cdot 2\ TBP_{(org)} \tag{5}$$

It is somewhat surprising that $TcO_4^-$ can substitute for the nitrate ion in the normal solvate, $UO_2(NO_3)_2 \cdot 2TBP$. More recent work[69] also supports the mechanism according to Equation 5. Variation in distribution ratios with nitric acid concentration for the extraction of $HTcO_4$ is shown in Figure 16. Dependence of the distribution ratios on process parameters for the extraction of $TcO_4^-$ as the $UO_2\ NO_3\ TcO_4 \cdot 2TBP$ solvate is shown in Figure 17. The magnitudes of the distribution ratios indicate that a few percent of the technetium may follow the uranium through Purex processing.

### E. Mass Action Effect of Uranium

Since uranium is present in overwhelming quantities relative to fission products and in addition is highly extractable, high solvent loading with uranium is used to suppress the extraction of fission product elements (Figure 18). The extraction of Pu (IV) is also suppressed[8,19,70,71] (Figure 19), but its extractability is sufficient at high uranium loadings to permit essentially quantitative recovery.[70] Nitric acid concentrations in the aqueous phase of about 2 to 3 *M* are about optimum with respect to reducing the coextraction of ruthenium and zirconium (Figure 20). However, in the use of backwashing or scrubbing to further purify actinide extracts it would appear to be beneficial[71] to use dual nitric acid scrubs: first an aqueous scrub with about 4 *M* $HNO_3$ to scrub out ruthenium and then a second scrub with about 1.5 *M* $HNO_3$ to scrub out zirconium.

## V. SEPARATIONS BY CONTINUOUS COUNTERCURRENT EXTRACTION

All versions of the Purex process are based on a sequence of countercurrent solvent

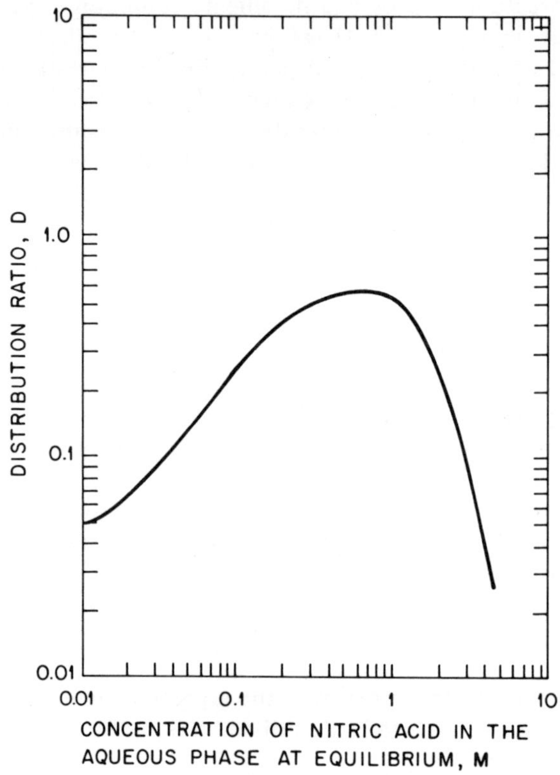


FIGURE 16. Distribution ratios of technetium as a function of the nitric acid concentration of the equilibrium aqueous phase at 30°C [1.09 *M* TBP in normal dodecane (30 vol %); tracer level Tc]. (Adapted from Siddall, T. H., Behavior of Technetium in the Purex Process, Rep. DP-364, Savannah River Laboratory, Aiken, S.C., 1959.)

extraction cycles. Solvent extraction cycles are defined in Purex processing as those operations in which separations are achieved by transferring the U and/or Pu from an aqueous phase to the organic phase and then recovering the U and/or Pu by back-extraction into an aqueous phase. The upper limit of decontamination of U and Pu from fission products and from transplutonium elements in one Purex cycle is of the order of $10^3$. At least three extraction cycles are required to permit direct handling of U and Pu products.

### A. Sequence of Solvent Extraction Cycles

The sequence in which separations and purifications are effected in the Purex process may be varied. In the more common version of the Purex process, U and Pu are codecontaminated from fission products and other actinides and then separated from each other (partitioned) in the first cycle (Figure 21). At the French LaHague Plant,[72] the U and Pu are first codecontaminated from fission products and other actinides in a coextraction-costripping cycle and then are partitioned in the second cycle. Koch[9,71] has reviewed the relative merits of the two different arrangements of the solvent extraction cycles and the choice would appear to depend on the methods used to insure that nuclear criticality is prevented.

Since both the arrangement of Purex cycles and the chemical conditions employed

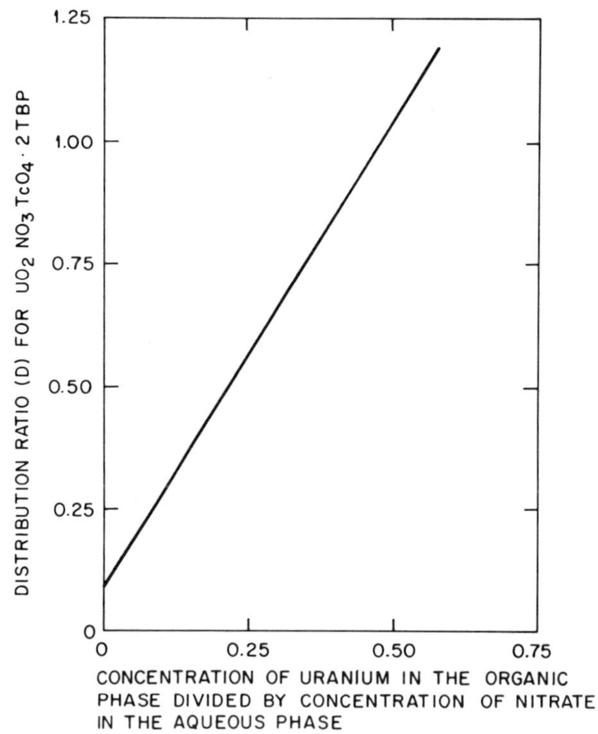


FIGURE 17. Equilibrium extraction of technetium as $UO_2NO_3TcO_4 \cdot 2$ TBP at 30°C [1.09 *M* TBP in normal dodecane (30 vol %); macro concentrations of uranium and tracer levels of technetium]. (Adapted from Siddall, T. H., Behavior of Technetium in the Purex Process, Rep. DP-364, Savannah River Laboratory, Aiken, S.C., 1959.)

in the cycles can be varied, further discussions on the details of the cycles are deferred to Section IX. Understanding of the chemistry of the reductants and oxidants which are utilized to effect actinide valence changes (Section VI) in the separations, and the effects of solvent degradation products on separation (Section VIII) are first necessary to understand the basis of the cycles.

## B. Countercurrent Extraction Principles Employed

To obtain concentrated products of U and Pu in both high yield and purity, judicious use of the principles of countercurrent liquid-liquid extraction is employed. Aqueous scrubbing (or back-washing) of organic phases containing actinides is extensively employed to improve separations from fission products (Figure 22). Similarly, the use of a pure extractant scrub in U-Pu partitioning minimizes the contamination of the aqueous Pu product by uranium. Simple equations cannot be used to calculate the extent of recoveries and purities of products obtained because the distribution ratios change from stage to stage in Purex separations. However, simple equations based on constant distribution ratios[73] can serve to illustrate the benefits of countercurrent extraction and scrubbing operations.

When the distribution ratio is constant, types 1 and 2 operations (Figure 22) are described by Equation 6 and types 3 and 4 operations are described by Equation 7,

$$\phi = \frac{(P_1 - 1)(P_2^m - 1)}{(P_1^{\,n+1} - 1)(P_2 - 1)\,P_2^{m-1} + (P_2^{m-1})(P_1 - 1)} \tag{6}$$

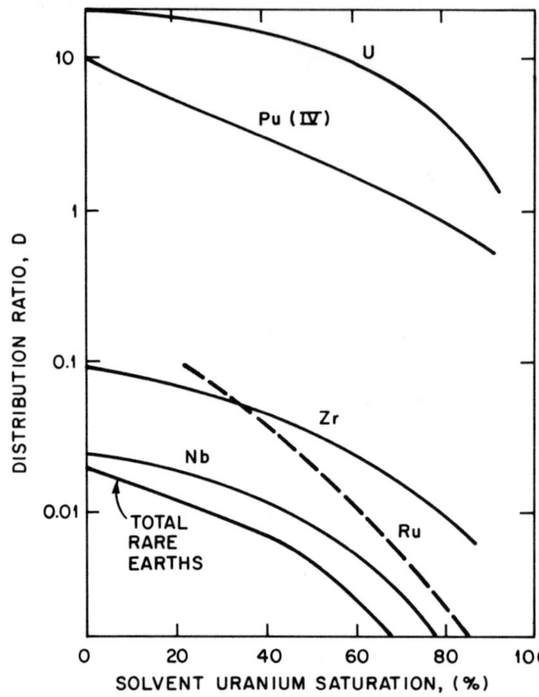


FIGURE 18. Effect of TBP saturation with uranium on the distribution ratio of uranium and of trace concentrations of Pu(IV), Zr(IV), Nb(V), rare earths, and ruthenium (mixture of various aqueous Ru species) (30% TBP; aqueous $HNO_3$ concentration of 2 *M* at 0% saturation increasing to about 3 *M* at 90% saturation). (Adapted from Stevenson, R. L. and Smith, P. E., *Reactor Handbook*, Vol. 2, 2nd ed., Stoller, S. M. and Richards, R. B., Eds., Interscience, New York, 1961, 107. With permission.)

$$\phi = \frac{P - 1}{P^{n+1} - 1} \tag{7}$$

where $\phi$ = fraction of a given component in the feed that remains unextracted (i.e., does not transfer to the counterphase), P = extraction factor, n = number of equilibrium extraction (or stripping) stages, and m = number of equilibrium scrubbing stages.

The extraction factor P is defined in terms of the direction of transfer that occurs during the particular type of operation. Definitions of P, $P_1$, and $P_2$ are as follows, when $V_o$ and $V_a$ are the volumes of the organic and aqueous phases, respectively.

$$\text{Type 1: } P_1 = D \cdot \frac{V_o}{V_a}; P_2 = D \cdot \frac{V_o}{V_a}$$

$$\text{Type 2: } P_1 = \frac{1}{D} \cdot \frac{V_a}{V_o}; P_2 = D \frac{V_o}{V_a}$$

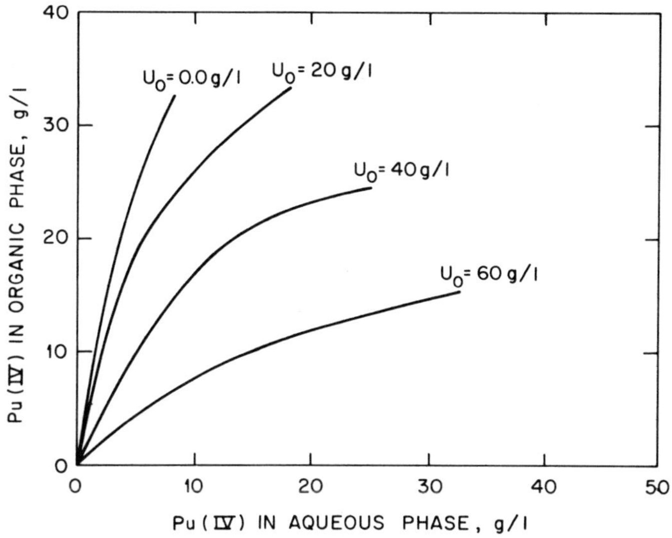


FIGURE 19. Equilibrium distribution of Pu(IV) between nitric acid and 20% TBP-dodecane for various organic phase uranium loadings ($U_o$) at 25°C (total $HNO_3$ = 3.25 *M*). (Modified from Koch, G., Ochsenfeld, W., and Schwind, E., Rep. KFK-440, *Kernforschungszentrum Karlsruhe Ber.*, 1969. With permission.)

$$\text{Type 3: } P = D \cdot \frac{V_o}{V_a}$$

$$\text{Type 4: } P = \frac{1}{D} \cdot \frac{V_a}{V_o}$$

Note that the ratio of $V_o/V_a$ can be varied in extraction and in scrubbing operations (Figure 22). The highest recovery and purity of products are attained by selecting $V_o/V_a$ ratios such that the extraction factors for the impurities are far below unity.

The relative benefits of aqueous scrubbing may be seen by consideration of the extraction of an aqueous solution containing elements A and B in equimolar amounts and where $D_A$ = 12 and $D_B$ = 0.12. Table 6 shows comparisons of the ratio of A/B in the extract and raffinate phase with and without the use of aqueous solution scrubbing. The aqueous scrub flow is 0.2 fraction of the feed flow. It is seen that as the number of scrub stages is increased, the composition of the raffinate proceeds toward pure B whereas the extract composition proceeds toward pure A. Without scrubbing, it is not possible to achieve both high recovery and high purity of A in the extract phase. High purity of B can be obtained in the aqueous phase but at the expense of a 12% loss of B to the extract phase. The use of an organic phase scrub is similarily beneficial when one component is being selectivity stripped from an organic phase while another component is being retained in that phase, as is the case in partitioning.

Sometimes more than one aqueous scrub is used in Purex extractions. In the case of dual acid scrubs, the additional scrub stream is introduced between the feed stream and the other scrub. It has also been found beneficial to carry out aqueous scrubbing at higher temperatures than those used in the extraction section. The higher temperatures suppress the extraction of the higher complexes of ruthenium (see Section III).

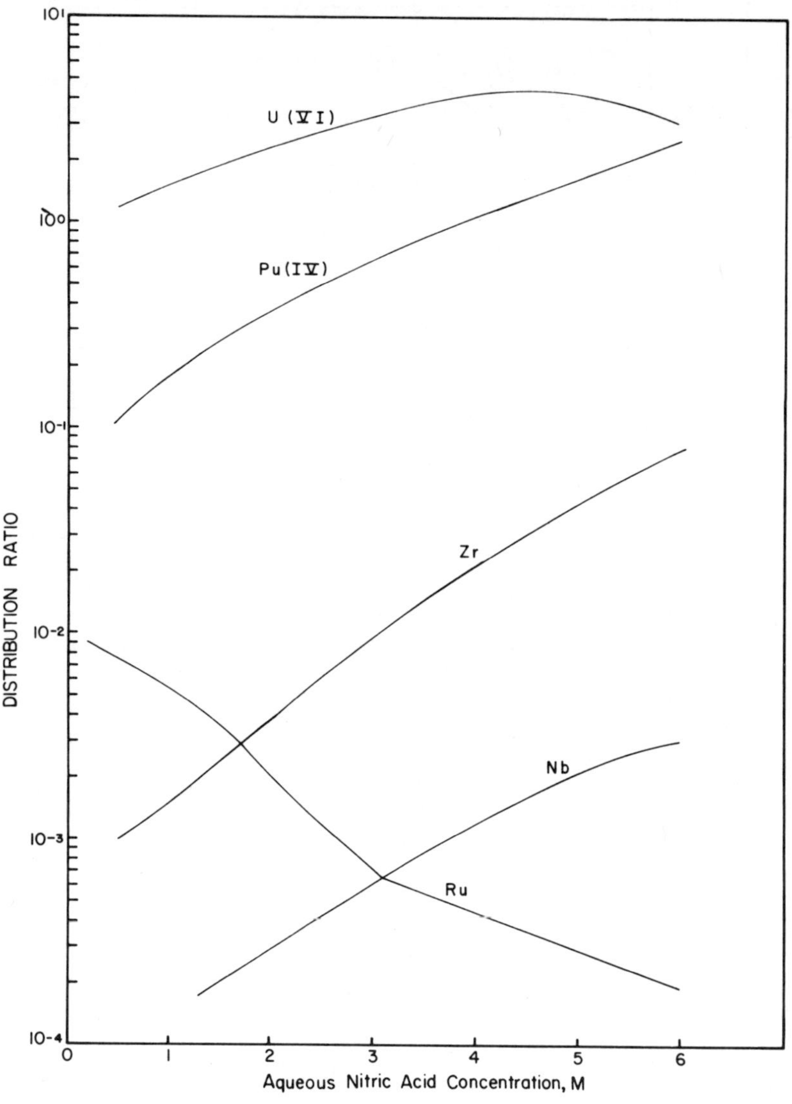


FIGURE 20. Effect of aqueous nitric acid concentration on the extraction of uranium (VI) and tracer concentrations of Pu(IV), and fission products [30% TBP phase loaded to 80% saturation with U(VI); initial aqueous U(VI) concentration = 0.840 *M*]. (Adapted from Stevenson, R. L. and Smith, P. E., *Reactor Handbook,* Vol. 2, 2nd ed., Stoller, S. M. and Richard, R. B., Eds., Interscience, New York, 1961, 107. With permission.)

## VI. VALENCE ADJUSTMENT IN SEPARATIONS OF NEPTUNIUM AND PLUTONIUM

Separations of uranium, neptunium, and plutonium are based on the inextractability of pentavalent neptunium and trivalent plutonium. The presence of nitrous acid in the initial feed which is formed both in the fuel dissolution process and by radiolysis complicates the redox chemistry that can be used to effect valence changes for neptunium and plutonium. Essentially complete separations of uranium, neptunium, and pluton-

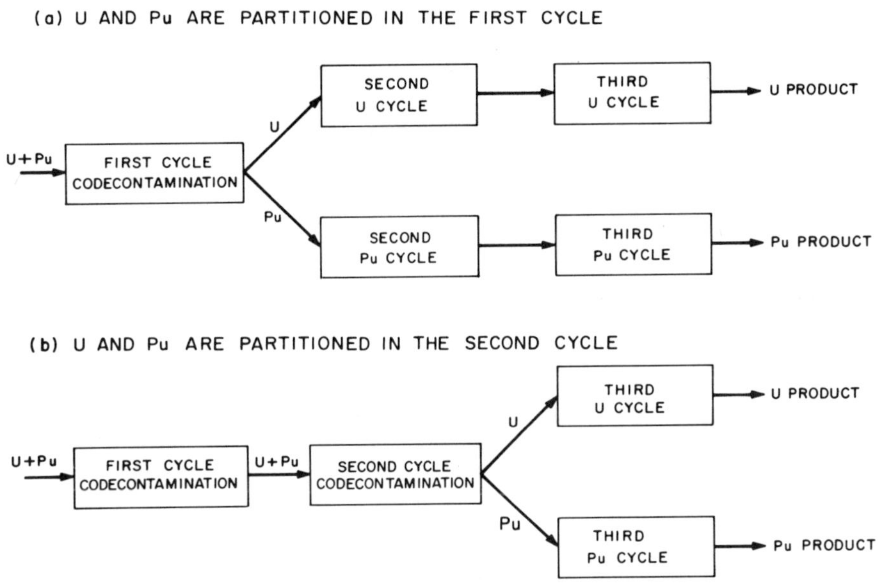


FIGURE 21. Alternative arrangements of the solvent extraction cycles in the Purex process.

ium cannot be achieved in one extraction cycle; multiple extraction cycles are required. In the more modern plants, redox reagents which do not increase quantities of salt wastes are preferred for valence adjustments. Methods used must effect valence changes rapidly. Newton[74] has extensively reviewed the literature on the kinetics of redox reactions of actinides.

## A. Neptunium

It is desirable that neptunium be rejected to the high activity waste, preferably in the first extraction cycle. However, it is not clear that the necessary extent of rejection can be accomplished in one cycle. It is not possible to control precisely the position of the equilibrium depicted in Equation 8 throughout all of the stages of extraction because of the extractability of $HNO_2$.

$$NpO^{2+} + \tfrac{1}{2}\,HNO_2 + \tfrac{1}{2}\,H_2O \rightleftarrows NpO_2^+ + {}^3/_2\,H^+ + \tfrac{1}{2}\,NO_3^- \qquad (8)$$

The oxidation of $NpO_2^+$ by nitrate ion is favored by high acidity and low nitrous acid concentrations.[21] The oxidation of $NpO_2^+$ proceeds relatively rapidly when the nitrous acid is in the $10^{-5}$ to $10^{-3}$ *M* range. At these low concentrations, nitrous acid acts as a catalyst for the oxidation reaction. Even at high acidities, nitrous acid in high concentrations ($10^{-2}$ *M*) reduces $NpO_2^{2+}$ to $NpO_2^+$. Maintenance of the desired inextractable pentavalent state thus depends on maintaining the nitrous acid concentration in the proper range.

The fraction of total neptunium present as $NpO_2^+$ is also governed by its disproportionation reaction.

$$2\,NpO_2^+ + 4\,H^+ \rightleftarrows Np^{4+} + NpO_2^{2+} + 2\,H_2O \qquad (9)$$

The relative proportion of $NpO_2^+$, $NpO_2^{2+}$, and $Np^{4+}$ is determined by the acidity, complexing anions, and the redox potential. $NpO_2^+$ is the most stable valence between

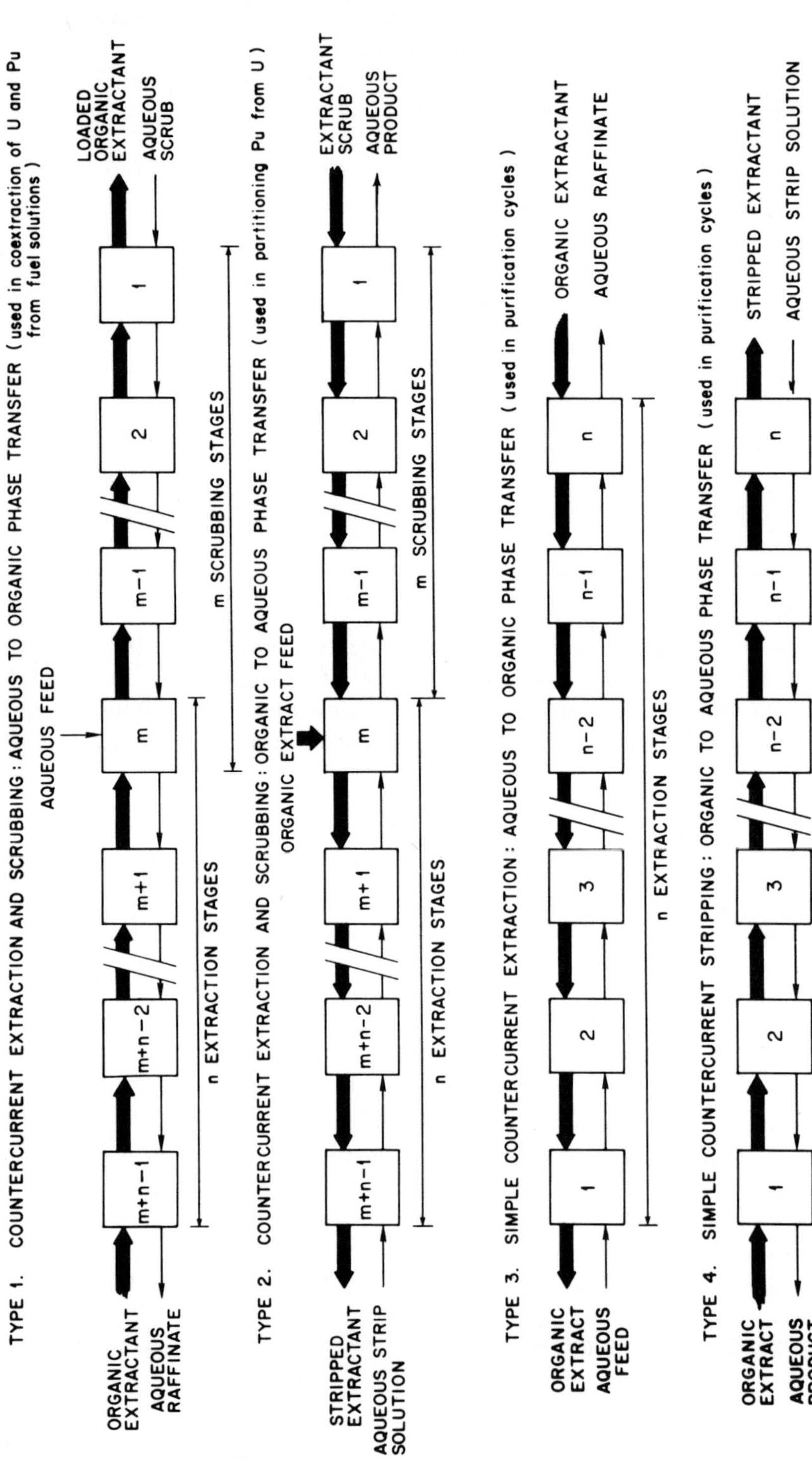


FIGURE 22. Types of countercurrent operations carried out in the Purex process.

### Table 6
### EFFECT OF AQUEOUS SCRUBBING ON THE PURITY OF AN EXTRACT PHASE

| Stages | | | % Unextracted | | A/B ratio | |
|---|---|---|---|---|---|---|
| Total | Extraction, n | Scrubbing, m | A | B | Aqueous raffinate | Organic extract |
| 4 | 4 | 0 | $4 \times 10^{-3}$ | 88.00 | $4.5 \times 10^{-5}$ | 8 |
| 5 | 4 | 1 | $9 \times 10^{-3}$ | 90.00 | $1 \times 10^{-4}$ | 9 |
| 6 | 4 | 2 | $9 \times 10^{-3}$ | 96.43 | $9.3 \times 10^{-5}$ | 28 |
| 7 | 4 | 3 | $9 \times 10^{-3}$ | 98.43 | $9.1 \times 10^{-5}$ | 64 |
| 8 | 4 | 4 | $9 \times 10^{-3}$ | 99.26 | $9.1 \times 10^{-5}$ | 137 |
| 10 | 4 | 6 | $9 \times 10^{-3}$ | 99.82 | $9.0 \times 10^{-5}$ | 555 |
| 14 | 4 | 10 | $9 \times 10^{-3}$ | 99.99 | $9.0 \times 10^{-5}$ | 9200 |

*Note:* Element A: D = 12, $P_1$ = 10, and $P_2$ = 60; Element B: D = 0.12, $P_1$ = 0.1, and $P_2$ = 0.5; Volumes: aqueous feed = 1, aqueous scrub = 0.2, and organic extractant = 1. A/B ratio in feed is unity.

about 0.1 and 2 *M* $HNO_3$. Above about 2 *M* $HNO_3$, the disproportionation reaction begins to proceed slowly, and in 6 to 8 *M* $HNO_3$ the $NpO_2^+$ disappears entirely. Rejection of neptunium to the aqueous phase is thus favored by $HNO_3$ concentration of <2 *M* if proper $HNO_2$ concentrations can be maintained.

The redox chemistry of neptunium in Pu partitioning and in purification cycles is discussed in the next section. After neptunium has been extracted into TBP, the valence forms are dictated by the redox methods used for controlling the valence of plutonium. Present specifications[75] for the return of uranium to gaseous diffusion plants set a maximum value of $^{237}Np$ of 1 ppm in the uranium. This means that an overall DF for Np of about 500 needs to be achieved in Purex processing and $UF_6$ conversion. Process details of achieving this objective have not been demonstrated although it would appear that chemical conditions could be worked out. In several tests, [72,76,77] results have shown that introduction of $HNO_2$ into the extraction stage that is one stage upstream (stage m + n − 2 in Figure 22) of the stage from which the aqueous waste emerges is effective in rejecting Np to the aqueous waste. About 99.7% of the Np was rejected[72] (DF = 330) and about 91% of the total Np rejected was $NpO_2^+$. High uranium saturation was also employed to minimize the amount $Np^{4+}$ or $NpO_2^{2+}$ in the extract.One can only conjecture that further studies will lead to improvements in controlling the concentration profile of $HNO_2$ in countercurrent extraction stages so that a greater rejection of Np can be obtained. Additional decontamination can of course be attained in the uranium purification cycles by employing repetitions of the extraction conditions used in the first cycle for the reextraction of the uranium in those cycles.

## B. Plutonium

Tetravalent plutonium is the normal stable valence in nitric acid solutions for most conditions in the Purex process. Only at $HNO_3$ concentrations of <0.5 *M* is $Pu^{4+}$ partially reduced to $Pu^{3+}$ by $HNO_2$.[78,79] For extractions, the plutonium valence is always adjusted to the tetravalent state, and in subsequent partitioning operations it is reduced to $Pu^{3+}$.

Fuel dissolution and feed adjustment steps insure that only tetravalent plutonium is present in the feed. After $Pu^{3+}$ is partitioned, it is converted again to $Pu^{4+}$ in the feed preparation steps for additional purification cycles. Oxidation of $Pu^{3+}$ can be accomplished readily using either nitrous acid[80-83] or an electrolytic oxidation process.[71,84-86] Mixtures of NO and $NO_2$ gases are used to generate nitrous acid.

$$2NO + HNO_3 + H_2O \rightleftarrows 3HNO_2 \quad (10)$$

$$3NO_2 + H_2O \rightleftarrows NO + 2HNO_3 \quad (11)$$

$$2NO_2 + H_2O \rightleftarrows HNO_2 + HNO_3 \quad (12)$$

In strong $HNO_3$, $HNO_2$ oxidizes $Pu^{3+}$ and reduces $PuO_2^{2+}$ to $Pu^{4+}$ quantitatively.[78]

$$Pu^{3+} + HNO_2 + H^+ \rightleftarrows Pu^{4+} + NO + H_2O \quad (13)$$

$$PuO_2^{2+} + HNO_2 + 2H^+ \rightleftarrows Pu^{4+} + HNO_3 + H_2O \quad (14)$$

At temperatures of 60 to 80°C, these reactions proceed rapidly. Concomitantly $Np^{4+}$ is also oxidized in this step and final adjustments of nitric and nitrous acid concentrations must be made to obtain $NpO_2^+$ when that form is desired.

Electrolytic oxidation[76-78] is not used for valence adjustment in higher activity feed but is used for preparation of feeds for subsequent purification of plutonium. The mechanism of the electrolytic oxidation process involves electrode reactions with hydrazine, nitrate ion, $U^{4+}$, and $Pu^{3+}$. The electrolytic cell compartments are not separated by a barrier diaphragm in this process. The hydrazine-electrode reaction is initiated by potentials of +800 mV at a Pt anode and reaction is completed at about +1600 mV, after which $Pu^{3+}$ and $U^{4+}$ are also oxidized. More detailed discussion of the electrode reactions is presented later in this section. The reductants that are employed in partitioning are the ferrous ion or the tetravalent uranium ion.

$$Pu^{4+} + Fe^{2+} \rightarrow Pu^{3+} + Fe^{3+} \quad (15)$$

$$2Pu^{4+} + U^{4+} + 2H_2O \rightarrow 2Pu^{3+} + UO_2^{2+} + 4H^+ \quad (16)$$

The reduction rates are rapid.[74] The reduction by ferrous ion is inversely proportional to the first power of the hydrogen ion concentration, whereas with $U^{4+}$ inverse dependence on the square of the hydrogen ion concentration is observed. The solution of $U^{4+}$ may be specially prepared or it may be generated *in situ* during the partitioning by electrolytic methods. Hydroxylamine nitrate (HAN) may be used as the reductant in the purification cycles for plutonium, but the reduction rate is too slow to be used in the first cycle where higher acidities are present.

$$2NH_3OH^+ + 4Pu^{4+} \rightarrow 4Pu^{3+} + N_2O + H_2O + 6H^+ \quad (17)$$

The reduction with HAN is inversely proportional to the fourth power of the hydrogen ion concentration.[87,88] The kinetics of the above reduction reactions have been studied extensively. The hydrogen ion dependencies reflect the role of hydrogen ions in the formation of activated complexes or rate-determining reactions.[74] Acidities of less than about 0.3 *M* are undesirable for $Pu^{4+}$ because of the tendency of $Pu^{4+}$ to undergo irreversible hydrolysis at low acidity. Reduction of $Pu^{4+}$ with HAN is very slow at acidities of about 1 *M*.

In addition to the primary reductants that are used to convert $Pu^{4+}$ to $Pu^{3+}$, holding reductants are utilized to prevent the reoxidation of $Pu^{3+}$ by nitrous acid. The principal function of the holding reductant is to rapidly destroy nitrous acid. In fact, holding reductants are necessary for maintaining the stability of $Fe^{2+}$, $U^{4+}$, or $Pu^{3+}$ in nitric acid solutions. This is because the $HNO_2$ formed in the slow oxidation of $Pu^{3+}$, $U^{4+}$, or $Fe^{2+}$ by $HNO_3$ (or by radiolysis) exerts a catalytic effect on the oxidation process. Hydrazine is used to stabilize uranous nitrate solutions because of the following reactions:[89]

$$U^{4+} + NO_3^- + H_2O \rightleftarrows UO_2^{2+} + HNO_2 + H^+ \text{ (slow)} \qquad (18)$$

$$U^{4+} + 2HNO_2 \rightarrow UO_2^{2+} + 2H^+ + 2NO \text{ (rapid)} \qquad (19)$$

$$2NO + HNO_3 + H_2O \rightleftarrows 3HNO_2 \text{(rapid)} \qquad (20)$$

Low acidity favors the oxidation of $U^{4+}$ by nitrate ion whereas $Pu^{3+}$ oxidation is favored by high acidity.

$$2Pu^{3+} + NO_3^- + 3H^+ \rightleftarrows 2\,Pu^{4+} + HNO_2 + H_2O \text{ (slow)} \qquad (21)$$

Sulfamate ($NH_2SO_3^-$) ions[8] or hydrazine[90,91] may be used to stabilize $Fe^{2+}$ solutions.

The mechanism of the rapid rate of oxidation of $Pu^{3+}$ by $HNO_2$ has been extensively investigated.[80-83] Two mechanisms were proposed, one involving the nitrosyl ion[80]

$$HNO_2 + H^+ \rightleftarrows NO^+ + H_2O \text{ (rapid)} \qquad (22)$$

$$NO^+ + Pu^{3+} \rightarrow NO + Pu^{4+} \text{ (rate controlling)} \qquad (23)$$

and the other[80] involving $N_2O_4$.

$$HNO_2 + HNO_3 \rightleftarrows N_2O_4 + H_2O \text{ (rapid)} \qquad (24)$$

$$N_2O_4 + Pu^{3+} \rightarrow NO_2^- + NO_2 + Pu^{4+} \text{ (rate controlling)} \qquad (25)$$

More recent work[82] proposes a mechanism involving $NO_2$ rather than its dimer, $N_2O_4$.

$$HNO_2 + HNO_3 \rightleftarrows 2NO_2 + H_2O \qquad (26)$$

$$Pu^{3+} + NO_2 \rightleftarrows Pu^{4+} + NO_2^- \text{ (rate controlling)} \qquad (27)$$

$$NO_2^- + H^+ \rightleftarrows HNO_2 \qquad (28)$$

Whatever the mechanisms of the oxidation reactions, nitrous acid must be destroyed rapidly to prevent its rapid reaction with $Pu^{3+}$, $Fe^{2+}$, or $U^{4+}$. The only reagents that react sufficiently rapidly with $HNO_2$ are hydrazine and sulfamic acid.[92-95] Although hydroxyline will reduce $Pu^{4+}$, it does not react sufficiently rapidly with $HNO_2$ to prevent reoxidation of $Pu^{3+}$ unless a large excess is employed. Sulfamic acid is generally considered an unsatisfactory holding reductant for use in modern plants because it is oxidized to sulfate ion which introduces complications in the recycle of nitric acid and in waste management. Also, $U^{4+}$ is currently preferred rather than $Fe^{2+}$ since uranium is already a constituent in the system and does not add to wastes. Ferrous sulfamate is the most commonly used reductant in older Purex plants. Sulfamate ions rapidly reduce nitrite ions.[94]

$$NH_2SO_3^- + NO_2^- \rightarrow N_2 + SO_4^{2+} + H_2O \quad (29)$$

The products of the reaction of nitrous acid and hydrazine depend on the ratio of the reactants.[96,97] With excess hydrazine, the following fast reaction occurs:

$$N_2H_4 + HNO_2 \rightarrow HN_3 + 2H_2O \quad (30)$$

On the other hand, with excess nitrous acid, a second fast reaction occurs yielding gaseous products:

$$HN_3 + HNO_2 \rightarrow N_2O + N_2 + H_2O \quad (31)$$

In addition, $N_2H_2$ slowly reacts with $HNO_3$. Koltunov and Marchenko[93] determined that the stoichiometry of the reaction in 0.1 to 0.5 *M* $HNO_3$ was described by the following equation:

$$7\,N_2H_4 + 12\,HNO_2 + H^+ \rightarrow NH_4^+ + HN_3 + 5N_2 + 6N_2O + 18\,H_2O \quad (32)$$

They propose a mechanism which involves the participation of the nitrosyl ion, $NO^+$. The mechanism of $N_2H_4$ reactions in the Purex process solutions is not completely understood. It would appear that the following competing reactions are involved:

1. Catalytic[98] decomposition of $N_2H_4$
2. Reaction of $N_2H_4$ with $HNO_2$
3. Reaction of $N_2H_4$ with $HNO_3$
4. Reaction of $HN_3$ with $HNO_2$
5. Reaction of $HN_3$ with $HNO_3$

The reduction of $Pu^{+4}$ to $Pu^{3+}$ or $NpO^{2+}$ to $Np^{4+}$ by hydrazine is too slow to be useful in partitioning by countercurrent separation. However, hydrazine does rapidly reduce $NpO_2^{2+}$ to $NpO_2^+$. Several studies have been made regarding the kinetics of these reactions.[48,99-102]

The relatively high extractability of $HNO_2$[103] (D values 5 to 18) under Purex conditions limits the effectiveness of hydrazine as a holding reductant for extractable $U^{4+}$ because hydrazine is virtually inextractable. Although $HN_3$ is highly extractable[104] (D $\sim$ 1 to 2) and is formed when excess $N_2H_4$ is present in the aqueous phase, the reaction rate of $HN_3$ with $HNO_2$ in the organic phase is too slow relative to the reaction rate of $HNO_2$ with trivalent plutonium. Therefore, it cannot satisfactorily protect against oxidation of $Pu^{3+}$ or the extractable reductant, $U^{4+}$, in the organic phase. $Pu^{3+}$ and $HNO_2$ in the organic phase exert a catalytic effect on the $HNO_3$ oxidation of $U^{4+}$ in that phase. Kinetic studies[89] show that the presence of plutonium added as either $Pu^{4+}$ or $Pu^{3+}$ rapidly accelerates (by order of magnitude) the oxidation rate of $U^{4+}$ by $HNO_3$ in the organic phase (Figure 23). Most of the reduction of $Pu^{4+}$ by $U^{4+}$ takes place in the organic phase,[104] whereas with inextractable $Fe^{2+}$ or HAN the reduction is accomplished in the aqueous phase. The reaction between $N_2H_4$ and $HNO_2$ has been assumed to occur principally at the aqueous-organic interface; nitrite destruction is increased by increasing the interfacial area of the mixed phases. Larger ratios of $U^{4+}$ to $Pu^{4+}$ are

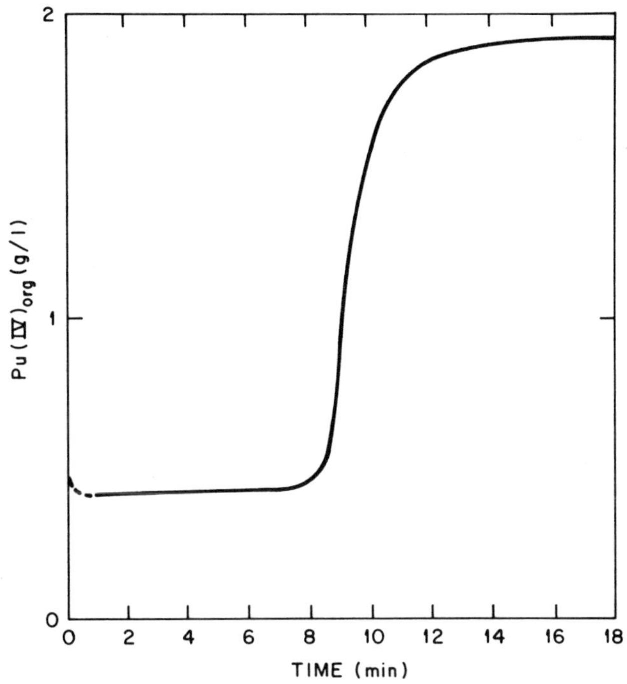


FIGURE 23. Oxidation of $Pu^{3+}$ in 30% TBP after reduction with excess U(IV) [total Pu = 1.9 g/ℓ; $HNO_3$ = 0.22 *M*; U(IV) = 1.45 g/ℓ]. (Adapted from Talmont, X. L. R., *Solvent Extraction Chemistry of Metals,* McKay, H. A. C., Ed., Macmillan, London, 1965, 103. By permission of Macmillan, London and Basingstoke.)

required to reduce the $Pu^{4+}$ than would be predicted by the values of the redox potential for the reduction reaction.

$$2\,Pu^{4+} + U^{4+} + 2\,H_2O \rightarrow UO_2^{2+} + 2\,Pu^{3+} + 4H^+ \quad E_o = 0.65V \tag{33}$$

The magnitude of the value of $E_o$ indicates a separation factor of Pu from U of about $10^9$ would be attained in Purex partitioning using equimolar amounts of $Pu^{4+}$ and $U^{4+}$ if $HNO_2$ could be rapidly destroyed.[104] In practice it has been found that about a five- to tenfold excess of $U^{4+}$ is required in partitioning to obtain separation factors of about $10^4$. The requirement of a five- to tenfold excess is a considerable disadvantage if $U^{4+}$ must be prepared separately and used as a reductive strip solution. Up to about 10% more uranium would have to be processed in solvent extraction operations. However, this disadvantage is overcome with the *in situ* generation of $U^{4+}$ by the electrolytic process. Satisfactory reductants for the destruction of $HNO_2$ in the organic phase are yet to be discovered. Biddle et al.[104] found phenylhydrazine, *o*-methylphenyl-hydrazine, and hydrazoic acid to be unsatisfactory.

Potential problems that may be caused by the presence of $HN_3$ in the Purex process have not been completely evaluated. This is largely due to the lack of knowledge regarding the $HNO_2$ contents of the Purex process streams and the amounts of $HN_3$ produced when $N_2H_4$ is used as a holding reductant. No difficulties in plant operations have been experienced to date that are attributable to $HN_3$. Aqueous solutions of $HN_3$ are stable up to concentrations of 20 to 30% and they present no safety problems.

However, because of its low boiling point (37°C), $HN_3$ in warm aqueous $HNO_3$ solution will tend to concentrate in vapor phases. Hydrazoic acid vapors are known to explode unpredictably. However, part of the $HN_3$ is extracted into the organic phase. Subsequently, some of it is stripped into actinide product solutions. Finally, the residual $HN_3$ in the solvent is removed as an azide salt in the cleanup step (see Section X) before recycle. Chemical methods are available for destruction of $HN_3$ in vapors or in process liquids if it should be warranted. An excess of $NO_2$ in tank vapor or acid solution can be destroyed by the reaction

$$2\,NO_2 + HN_3 \rightarrow HNO_3 + N_2 + N_2O \qquad (34)$$

In alkaline media such as caustic scrubber wastes (see Section X), it can be destroyed by the azide-nitrite reaction:

$$NaNO_2 + NaN_3 + H_2O \rightarrow N_2 + N_2O + 2\,NaOH \qquad (35)$$

Since $U^{4+}$ and hydrazine have been used under different chemical conditions in different Purex plants in partitioning of Pu and in purification cycles, further discussion is reserved for Section IX where those cycles are discussed. It would appear that the electrolytic reduction process[74,86,105,106] where $U^{4+}$ is produced *in situ* in the mixed phases is the most advantageous method to partition plutonium.

The chemical reactions involved in the *in situ* electrolytic reduction process have been studied[71,84-86] extensively. Tetravalent plutonium is reduced both directly at the cathode and by the $U^{4+}$ that is simultaneously generated at the cathode.

$$Pu^{4+} + e^- \rightarrow Pu^{3+} \qquad E_o = 0.92\ V \qquad (36)$$

$$UO_2^{2+} + 4H^+ + 2e^- \rightarrow U^{4+} + 2H_2O \qquad E_o = 0.33\ V \qquad (37)$$

The first reaction is reversible whereas the second is irreversible. The mechanism involves the following equation:

$$2UO_2 + 2e^- \rightleftarrows 2UO_2^+ \text{ (at the electrode)} \qquad E_o = 0.06\ V \qquad (38)$$

The $UO_2^+$ then rapidly disproportionates according to

$$2UO_2^+ + 4H^+ + 4H_2O \rightarrow UO_2^{2+} + U^{4+} + 6H_2O \qquad (39)$$

The $U^{4+}$ formed in turn reduces $Pu^{4+}$.

$$2Pu^{4+} + U^{4+} + 2H_2O \rightarrow 2Pu^{3+} + UO_2^{2+} + 4H^+ \qquad (40)$$

Since $U^{4+}$ is extractable, this reaction takes place in both phases during the partitioning process.

The presence of hydrazine in cathode solutions effectively prevents the formation of nitrite ions.

$$NO_3^- + 3H^+ + 2e^- \rightleftarrows HNO_2 + H_2O \qquad E_o = 0.94\ V \qquad (41)$$

In addition to its reaction with nitrous acid, the hydrazinium ion, $N_2H_5^+$, is oxidized at the anode.

$$N_2H_5^+ \rightarrow N_2 + 5H^+ + 4e^- \qquad E_o = 0.23\ V \qquad (42)$$

Any neptunium present as $NpO_2^+$ is reduced at the cathode according to

$$NpO_2^{2+} + e^- \rightarrow NpO_2^+ \qquad E_o = 1.15\ V \qquad (43)$$

$$NpO_2^+ + e^- \rightarrow Np^{4+} \qquad E_o = 0.75\ V \qquad (44)$$

Platinum and titanium are suitable for anode and cathode materials, respectively, in regard to corrosion. The gaseous products produced at the electrodes contain about equal volumes of $H_2$ and $N_2$ and about 1 vol % NO. The anode gas contains about 95% $N_2$, 4% $O_2$, and 1% NO. The oxygen may have originated from an air impurity. However, the possibility of producing $O_2$ by the following electrode reaction cannot be ruled out:

$$2H_2O \rightarrow O_2 + 4H^+ + 4e^- \qquad E_o = -1.23\ V \qquad (45)$$

Electrolytic oxidation and reduction processes have been shown to be technically feasible on the small pilot plant scale [71,84-86] but it remains to be seen if these methods perform satisfactorily in large-scale plants. The electrolytic reduction process has been installed for U-Pu partitioning in the Barnwell Nuclear Fuel Plant (BNFP)[105,106] at Barnwell, S. C., which is yet to operate with irradiated fuel.

## VII. SOLVENT DEGRADATION

TBP and diluents undergo both radiolytic and chemical attack in Purex processing. Radiolysis is by far the greatest contributor to solvent degradation and some of the products formed reduce decontaminations from zirconium, niobium, and ruthenium and the recovery of uranium and plutonium. During simultaneous radiolytic and chemical attack, there are produced a variety of gases which are essentially insoluble and numerous soluble compounds which are retained by the solvent. The soluble degradation products may be placed in the following general categories:

1. The primary products of the radiolysis and chemical hydrolysis of TBP. These products consist of dibutylphosphoric acid (HDBP), monobutylphosphoric acid ($H_2MBP$), and phosphoric acid.
2. Diluent degradation products which result primarily from nitration and oxidation reactions.
3. Secondary reaction products which are formed from intermediates of TBP and diluent molecules. In these secondary reactions, alkyl phosphoric acids are produced which contain long-chain alkyl groups derived from the diluent.

The quantities of degradation products formed are dependent on the radiation intensity, exposure time, process reagent concentrations, and temperature. The primary products formed from TBP exert an immediate but temporary effect on separations because they are effectively removed in solvent purification steps prior to recycle (see Section X). However, formation of degradation products from secondary reactions among intermediate species or molecular fragments of TBP and diluents occurs slowly and exerts a long-term effect on separation performance. These compounds are not effectively removed by simple solvent recycle methods such as sodium carbonate scrubbing. Radiolytic decomposition of the solvent is orders of magnitude greater than is its decomposition by chemical reactions with other process chemicals.

The units used to discuss the radiation effects on solvents throughout subsequent discussions are

1. Radiation yield: a unit G value is one molecule destroyed or produced per 100 electron volts (eV) of absorbed energy.
2. Radiation dose: Watt · hour · liter$^{-1}$ (Wh$\ell^{-1}$) — For reference purposes, an absorbed dose of 1 Wh$\ell^{-1}$ ($2.25 \times 10^{22}$ eV$\ell^{-1}$) and a unit G value produces $3.74 \times 10^{-4}$ g mol or 8.3 m$\ell$ of gas at STP.

### A. TBP Degradation

It has been established that TBP decomposes both by radiolysis and by chemical hydrolysis according to the following reactions:[8,107,108]

$$(C_4H_9O)_3PO + HOH \rightarrow (C_4H_9O)_2POOH + C_4H_9OH \quad (46)$$

$$(C_4H_9O)_2POOH + HOH \rightarrow (C_4H_9O)PO(OH)_2 + C_4H_9OH \quad (47)$$

$$(C_4H_9O)PO(OH)_2 + HOH \rightarrow H_3PO_4 + C_4H_9OH \quad (48)$$

Actually, the reactions can be considered as hydrolysis or deaklylation.[109] The butanol ($C_4H_9OH$) formed in the reaction is aqueous-soluble and can undergo nitration and oxidation reactions at an appreciable rate at temperatures of about 100°C and greater. Under usual solvent extraction conditions (<5 *M* $HNO_3$ and <75°C) these reactions are not significant because of the short contact times. Appreciable quantities of radiolytic gases are also produced (Table 7)[110] but these are of little concern in solvent extraction operation.

In thermal hydrolysis, the reaction rates increase with increasing acid concentration, temperature, and TBP concentration.[8,107,108] The effect of nitrate ion concentration is small. The aqueous phase thermal hydrolysis reactions are first order with respect to nitric acid concentration. The first order rate constants for the first hydrolysis reaction of TBP in the aqueous and organic phases at 25°C are of the order of $10^{-5}$ and $10^{-7}$ hr$^{-1}$, respectively. The energy of activation for the aqueous phase hydrolysis is estimated to be 20 to 25 kcal mol$^{-1}$ whereas it is estimated to be 15 to 20 kcal mol$^{-1}$ for the organic phase hydrolysis. The organic phase reaction is the more important reaction since the solubility of TBP in aqueous phases is very low ($\sim 1 \times 10^{-3}$ *M*). The hydrolysis rates of HDBP and $H_2MBP$ are not accurately known except that HDBP hydrolyzes more rapidly than $H_2MBP$. In the aqueous phase hydrolysis,[8] the relative rates are in the approximate ratio of 4:2:1 for the first, second, and third hydrolysis steps, respectively. Rates for hydrolysis of HDBP and $H_2MBP$ have not been reported for the organic phase reactions. The effects of temperature and nitric acid concentration on the rate constants for the formation of HDBP by the first hydrolysis reaction are shown in Table 8 for organic phases in equilibrium with their aqueous nitric acid phases.[107] The reaction appears to be first order dependent with respect to $HNO_3$ in 100% TBP phases, whereas it is a substantially lesser power wth 20% TBP. The effect of the presence of uranyl nitrate is uncertain. There were disagreements among early workers whether uranyl nitrate accelerated the hydrolysis. It has recently been shown

## Table 7
## G VALUES FOR GASEOUS PRODUCTS FROM WET SOLVENTS[110]

| Organic phases[a] | Gas evolved | $H_2$ | $CH_4$ | CO | $N_2$ | $CO_2$ | $N_2O$ | NO | Total hydrocarbons |
|---|---|---|---|---|---|---|---|---|---|
| TBP-$H_2O$-$HNO_3$ (2 *M*) | 0.85 (max) | 0.5 | 0.016 | 0.074 | 0.155 | 0.059 | 0.113 | 0.118 | 0.13 |
| 20.8% TBP/OK (dry) | 1.97 | 1.81 | 0.10 | — | — | — | — | — | — |
| 19.5% TBP/OK-$H_2O$-$HNO_3$ (0.6 *M*) | 1.36 | 1.18 | 0.094 | 0.031 | 0.267 | 0.037 | 0.39 | 0.58 | 0.33 |
| 20.7% TBP/OK-$H_2O$-$HNO_3$ (0.15 *M*)-$UO_2(NO_3)_2$ (0.3 *M*) | 1.16 | 1.03 | 0.036 | — | 0.145 | 0.012 | 0.028 | — | 0.16 |

[a] OK = odorless kerosine.

Table 8
ORGANIC PHASE HYDROLYSIS OF TBP[8]

| Equilibrium composition of phases | | | |
|---|---|---|---|
| Aqueous | Organic | Temperature (°C) | Rate constant ($k_1$ ($hr^{-1}$) × $10^5$) |
| 1 $M HNO_3$ | 20% TBP in kerosine | 70 | 2.3 |
| | 0.14 $M HNO_3$ | 50 | 0.48 |
| | | 30 | 0.11 |
| | | 21 | 0.058 |
| 5 $M HNO_3$ | 20% TBP in kerosine | 60 | 3.2 |
| | 0.67 $M HNO_3$ | 50 | 0.84 |
| | | 30 | 0.17 |
| | | 21 | 0.10 |
| 1 $M HNO_3$ | 100% TBP | 70 | 10 |
| | 0.86 $M HNO_3$ | 50 | 1.2 |
| | | 30 | 0.14 |
| 5 $M HNO_3$ | 100% TBP | 70 | 42 |
| | | 50 | 5.5 |
| | | 30 | 0.71 |
| | | 21 | 0.25 |

that photochemical reactions in the $UO_2(NO_3)_2 \cdot 2TBP$ phase initiated by visible light sources at 400- to 500-nm wavelengths also produce HDBP.[111] Failure to take this photochemical reaction into account may have been the reason for earlier discrepancies.

Studies[110] of radiolysis of TBP using $\gamma$ rays from $^{60}Co$ (1.6 MeV) or 1 MeV electrons have been used to determine G values for gaseous products and for the soluble products: HDBP, $H_2MBP$, and butanol (Tables 7 and 9). Organic phases containing dissolved water and $HNO_3$ release less gases than do the dry solvents because of the reduction of yields of $H_2$ and $CH_4$. The presence of $UO_2(NO_3)_2$ and $HNO_3$ causes G values of gaseous products to be diminished. Hydrocarbons of chain length $C_2$ to $C_4$ contain both the saturated and unsaturated species. HDBP is produced in a greater yield than is $H_2MBP$. The ratio of HDBP to $H_2MBP$ is seen to vary with the type of radiation. Typical yields[110] obtained by irradiation of organic phases containing nitric acid and uranyl nitrate are given in Table 10. Nitric acid increases the yield of HDBP and $H_2MBP$ whereas water, uranyl nitrate, and TBP decrease it.

Aqueous phases saturated with TBP evolve less gases during irradiation than do organic phases.[110] Carbon dioxide and nitrous oxide are major constituents with nitric oxide, nitrogen, and hydrogen as minor constituents. Yields of alkylphosphoric acids are comparable to the values given in Table 10 for the irradiation of dry TBP. In oxygenated solutions[112] yields decrease with increasing nitrc acid concentration. Values of G(HDBP) decrease from 1.87 to 0.49 as the nitric acid concentration is increased from 0.5 to 4 *M*. Free hydroxyl radicals are postulated to explain the degradation mechanism through their reactions with TBP and $HNO_3$.

Irradiation of mixed TBP and aqueous nitric acid phases shows the quantities of soluble products depend on the type and concentration of the diluent, the nitric acid concentration, and the presence or absence of oxygen. The effects of diluent type and nitric acid concentrations[113] are shown in Table 11. Some diluents such as $CCl_4$ are more effective than hydrocarbon diluents in their ability to transfer energy to the TBP. For any given diluent G(HDBP + $H_2MBP$) increases linearly with TBP concentrations except at low (<10 vol %) TBP concentrations where energy transfer is enhanced. The presence of $O_2$ inhibits the formation of HDBP. The results of Barelko and co-work-

## Table 9
## DIFFERENCES IN YIELDS OF SOLUBLE PRODUCTS FOR ORGANIC PHASE IRRADIATIONS

| | $^{60}Co$-γ's[107] | 1-MeV electrons[110] |
|---|---|---|
| Radiolysis of dry TBP | | |
| G values for HDBP | 1.9 | 1.52 |
| G values for $H_2MBP$ | 0.3 | 0.17 |
| Molar ratio HDBP/$H_2MBP$ | 6:1 | 9:1 |
| G value for butanol | 0.65 | — |
| Radiolysis of wet TBP (6.4%) | | |
| G value for (HDBP + $H_2MBP$) | 1.3 | 0.90 |
| G value for butanol | 0.3 | Nil |

## Table 10
## COMBINED YIELD OF HDBP AND $H_2MBP$, G(HDBP + $H_2MBP$), FOR DIFFERENT SOLVENT PHASE COMPOSITIONS IN ORGANIC PHASE IRRADIATIONS[110]

| System | G(HDBP + $H_2MBP$) |
|---|---|
| TBP (dry) | 1.8 |
| TBP-$H_2O$ | 0.9 |
| TBP-$H_2O$-$HNO_3$ (1.5—2.5 *M*) | 2.2 |
| TBP-$H_2O$-$HNO_3$-uranium | 1.6—1.7 |
| 20% TBP/OK (dry) | 0.55 |
| 20% TBP/OK-$H_2O$-$HNO_3$ (0.5—0.6 *M*) | 0.43 |
| 20% TBP/OK-$H_2O$-$HNO_3$-uranium | 0.2 |

## Table 11
## YIELD OF DEGRADATION PRODUCTS OF TBP IN MIXED PHASE IRRADIATIONS: DOSE = 160 Wh$\ell$[113]

| Composition of initial phases | | Concentration of degradation products ($M \times 10^2$) | | |
|---|---|---|---|---|
| | | Organic phase | | Total phosphorus[a] |
| Organic | Aqueous $HNO_3$ (*M*) | HDBP | $H_2MBP$ | Aqueous phases |
| 100% TBP | 0.5 | 13.8 | 5.0 | 1.4 |
| | 3 | 13.9 | 4.9 | 2.0 |
| 30% TBP-(n-dodecane) | 0.5 | 3.2 | 0.8 | 1.7 |
| | 3 | 3.7 | 0.9 | 1.4 |
| 30% TBP-mesitylene | 0.5 | 1.5 | 0.4 | 0.4 |
| | 3 | 2.2 | 0.6 | 0.4 |
| 30% TBP-$CCl_4$ | 0.5 | 2.1 | 0.7 | 0.6 |
| | 3 | 13.6 | 3.5 | 0.6 |

[a] Mainly $H_2MBP$.

ers[114,115] typify this effect. In the absence of $O_2$, G(HDBP) was determined to be 1.2, whereas it was 0.5 in the mixed phases that were sparged by $O_2$ at 1 atm pressure. The

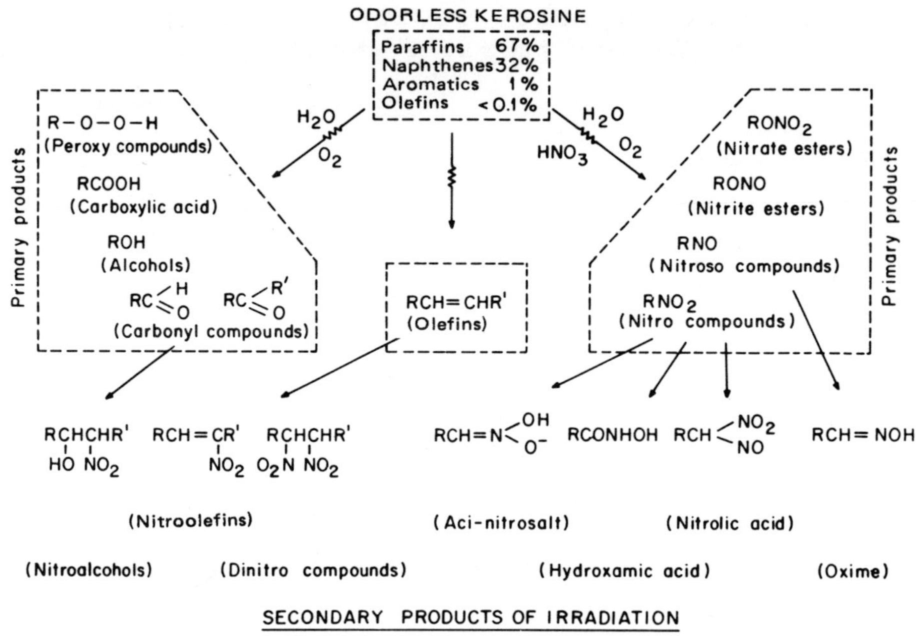


FIGURE 24. Some possible primary and secondary reaction products of diluent degradation.[117,118]

ratios of G(HDBP)/G($H_2$MBP) were not constant with increasing dosages although total yields are constant for the both systems. The ratio of G(HDBP)/G($H_2$MBP) decreased linearly from an initial ratio of about 7 at 10 $Wh\ell^{-1}$ to about 2 at 100 $Wh\ell^{-1}$.

## B. Diluent Degradation

Normal dodecane (NDD), odorless kerosine (OK), and other diluents are attacked either by radiation or by nitrous-acid-catalyzed chemical reactions with nitric acid.[113,116-120] The formation of a variety of soluble chemical species is possible by several primary reactions, which in turn can undergo slow secondary reactions[117,118] (Figure 24). Radiation, as previously noted in Table 7, also produces gaseous species. A relatively pure normal hydrocarbon fraction consisting of about 98% n-dodecane is the most resistant hydrocarbon diluent and is presently used in the Savannah River Plant.[120] It is well established that the order of susceptibility of hydrocarbons isolefins > naphthenes > isoparaffins > n-paraffins. Certain aromatic type diluents (e.g., ethylbenzenes) show good stability in general[121] but have yet to find applicability in Purex processing.

Since $HNO_2$ is extracted by TBP and also catalyzes nitric acid reactions, it is effective in promoting nitration and oxidation reactions in the organic phase. The G values for the production of nitrite ion in the aqueous phase by $\alpha$, $\beta$, or $\gamma$ irradiation[122,123] are of the order of 2 in 1 to 5 $M\,HNO_3$.

Only a few of the primary products of degradation in Figure 24 have been identified and their G values measured. The primary compound types that have been identified by functional group analysis[113,117] are R-$NO_2$, $RONO_2$, $RCO_2H$, and RCOR. The G values for their formation depend on irradiation conditions with respect to presence or absence of air, the degree of mixing of phases, and the $HNO_2$ concentration.[117] The

**Table 12**
**CONCENTRATIONS OF PRIMARY PRODUCTS OF DILUENT DEGRADATION IN THE IRRADIATION OF 20% TBP-OK**

| Structure type | R-$NO_2$ | $RONO_2$ | $RCO_2H$ | RCOR |
|---|---|---|---|---|
| Concentration, *M* | $5 \times 10^{-3}$ | $2.8 \times 10^{-2}$ | $1 \times 10^{-2}$ | $7.4 \times 10^{-2}$ |

*Note:* Dose = 7.5 Wh$\ell^{-1}$. 2.7 volumes of an aqueous phase [3 *M* $HNO_3$, 0.71 *M* $UO_2(NO_3)_2$, $10^{-3}$ *M* $Zr(NO_3)_4$ $10^{-3}$ *M* $NaNO_2$] continuously stirred with 1.0 volume of the organic phase.

From Huggard, A. J. and Warner, B. F., *Nucl. Sci. Eng.*, 17, 638, 1963. With permission.

concentrations of these compounds after a 7.5-Wh$\ell^{-1}$ radiation dose are given in Table 12 for a 20% TBP-OK phase that was stirred with the aqueous phase during the irradiation.[117] The amounts of these compounds were observed to increase linearly with total dose to about Wh$\ell^{-1}$.

Identification of secondary reaction products by direct methods is not possible because of the slowness of the reactions and the resultant low concentrations. Indirect methods such as the known extraction behavior of candidate compounds with respect to test cations have been used in attempts to explain the extraction behavior of irradiated solvents (Section VIII). However, these indirect tests[117-119,124] have not shown that any of the secondary products are responsible for retention of fission products, uranium, or plutonium by irradiated solvents during the irradiation.

### C. Secondary Degradation Compounds

On irradiation, higher alkylphosphoric acids and esters are formed by slow secondary reactions between the primary degradation products of TBP and those of the diluent.[125-128] The production rate of these secondary products is extremely slow. Total concentration of the species is $10^{-5}$ *M* when a 20% TBP-NDD phase previously equilibrated with 0.5 *M* $HNO_3$ is irradiated to a $^{60}Co$ $\gamma$ dose of 300 Wh$\ell^{-1}$. Nevertheless, Karlsruhe workers have recently been able to concentrate these species by distillation and identify many of them after fractional separation by a combination of solvent extraction, liquid and gas chromatography, and mass spectrographic techniques.[126-128] The following type compounds were identified:

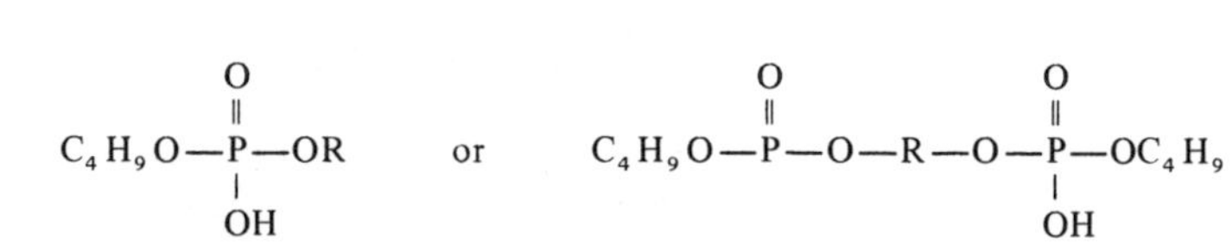

where R contains $C_5$ to $C_{14}$ alkyl chains.

In addition to the higher phosphoric acids and diesters, a fraction was isolated which contained a small quantity of polycarbonyl compounds. It is not known whether the polycarbonyl compounds are derived primarily from the diluent or also involve participation of TBP degradation products. Because of their lower concentration, the effect of the carbonyl compounds in solvent extraction is small in comparison to the acid phosphate esters.

## VIII. EFFECTS OF SOLVENT DEGRADATION PRODUCTS ON SEPARATIONS

Alkylphosphoric acids and their diesters are the only compounds formed by either radiolytic or chemical attack of the solvent that definitely have been shown to exert appreciable chemical effects on Purex separations. If concentrations of these compounds are allowed to build up, retention of U, Pu, Zr-Nb, and Ru by the solvent eventually results and then continues to increase with additional build-up. At very high doses changes in physical properties such as viscosity affect phase disengagement times.[113,117] Physical property changes are related to degradation products derived from the diluent and TBP. The compounds formed with the lower molecular weight alkylphosphoric acids are of low solubility, especially those formed with zirconium, and interfacial precipitates or third phases result at relatively low concentrations of HDBP and $H_2MBP$. Polycarbonyl compounds are formed at rates that are slower than the formation rates of compounds with phosphoric acid groups and have essentially no effect relative to these phosphoric acid compounds. Compounds of the structure types $RNO_2$, $RONO_2$, RCOOH, and RCOR have been shown to have insignificant effects. Furthermore, secondary products of diluent degradation that have been postulated have no effect. Aci-forms of nitro compounds[119] or hydroxamic acids[118,124] have not been detected in Purex streams. Even if hydroxamic acids are formed, they would be rapidly destroyed by nitrous acid.[124]

### A. Extractions with Alkylphosphoric Acids

Alkylphosphoric acids are powerful extractants[24,129,130] and extractable cations are not readily stripped from the organic phase with $HNO_3$. In low-polarity diluents, like kerosine or dodecane, cation exchange is the principal mechanism of extraction at nitric acid concentrations employed (<3 *M*) in the Purex process. At high $HNO_3$ concentrations, neutral salt solvates are also formed. Reactions or the extraction of $UO_2^{2+}$ are[131]

$$UO_2^{2+} + (HDBP)_2 \rightleftarrows UO_2(DBP)_2 + 2H^+ \quad (49)$$

$$UO_2^{2+} + 2(HDBP)_2 \rightleftarrows UO_2(DBP)_2 \cdot (HDBP)_2 + 2H^+ \quad (50)$$

$$UO_2^{2+} + 2NO_3^- + (HDBP)_2 \rightleftarrows UO_2(NO_3)_2 \cdot (HDBP)_2 \quad (51)$$

Because of the presence of a large excess of TBP in the Purex process, the effects of alkylphosphoric acids are reduced by hydrogen bonding with TBP according to the reaction[132]

$$(HDBP)_2 + 2TBP \rightleftarrows 2HDBP \cdot TBP \quad (52)$$

Numerous studies have shown the distribution ratios for the forward and reverse extractions of tracer level concentrations of $UO_2^{2+}$, $Pu^{4+}$, and $Zr^{4+}$ are unaffected by the presence of alkylphosphoric acid in concentrations of about $10^{-4}$ *M* in 20 to 30% TBP.[8,110,113] Above this concentration, however, the distribution ratios markedly increase with increasing concentration and are about second-power dependent on the alkylphosphoric acid concentration (Figure 25). Most trivalent species are only appreciably extracted at relatively low (<0.5 *M*) nitric acid concentrations. Plutonium can be stripped from solvents containing HDBP with strip solutions of about 1 *M* $HNO_3$ by reducing it to $Pu^{3+}$.

The extractable compounds formed with alkylphosphoric acids in the presence of excess TBP and $HNO_3$ concentrations of 3 *M* or less are typified by the dibutylphosphate compounds, $Zr(OH)(NO_3)(DBP)_2$, $Pu(NO_3)_2(DBP)_2$. These formulas are based

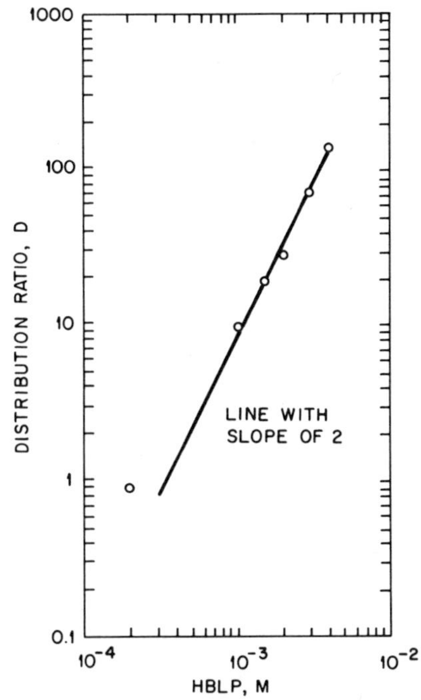


FIGURE 25. Distribution of zirconium between HBLP-30% TBP-dodecane and 3 *M* $HNO_3$ at 20°C (tracer level Zr concentrations). (Adapted from Maya, L. and Bopp, C. D., *J. Inorg. Nucl. Chem.*, 40, 1147, 1978. With permission.)

on the observed power dependence of distribution ratios on $H^+$, $NO_3^-$, and alkylphosphoric acid concentrations and on the analyses of precipitates of these compounds.[131,134-137] Structure considerations indicate the zirconium and plutonium compounds may exist as polymeric species in the organic phase through phosphoryl oxygen and free oxygen bridges. In addition, higher solvates formed by hydrogen bonding with alkylphosphoric acids (HA) may be formed such as $Pu(NO_3)_2A_2(HA)_2$ and $Pu(NO_3)_2A_2(HA)_4$. Wallace[138] has also proposed polymeric units of the general formula,

$$(RuNO)_n(DBP)_{2n-2}X_{n+2}(H_2O)_2$$

where X is $OH^-$ or $NO_3^-$, and n is the number of polymer units. These higher solvates and polymeric species of zirconium and ruthenium are suppressed in the presence of large excesses of TBP because of the formation of HDBP·TBP.

Relatively little is known regarding the distribution ratios for the extraction of metal cations by $H_2MBP$ except that the tendency for formation of alkylphosphoric acid compounds is considerably less than with HDBP. Studies have been largely qualitative because of the high aqueous phase of solubility of $H_2MBP$ and the insolubility of metal compounds in either organic or aqueous phases. The relative order of compound formation with alkylphosphoric acids[58] and metal cation extractability is

$$Zr^{4+} > Pu^{4+} > UO_2^{2+} > Ru(NO)3^+ > Nb^{5+}$$

In the bulk separation of fission products, uranium is present in a large excess and practically all of the HDBP and some of the $H_2MBP$ are used in forming the uranium compounds. The residual HDBP and $H_2MBP$ are utilized in the formation of the zirconium and plutonium compounds.

The radiation dose resulting in a single pass of solvent through pulse column equipment employed at Eurochemic[139] is reported to be as much as 0.5 $Wh\ell^{-1}$ which results in about $10^{-5}$ *M* concentrations of total alkylphosphoric acids. Essentially no problems are encountered with fission product decontamination in this operation.

### B. Solubilities of Alkylphosphoric Acids and Their Compounds

In the TBP-hydrocarbon diluent-$HNO_3$ system, HDBP and the higher alkylphosphoric acids favor the organic phase whereas $H_2MBP$ favors the aqueous phase.[132] The distribution ratios of HDBP lie between 5 and 20 for 20% TBP over concentration ranges of 0.1 to 10 *M* $HNO_3$ and $10^{-5}$ and 0.1 *M* HDBP. On the other hand, the distribution ratios of $H_2MBP$ are all below 0.3 over the above concentration range. The compounds formed by reaction of HDBP or $H_2MBP$ are of limited solubility in both the aqueous and organic phases (Figure 26).[140] Data on quantitative measurements of the solubility of plutonium compounds are lacking in the literature. Only the organic phase solubility can be estimated for $Pu(NO_3)_2(DBP)_2$ and its value is about $10^{-3}$ *M* in 10% TBP.[136] Solubilities generally increase with increasing nitric acid concentrations. There is generally good agreement among the investigators[140-143] in regard to the solubilities of the uranium and zirconium compounds. The increase in solubility of the zirconium compound with nitric acid concentration has been utilized to suppress precipitation in mixer-settlers at Karlsruhe.[71,141] The much higher solubility of the uranium compounds permits the use of high uranium loading of the solvent for the suppression of $Zr(OH)(NO_3)(DBP)_2$ precipitation.

The approach to equilibrium in solubility studies is slow and requires several days. The actual time would appear to depend on the aging conditions and method of compound preparation.[136,140-143] Increasing acidity slows the rate of precipitation. The solids are highly voluminous, settle slowly, and are apparently surface active.[140] Considerable foaming was noted when these solid compounds are agitated with the liquid phases. Barelko et al.[115] describe the zirconium precipitate as a slowly settling sol when formed in the irradiation of mixed TBP and aqueous phases. This precipitate was determined to be a compound containing monobutyl phosphate, dibutyl phosphate, and possibly some $PO_4^{3-}$. Perhaps the term "interfacial crud" that has been applied when these compounds precipitate and accumulate at the interface during Purex separations is appropriate.

### C. Effect of Phosphoric Acid

There is quantitative information in the literature on the formation of phosphate precipitates in solvent extraction operations. As Faugeras and Talmont[144] have pointed out, zirconium phosphate, even in nitric acid, is the most insoluble of phosphates. Analyses of aqueous phases from mixed phase irradiations usually are based on total phosphorus analyses and are often reported as the aqueous soluble $H_2MBP$, which is formed in greater abundance. In addition, phosphoric acid concentrations of $PO_4^{3-}$ do not appreciably affect distribution ratios. Distributions rations of Pu(IV) between 3 *M* $HNO_3$ and 30% TBP are only reduced by a factor of 2 by the presence of $PO_4^{3-}$ in 0.025 *M* concentrations.[8,15]

It is normal practice to remove TBP and its degradation products from aqueous high-activity waste by steam stripping[8,117] or diluent washes[142] prior to its concentration in evaporators. The relatively long residence times in the evaporators and the intense

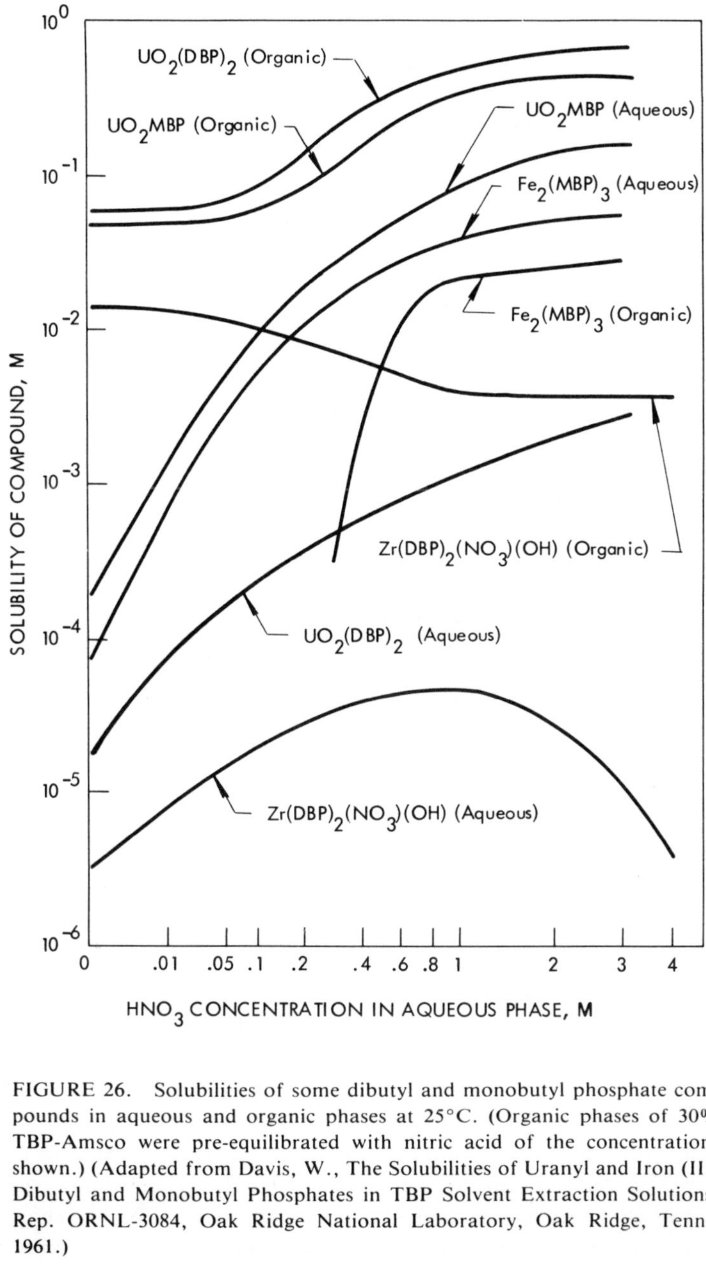


FIGURE 26. Solubilities of some dibutyl and monobutyl phosphate compounds in aqueous and organic phases at 25°C. (Organic phases of 30% TBP-Amsco were pre-equilibrated with nitric acid of the concentrations shown.) (Adapted from Davis, W., The Solubilities of Uranyl and Iron (III) Dibutyl and Monobutyl Phosphates in TBP Solvent Extraction Solutions, Rep. ORNL-3084, Oak Ridge National Laboratory, Oak Ridge, Tenn., 1961.)

radiation would convert essentially all these organophosphorous compounds to phosphoric acid. Greater concentration factors are possible because insoluble phosphates and phosphomolybdates are avoided.[117]

### D. Collective Effects

Although primary degradation product concentrations increase approximately linearly with dose, retention of fission products increases in a nonlinear fashion[117] after exposures of about 20 $Wh\ell^{-1}$ (Figure 27). This observation of a threshold value of dose indicates the primary products are not responsible for fission product retention. The shape of the zirconium retention curve suggests that secondary chemical and ra-

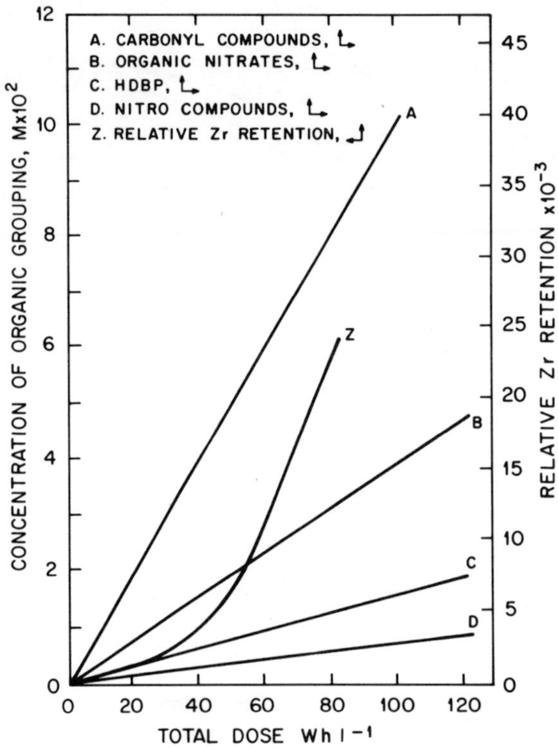


FIGURE 27. Effect of irradiation dose on the radiation yields of degradation products and the relative retention of zirconium (20% TBP-in odorless kerosine). (Adapted from Huggard, A. J. and Warner, B. F., *Nucl. Sci. Eng.*, 17, 638, 1963. With permission.)

diolytic reaction products are formed which are powerful extractants. Zirconium retention is measured by the Windscale Z number method and is a relative measurement. The Z number is determined by a standard test in which the zirconium retention is measured in a certain prescribed back-extraction procedure. Results are then calculated on the basis of moles of zirconium retained by $10^9$ ℓ of solvent.

Because the Z number does not measure the distribution ratio directly and the effects of the ratio of extractant to metal ion are unknown, quantitative interpretation of curve Z, Figure 27 is not possible. However, tests of compounds representing the primary and the possible secondary degradation products of diluents (Figure 24) show that none of the primary compounds, either individually or collectively, cause appreciable retention of zirconium.[117] Of the secondary products, only the hydroxamic acids gave high Z numbers. Hydroxamic acids, however, are rapidly destroyed by nitrous acid and have not been shown to be present in irradiated solvents or in Purex process streams.[118,124] Nitro compounds are not converted to their aci-forms under Purex conditions and do not contribute to fission product retention as once suggested.[119] Therefore it appears that the higher alkylphosphoric acids previously noted in Section VII are principally responsible for the retention of fission products after long-term exposure of the solvent.

## IX. CHEMICAL ASPECTS OF EXTRACTION CYCLES

It is desirable in large commercial plants to achieve the necessary decontamination of U and Pu using TBP extraction chemistry for all of the process cycles. Uranium purifications have usually been carried out in three TBP extraction cycles and, in some cases, with additional purification from $^{95}Zr$-$^{95}Nb$ using silica gel columns. In the third Pu purification cycle, however, older plants have used either cation exchange, anion exchange, or amine extraction to obtain concentrated plutonium products and thereby eliminate the need for plutonium solution evaporators. Even stripping Pu from TBP as its sulfate complexes[145] has been used to obtain concentrated products. All of these alternatives that have been employed for the third Pu cycle present many difficulties from either a plant operability, nitric aid recycle, or waste management point of view.

The use of complexing agents such as fluoride[146,147] to enhance fission product decontamination, and tartrate,[148] or sulfate[149] complexing rather than a reductant to effect partitioning has been shown to be chemically feasible. Formic acid complexing has been utilized to assist in stripping.[150] However, the use of such reagents would complicate both the recycle of nitric acid and waste management, and, furthermore, they have not been evaluated in this respect. Further evaluation of the use of organic complexing agents is probably warranted since effective methods for conversion of such compounds to highly volatile species (e.g., $CO_2$) could perhaps be developed.

### A. The First (High Activity) Cycle

In the first cycle, it is desirable to achieve a high decontamination from fission product activity and to minimize degradation of the solvent by limiting the contact time of the solvent. High fission product decontamination is achieved by using a high loading of uranium in the extractant and back-scrubbing of the U-Pu extract with aqueous $HNO_3$ scrub solutions. Centrifugal extractors or pulse columns are used to minimize the contact time of the solvent phase with the highly radioactive aqueous phase, thus minimizing solvent degradation. Comparisons of solvent extraction contactors used in radiochemical separations have been reviewed elsewhere.[150-153] Estimations of the solvent degradation and the resultant decontamination factors for Zr-Nb have been made for various type contactors[153] (Table 13). These predicted values are in good agreement with experimental values.[152] Methods of estimating solvent doses have been discussed by several authors.[116,139,153]

High acidity (2 to 4 $M$ $HNO_3$) in the feed and scrubbing stages suppresses interfacial precipitates of zirconium-alkylphosphoric acid compounds. From previously cited chemistry (see Section IV) it would appear beneficial to carry out two acid scrubbing operations. Koch[71] recommends first scrubbing the extract phase with 4 $M$ $HNO_3$ to enhance the back-washing of ruthenium and then scrubbing with 1.5 $M$ $HNO_2$ to improve back-washing of zirconium.

The use of two scrubbing temperatures for scrubbing Zr-Nb and Ru from the extract phase has also been shown to be beneficial in continuous countercurrent mixer-settler tests.[154] The bulk of the zirconium is first removed using a multistage scrubbing at 30°C. Subsequently, enhanced removal of ruthenium and highly extractable form of zirconium is accomplished by a multistage scrub at 70°C. Improved ruthenium removal results from more rapid dissociation of the highly extractable trinitrato complex to the relatively inextractable mono and dinitrato species at elevated temperatures. The reason for the beneficial effect of higher scrubbing temperatures on Zr-Nb removal is unknown. The species are unidentified. Since only the aqueous products were analyzed after U-Pu partitioning, it is not known whether improved decontamination was accomplished by retention of Zr-Nb as alkylphosphoric compounds by the organic phase or by thermal conversions to less extractable aqueous zirconium species. It should be borne in mind, however, that the acid hydrolysis of TBP is accelerated by increasing temperatures for either organic or aqueous phase reactions.

**Table 13**
**ESTIMATED SOLVENT DEGRADATION AND DECONTAMINATION FACTOR (DF) FOR Zr-Nb USING VARIOUS TYPES OF CONTACTORS**

| Contactor type | Dose to solvent ($Wh\ell^{-1}$) | DBP concentration ($M$) | DF(Zr-Nb) |
|---|---|---|---|
| Mixer-settler | 2.84 | $5.3 \times 10^{-4}$ | 10 |
| Pulse column | 0.19 | $3.5 \times 10^{-5}$ | 300 |
| Centrifugal contactor | 0.064 | $1.2 \times 10^{-5}$ | 500 |

*Note:* Solvent: 20% TBP — odorless kerosine. Fuel: 43,000 Mwd/t; 50 day decayed.

From Warner, B. F., Naylor, A., Duncan, A., and Wilson, P. D., *Proc. Int. Solvent Extraction Conf.*, Vol. 2, Jeffreys, G. V., Ed., Society of Chemical Industry, London, 1974, 1481. With permission.

Increased uranium loading of the solvent phase has also been shown to suppress interfacial precipitation of Zr. However, the upper limit of operability with respect to phase separation time is about 80% saturation of the solvent by uranium. This is because the densities of the aqueous and organic phases are nearly equal at about 80% uranium saturation.[8]

### B. U-Pu Partitioning

Partitioning of U-Pu using $U^{4+}$ as the reductant for $Pu^{4+}$ has been demonstrated at the Eurochemic plant[139] at Mol, Belgium, the La Hague plant[72] in France, and the Milliplant [71,86,141] at Karlsruhe, West Germany. Adequate separations are obtained in overall processing either using the $U^{4+}$ stabilized by $N_2H_2$ as a strip solution or by generating $U^{4+}$ *in situ* by the electrolytic process. Separation factors attained using $U^{4+}$ in the reductive strip solution have generally been an order of magnitude less than those achieved with ferrous sulfamate (Table 14). However, further separations of Pu from uranium can be achieved in an additional uranium purification cycle by using $U^{4+}$ as the reductant. Eurochemic results showed the total $\alpha$ specification[75] (<1500 dis $min^{-1}$ $g^{-1}$) on uranium products was met by repeating the $U^{4+}$ reduction in the second U cycle. Little additional decontamination of U from Np is accomplished when $U^{4+}$ is used as a reductant in the second cycle due to the reduction of Np(V) to Np(IV). However, it can be removed satisfactorily in the third U cycle.

Tests of the *in situ* electrolytic process at Karlsruhe continue to show effective separations of U and Pu.[71,86,141] Separation factors of plutonium from uranium are about an order of magnitude higher using this method than when reductive strip solutions of $U^{4+}$ are employed.

### C. Uranium Purification Cycles

Further fission product decontamination is accomplished by employing the same chemical equilibrium principles that were employed in the first cycle. Further decontamination of U from Pu can be achieved by reductively stripping the plutonium with a $U^{4+}$ strip solution or by the electroreduction method in the second cycle. Final Np decontamination can probably be achieved most conveniently in the third cycle by feeding a nitrous acid scrub stream into an intermediate extraction stage near the raffinate exit in the countercurrent extraction step of the cycle. Here a concentration of $>10^{-2}$ $M$ $HNO_2$ effectively scrubs neptunium into the raffinate by reducing it to the pentavalent state. The nitric acid is recovered by distillation and the small volume of

**Table 14**
**COMPARISON OF THE USE OF URANOUS NITRATE-HYDRAZINE AND FERROUS SULFAMATE IN U-Pu PARTITIONING AT EUROCHEMIC**

| | Ferrous sulfamate | Uranous nitrate-hydrazine |
|---|---|---|
| Mole ratio | Fe/Pu∿ 2 | U/Pu ∿ 8 |
| Separation factors | | |
| U product | $2 \times 10^3$ | $3 \times 10^2$ |
| Pu product | ∿ $10^5$ | $6 \times 10^3$ |

*Note:* Fuel characteristics: burn-up ∿ 15,000 Mwd/t; cooling time ∿ 150 days; Pu content ∿ 5 g/kg U; and specific activity ∿ 300 to 600 βCi/kg

From Van Geel, J., Joseph, C., Detilleux, E., Heinz, W., Cento, J., and Gustafsson, B., *Proc. Int. Solvent Extraction Conf.*, Vol. 1, Gregory, J. G., Evans, B., and Weston, P. C., Eds., Society of Chemical Industry, London, 1971, 577. With permission.

still bottoms containing small amounts of Np and fission products is combined with the high-level waste.

Intercycle evaporators are utilized to produce concentrated aqueous feeds of uranyl nitrate solutions so that extraction volumes are minimized. The aqueous phases contain small quantities of dissolved and entrained TBP and its degradation products which can undergo highly exothermic reactions[8,155,156] with hot nitric acid or uranyl nitrate if allowed to concentrate in the evaporation. Care is exercised in carrying out the evaporations. A pure diluent wash[134] or steam stripping [8,117] is used to remove most of the TBP and its degradation products before it is routed to the evaorators. A further precaution is taken by limiting the temperature used in evaporators. The highly exothermic and autocatalytic reactions with nitrate ion can be prevented[8,145] by maintaining temperatures below 130°C.

Workers at Karlsruhe[71] have proposed and studied process modifications that do not utilize intercycle evaporators. The total liquid throughput in solvent extraction steps is only increased by about 30%. In their process modification, concentrated nitric acid is added to the feed solutions so that high uranium loadings can be achieved in the extraction step, thus permitting a high uranium product concentration on stripping. The advantages and disadvantages of the Karlsruhe method compared to methods which employ intercycle evaporators have been reviewed[71] recently. The use of elevated temperatures (60°C) has been found to considerably improve stripping.

### D. Plutonium Purification Cycles

After partitioning, the plutonium is obtained as $Pu^{3+}$ in the product solution and the $Pu^{3+}$, hydrazine, and reacted $U^{4+}$ must be reoxidized for the subsequent purification cycles. The oxidation may be accomplished by sparging feed solutions with gaseous oxides of nitrogen or by electrolytic methods. Since lower loading of the extractant must be employed because of the limits of third phase formation in the $Pu(NO_3)_4$-$HNO_3$-$H_2O$-TBP-dodecane system (∿40 g/ℓ of Pu in 30% TBP), decontamination factors from fission products are less than those obtained in the uranium purification cycles. Also, lower loadings in extraction produce dilute products on stripping. Concentrated products can be produced only by reductively stripping the plutonium as $Pu^{3+}$.

Reduction of $Pu^{4+}$ is accomplished either by the electrolytic plutonium reduction process or by the use of hydroxylamine nitrate. Neither of thee processes adds to the volume of wastes since only gaseous species are generated. Product concentrations of about 60 g/ℓ have been demonstrated using reductive stripping. Such products are suitable for direct oxalate precipitation.

Plutonium can of course be stripped in its tetravalent state but dilute products would result, and intercycle evaporators would be required to avoid excessive liquid volumes in subsequent operations. In the evaporation of $Pu(NO_3)_4$ solutions conditions must be avoided which lead to irreversible formation of plutonium polymer and deposition of solid plutonium compounds. Partial reduction of $Pu^{4+}$ by solutions of low concentrations of $HNO_2$ (0.01 to 0.1 *M*) and about 0.3 to 0.5 *M* $HNO_3$ has been shown to improve product concentrations.[78,79]

## X. RECYCLE OF NITRIC ACID AND THE SOLVENT

Nitric acid is recovered by the physical process of distillation and is recycled.[8] For solvent purification, however, chemical scrubbing methods have been employed in existing Purex plants. The most widely used scrubbing process[8,118,157] is based on the use of an aqueous solution of sodium carbonate which is very effective in the removal of $H_2MBP$ and HDBP as their aqueous-soluble sodium salts. Sodium carbonate or other aqueous solutions of basic chemical salts are ineffective for removal of the higher alkylphosphoric acids[128,158] and diluent degradation products. Sodium hydroxide is not used because precipitates containing uranium and zirconium can be formed under some conditions.[128] The higher alkylphosphoric acids can be removed by sorption on macroreticular resins[159] or on solids such as $PbO_2$.[128] Distillation is the only method that has been shown to be effective for the removal of radiolysis products from both the diluent and TBP. It has been shown that distillation processes[117,119] are capable of producing purified solvent that is of the same quality as unirradiated solvent; thus it is a total purification method.

Total or more complete purification methods have not been required in the past because the fuels that were being processed were of relatively low burnup and removal of the primary radiolysis products, $H_2MBP$ and HDBP, gave adequate solvent performance. This was because concentrations of secondary products which are not removed by sodium carbonate scrubbing increase very slowly on repeated recycle of the solvent. The requirement of fresh solvent make-up to allow for the relatively small in-process losses (0.5% or less) of solvent due to solvent solubility, entrainment, and evaporation also helps in the retardation of the build-up of impurities in the recycled solvent.[117] Even with the higher burnup LWR fuels, aqueous scrubbing methods are adequate if solvent exposure times to irradiation are minimized by employing pulse columns or centrifugal contactors in the main-line separations.

Hydrazine carbonate[128] scrubbing appears to be preferable to sodium carbonate scrubbing from a waste management point of view. The spent hydrazine carbonate solution is treated by an anodic oxidation process which destroys all of the $N_2H_5^+$ and some of the $NO_3^-$ and permits the wastes to be highly concentrated by evaporation. The $N_2H_5^+$ is completely converted to $N_2$. The anodic gases are 95% $N_2$ with the remainder being a mixture of $NO_x$ and $O_2$. The electrode reactions of these ions were described in Section VI.

The rates of removal of $H_2MBP$ and HDBP are quite rapid in the scrubbing processes with alkaline carbonate solutions.[128] However, the compounds that are formed with cations are removed somewhat more slowly, particularly zirconium and ruthenium compounds. Increasing scrubbing temperatures improves the rate of removal. Allowing the solvent containing the ruthenium compounds to age drastically increases the time required to remove them by alkaline scrubbing.[160] For unaged solvents containing HDBP, Zr, and Ru, decontamination factors of 100 were obtained for both Zr and Ru in 5-min contacts using either $Na_2CO_3$ or hydrazine carbonate scrubbing solutions.[128] Using the falling drop method, mass transfer rates for both the zirconium compound and for unbound HDBP were found to be first order with respect to initial concentrations. The rate of removal of the Zr compound was found to be about 2.5 times slower than that of unbound HDBP. Aging the solvents containing nitrosyl ruthenium, $H_2MBP$, and HDBP has been shown to have a profound effect on ruthenium retention.[160] After 2 days of aging, a contact time of 200 min was required to remove 98% of the Ru from the solvent. This washing behavior suggests that the responsive species is converted to a nonresponsive species [although the 20% TBP initally contained unrealistic amounts of HDBP (0.1 *M*) and $H_2MBP$ (0.012 *M*)].

Degradation compounds derived solely from the diluent and the higher HA compounds are not effectively removed by alkaline scrubbing solutions.[128,158] Orth[157] reports very small decontamination factors for unknown species whih retain Zr and Ru. The compounds solely derived from the diluent and apparenly the higher alkylphosphoric acids can be removed using alkanol amine scrubbing.[118] However, this method has not found applicability because of a variety of process difficulties in its application and of waste management considerations. For similar reasons, other scrubbing solutions found beneficial in regard to lessening ruthenium retention, such as $Na_2CO_3$ — $KMnO_4$ and hydroxylamine nitrate solutions have not found application.[157]

## ACKNOWLEDGMENTS

The author is indebted to several people for valuable discussions on selected topics in Purex solvent extraction chemistry and, in some cases, for providing copies of their personal literature searches in certain topics. I am especially thankful for many valuable discussions with Dr. D. O. Campbell, Dr. A. D. Kelmers, and Dr. L. Maya of Oak Ridge National Laboratory (ORNL). Dr. Campbell (author of Chapter 3 of this volume) and I also freely exchanged literature source materials, since our subject matter was interrelated. Dr. Kelmers kindly provided a bibliography and an evaluation of the literature on certain aspects of the chemistry of hydroxylamine, hydrazine, and nitrous acid which he and D. Y. Valentine of ORNL had assembled. Discussions with him regarding this subject matter were invaluable. Special thanks are also due Dr. O. L. Keller, ORNL and to Dr. W. W. Schulz of the Rockwell International Corporation, Hanford, Wash., for valuable discussions on the chemistry of neptunium and transplutonium elements. Gratitude is also expressed to Dr. E. P. Horwitz and Dr. G. W. Mason of Argonne National Laboratory for valuable discussions on the chemistry of alkylphosphoric acids.

## REFERENCES


1. **Koch, G.**, Reprocessing of nuclear fuels, *Kerntechnik,* 20, 363, 1978.
2. **Bebbington, W. P.**, The reprocessing of nuclear fuels, *Sci. Am.*, 235, 30, 1976.
3. **Shelton, E. B.**, Coordinator, Experience with Processing Irradiated Fuel at the Savannah River Plant (1954 to 1976), Rep. DP-1467, Savannah River Laboratory, Aiken, S.C., 1977.
4. **Alcock, K., Grimley, S. S., Healy, T. V., Kennedy, J., and McKay, H. A. C.**, The extraction of nitrates by tri-n-butyl phosphate (TBP). I. The system TBP + diluent + $H_2O$ + $HNO_3$, *Trans. Faraday Soc.*, 52, 39, 1959.
5. **Healy, T. V. and McKay, H. A. C.**, The extraction of nitrates by tri-n-butyl phosphate (TBP). II. The nature of the TBP phase, *Trans. Faraday Soc.*, 52, 633, 1959.
6. **McKay, H. A. C. and Healy, T. V.**, Tri-n-butyl orthophosphate (TBP) as an extracting solvent, *Prog. Nucl. Energy Ser. 3,* 2. 546, 1958.
7. **Leroy, P.**, Study of the Solvent 30% Tributyl Phosphate in Dodecane, CEA-3207, Fontenay-aux-Roses Nuclear Research Center, France, 1967; ORNL-tr-4344, Oak Ridge National Laboratory, Oak Ridge, Tenn., 1967.
8. **Stevenson, R. L. and Smith, P. E.**, Aqueous separation, in *Reactor Handbook,* Vol. 2, 2nd ed., Stoller, S. M. and Richards, R. B., Eds., Interscience, New York, 1961, 107.
9. **Koch, G.**, The chemical reprocessing of irradiated nuclear fuels, *Chem. Ztg.*, 101, 64, 1977.
10. **Bagawde, S. V., Vasudava Rao, P. R., Ramakrishna, V. V., and Patil, S. K.**, The effect of temperature on the extraction of uranium (VI) from nitric acid by tri-n-butyl-phosphate, *J. Inorg. Nucl. Chem.*, 40, 1913, 1978.
11. **Davis, W., Jr.**, Thermodynamics of extraction of nitric acid by tri-n-butyl-phosphate-hydrocarbon solutions. I. Distribution studies with TBP in Amsco 125-82 at intermediate and low acidities, *Nucl. Sci. Eng.*, 14, 159, 1962.
12. **Davis, W., Jr.**, Thermodynamics of extraction of nitric acid by tri-n-butyl-phosphate-hydrocarbon diluent solutions. II. Densities, molar volumes, and water solubilities of TBP-Amsco 125-82-$HNO_3$-$H_2O$ solutions, *Nucl. Sci. Eng.*, 14, 169, 1962.
13. **Davis, W., Jr.**, Thermodynamics of extraction of nitric acid by tri-n-butyl-phosphate-hydrocarbon diluent solutions. III. Comparison of literature data, *Nucl. Sci. Eng.*, 14, 174, 1962.
14. **Davis, W., Jr., Mrochek, J., and Hardy, C. J.**, The system: tri-n-butyl-phosphate (TBP)-nitric acid-water, I. Activities of TBP in equilibrium with aqueous nitric acid and partial molal volumes of the three components in the TBP phase, *J. Inorg. Nucl. Chem.*, 28, 2001, 1966.
15. **McKay, H. A. C., Scargill, D., and Wain, A. G.**, Extraction by neutral organophosphorous compounds, in *Gmelin Handbuch der Anorganischen Chemie,* Vol. 21, Koch G., Ed., Springer-Verlag, Berlin, 1975, 177.
16. **Petrich, G. and Zolarik, A.**, Distribution of U(VI), Pu(IV), and nitric acid in the system uranyl nitrate-plutonium (IV) nitrate-nitric acid-water/30% TBP in aliphatic diluents: compilation and critical evaluation of equilibrium data, Rep. KFK-2576, *Kernforschungszentrum Karlsruhe Ber.*, 1977.
17. **Watson, S. B. and Rainey, R. H.**, Modifications of the Sephis Code for Calculating the Purex Solvent Extraction System, Rep. ORNL/ TM-5123, Oak Ridge National Laboratory, Oak Ridge, Tenn., 1975.
18. **Scotten, W. C.**, Solvex — A Computer Program for Simulation of Solvent Extraction Processes, Rep. DP-1391, Savannah River Laboratory, Aiken, S.C., 1975.
19. **Ochsenfeld, W., Baumgartel, G., and Schmieder, H.**, Empirical functions for the calculation of simultaneous Pu(IV) and U(VI) distribution in the system metal nitrates — nitric acid/tributyl phosphate-n-dodecane, in *Solvent Extraction Research,* Kertes, A. S. and Marcus, Y., Eds., New York, 1969, 373.
20. **Richardson, G. L.**, The effect of High Solvent Radiation Exposures on TBP Processing of Spent LMFBR Fuels, Rep. HEDL-TME 73-51, Hanford Engineering and Development Laboratory, Richland, Wash., 1973.
21. **Aochi, T., Tsujino, T., Kohsaka, A., and Hoshino, T.**, Computer evaluation of chemical flowsheets for plutonium purification with TBP, in *Proc. Int. Solvent Extraction Conf.*, Vol. 1, Gregory, J. G., Evans, B., and Weston, P. C., Eds., Society of Chemical Industry, London, 1971, 138.
22. **Shulz, W. W. and Benedict, G. E.**, Neptunium-237 production and recovery, in AEC Critical Review Series, TID-25955, U.S. Atomic Energy Commission Technical Information Center, Oak Ridge, Tenn., 1972, Chap. 2.
23. **Usachev, V. N. and Chaikhorskii, A. A.**, Investigation of complex formation in nonaqueous solutions. VII. On the extraction of neptunium (V) by tributylphosphate, *Sov. Radiochem.*, 8, 45, 1966.
24. **Ahrland, S., Liljenzin, J. O., and Rydberg, J.**, Actinides: solution chemistry, in *Comprehensive Inorganic Chemistry,* Vol. 5, Bailar, J. C., Jr., Emeleus, H. J., Nyholm, R., and Trotman-Dickenson, A. F., Eds., Pergamon Press, Oxford, 1973, 465.

25. **Patil, S. K., Ramakrishna, V. V., Avadhany, G. V. N., and Ramaniah, N. V.,** Some studies of the TBP extraction of actinides, *J. Inorg. Nucl. Chem.*, 35, 2357, 1973.
26. **Best, G. F., Hesford, E., and McKay, H. A. C.,** Tri-n-butyl phosphate as an extracting agent for inorganic nitrates. VII. The trivalent act inide nitrates, *J. Inorg. Nucl. Chem.*, 12, 136, 1959.
27. **Alcock, K., Best, G. F., Hesford, E., and McKay, H.A.C.,** Tri-n-butyl phosphate as an extracting solvent for inorganic nitrates. V. Furth er results for the tetra- and hexavalent actinide nitrates, *J. Inorg. Nucl. Chem.*, 6, 328, 1958.
28. **McKay, H. A. C. and Streeton, R. J. W.,** Extraction of uranium (IV) nitrate by tri-n-butyl phosphate and by dibutyl carbitol, *J. Inorg. Nucl. Chem.*, 27, 879, 1965.
29. **Hesford, E., McKay, H. A. C., and Scargill, D.,** Tri-n-butyl phosphate as an extracting solvent for inorganic nitrates. IV. Thorium nitrate, *J. Inorg. Nucl. Chem.*, 4, 321, 1957.
30. **Best, G. F., McKay, H. A. C., and Woodgate, P. R.,** Tri-n-butyl phosphate as an extracting solvent for inorganic nitrates. III. The plutonium nitrates, *J. Inorg. Nucl. Chem.*, 4, 315, 1957.
31. **Scargill, D., Alcock, K., Fletcher, J. M., Hesford, E., and McKay, H. A. C.,** Tri-n-butyl phosphate as an extracting solvent for inorganic nitrates. II. Yttrium and the lower lanthanide nitrates, *J. Inorg. Nucl. Chem.*, 4, 304, 1957.
32. **Hesford, E., Jackson, E. E., and McKay, H. A. C.,** Tri-n-butyl phosphate as an extracting solvent for inorganic nitrates. VI. Further results for the rare earth nitrates, *J. Inorg. Nucl. Chem.*, 9, 279, 1958.
33. **Mills, A. L. and Lillyman, E.,** A review of metallic fast reactor fuel reprocessing at Dounreay, in *Proc. Int. Solvent Extraction Conf.*, Vol. 2, Jeffreys, G. V., Ed., Society of Chemical Industry, London, 1974, 1499.
34. **Codding, J. W., Haas, W. O., Jr., and Heumann, F. K.,** Tributyl phosphate-hydrocarbon system: organizing equilibrium data, *Nucl. Technol.*, 50, 145, 1958.
35. **Kertes, A. S.,** The chemistry of the formation and elimination of a third phase in organophosphorous and amine extraction systems, in *Solvent Extraction Chemistry of Metals*, McKay, H. A. C., Ed., Macmillan, London, 1965, 377.
36. **Donaldson, D. M., Hartley, K., Lees, P., and Parkinson, N.,** Reprocessing fast reactor fuels at Dounreay, *Trans. Metall. Soc. AIME*, 227, 191, 1963.
37. **Komorav, E. V., Shput, L. B., and Pushlenkov, M. F.,** Thermodynamic equilibrium constants and thermal effects of reactions of extraction of uranyl nitrates by trialkyl phosphates in inert solvents, *Sov. Radiochem.*, 13, 395, 1971.
38. **Horner, D. E.,** Formation of Third Phases and the Effect of Temperature on the Distribution of Plutonium and Uranium in Extractions by Tri-n-Butyl Phosphate, Rep. ORNL-4724, Oak Ridge National Laboratory, Oak Ridge, Tenn., 1971.
39. **Adamskii, N. M., Karpacheva, S. M., Mel'nikov, I. N., and Rozen, A. M.,** The effect of temperature on the extraction of nitric acid with tributyl phosphate, *Sov. Radiochem.*, 2, 12, 1960.
40. **Shevchenko, V. B. and Fedorov, I. A.,** The effect of temperature on the extraction of uranyl-, plutonium-, ruthenium- and zirconium-nitrates with tributyl phosphate, *Sov. Radiochem.*, 2, 4, 1960.
41. **Baumgartner, F. and Finsterwalder, L.,** Kinetics of extraction of uranium (VI) and plutonium (IV) with n-tributylphosphate, in *Solvent Extraction Research*, Kertes, A. S. and Marcus, Y., Eds., Interscience, New York, 1969, 313.
42. **Baumgartner, F. and Finsterwalder, L.,** On the transfer mechanism of uranium (VI) and plutonium (IV) nitrate in the system nitric acid-water/tributylphosphate-dodecane, *J. Phys. Chem.*, 74, 108, 1970.
43. **Pushlenkov, M. F., Shchepetil'nikov, N. N., Kuznetsov, G. I., Kasimov, F. D., Yasnovitskays, A. L., and Yakovlev, G. N.,** The kinetics of uranium, plutonium, ruthenium and zirconium extraction with tributylphosphate, in *Proc. Int. Solvent Extraction Conf.*, Vol. 1, Jeffreys, G. V., Ed., Society of Chemical Industry, London, 1974, 493.
44. **Shchepetil'nikov, N. N., Kuznetsov, G. I., Pushlenkov, M. F., and Yakovlev, G. N.,** Study of the rate of extraction in systems with tributyl phosphate. V. Kinetics of the mass transport of uranium and plutonium during mixing in a centrifugal field, *Sov. Radiochem.*, 14, 356, 1972.
45. **Yagodin, G. A., Sinefribova, O. A., and Chekmarev, A. M.,** The effect of the polymerization of zirconium and hafnium on their extraction by different classes of organic compounds, in *Proc. Int. Solvent Extraction Conf.*, Vol. 3, Jeffreys, G. V., Ed., Society of Chemical Industry, London, 1974, 2209.
46. **Yagodin, G. A., Tarasov, V. V., and Kazak, V. G.,** Kinetics of thermal decomposition of unextractible compounds of zirconium in nitric acid solutions, *Sov. Radiochem.*, 15, 117, 1973.
47. **Chetverina, R. B., Boichinova, E. S., Vinter, I. K., and Benderskaya, L. A.,** Dependence of the ion-exchange properties of modified zirconium phosphate on the quality of the original zirconium oxynitrate, *J. Appl. Chem. USSR*, 49, 2637, 1976.

48. **Orth, D. A., McKibben, J. M., and Scotten, W. C.,** Progress in tributyl phosphate technology, in *Proc. Int. Solvent Extraction Conf.,* Vol. 1, Gregory, J. G., Evans, B., and Weston, P. C., Eds., Society of Chemical Industry, London, 1971, 514.
49. **Alcock, K., Bedford, F. C., Hardwick, W. H., and McKay, H. A. C.,** Tri-n-butyl phosphate as an extracting solvent for inorganic nitrates. I. Zirconium nitrate, *J. Inorg. Nucl. Chem.,* 4, 100, 1957.
50. **Klyuchnikov, V. M., Zaitsev, L. M., Korovin, S. S., Solov'eva, E. S., and Apraskin, I. A.,** Extraction of zirconium and hafnium hydroxide nitrates by tributyl phosphate, *Russ. J. Inorg. Chem.,* 17, 408, 1972.
51. **Ergov, G. F., Fomin, V. V., Frolov, Yu. G., and Yagodin, G. A.,** Tributyl phosphate solvates of zirconium and hafnium nitrates, *Russ. J. Inorg. Chem.,* 5, 503, 1960.
52. **Solovkin, A. S.,** Thermodynamics of the extraction of zirconium, present in the monomeric state, from nitric acid solutions by tri-n-butyl phosphate (TBP), *Russ. J. Inorg. Chem.,* 15, 983, 1970.
53. **Brown, P. M.,** Nitro complexes of nitrosyl ruthenium, *J. Inorg. Nucl. Chem.,* 13, 73, 1960.
54. **Scargill, D., Lyon, C. E., Large, N. R., and Fletcher, J. M.,** Nitratoaquo complexes of nitrosylruthenium III, *J. Inorg. Nucl. Chem.,* 27, 161, 1965.
55. **Fletcher, J. M., Lyon, C. E., and Wain, A. G.,** The partition coefficients of nitratonitrosylruthenium between nitric acid and TBP phases, *J. Inorg. Nucl. Chem.,* 27, 1841, 1965.
56. **Fletcher, J. M., Brown, P. G., Gardner, E. R., Hardy, C. J., Wain, A. G., and Woodhead, J. L.,** Nitrosyl ruthenium nitrato complexes in aqueous nitric acid, *J. Inorg. Nucl. Chem.,* 12, 154, 1959.
57. **Fletcher, J. M., Jenkins, I. L., Lever, F. M., Martin, F. S., Powell, A. R., and Todd, R.,** Nitrato and nitro complexes of nitrosylruthenium, *J. Inorg. Nucl. Chem.,* 1, 378, 1955.
58. **Brown, P. G. M., Fletcher, J. M., Hardy, C. J., Kennedy, J., Scargill, D., Wain, A. G., and Woodhead, J. L.,** The significance of certain complexes of ruthenium, niobium, zirconium and uranium in plant processes, *Prog. Nucl. Energy Ser. 3,* 2, 129, 1961.
59. **Fletcher, J. M. and Scargill, D.,** Factors influencing the variability of ruthenium decontamination in TBP processes, in *Solvent Extraction Chemistry of Metals,* McKay, H. A. C., Ed., Macmillan, London, 1965, 27.
60. **Duncan, A., Naylor, A., and Warner, B. F.,** Fission product behavior in the first contactor of a TBP process, in *Solvent Extraction Chemistry of Metals,* McKay, H. A. C., Ed., Macmillan, London, 1965, 3.
61. **Joon, K.,** RuNO-nitro complexes, conversion rates and extraction coefficients, in *Solvent Extraction Research,* Kertes, A. S. and Marcus, Y., Eds., Interscience, New York, 1969, 323.
62. **Rudstam, G.,** Studies of nitrosylruthenium complexes in nitric acid solutions using repeated extractions, *Acta Chem. Scand.,* 13, 1481, 1959.
63. **Pushlenkov, M. E., Shchepstil'nikov, N. N., Kuznetsov, G. I., Kasimov, F. D., Yasnovitskaya, A. L., and Yakovlev, G. N.,** Kinetics of extraction of nitrosoruthenium from nitric acid solutions by tri-n-butyl phosphate with mixing of the liquids in a centrifugal field, *Sov. Radiochem.,* 15, 252, 1973.
64. **McKay, H. A. C.,** An obstinate fission product, *Chem. Br.,* 11, 257, 1975.
65. **Hardy, C. J. and Scargill, D.,** The extraction of niobium (V) from nitric acid solution by tri-n-butyl phosphate (TBP), *J. Inorg. Nucl. Chem.,* 13, 174, 1960.
66. **Moore, J. G. and Rainey, R. H.,** Extraction of niobium-95 from nitric acid solutions with tri-n-butyl phosphate, Rep. ORNL-3285, Oak Ridge National Laboratory, Oak Ridge, Tenn., 1962.
67. **Moore J. G.,** Removal of niobium-95 from zirconium solutions with Vycor glass, *J. Inorg. Nucl. Chem.,* 20, 116, 1961.
68. **Siddall, T. H.,** Behavior of Technetium in the Purex Process, Rep. DP-364, Savannah River Laboratory, Aiken, S.C., 1959.
69. **Macasek, F.,** Co-extraction of technetium with uranyl nitrate tributylphosphate complex, *Radiochem. Radioanal. Lett.,* 22, 175, 1975.
70. **Koch, G., Ochsenfeld, W., and Schwind, E.,** Flowsheet studies on processing of plutonium fuels by solvent extraction, Rep. KFK-990, *Kernforschungszentrum Karlsruhe Ber.,* 1969.
71. **Koch, G., Ochsenfeld, W., Schmieder, H., and Weinlander, W.,** Chemical aspects in the design of a flowsheet for a large-scale LWR reprocessing plant, *Kerntechnik,* 18, 253, 1976.
72. **Bathellier, A.,** Transuranium elements in the reprocessing of nuclear fuels, *Bull. Inf. Sci.,* 217, 7, 1977.
73. **Alders, L.,** *Liquid-Liquid Extraction,* 2nd ed., Elsevier, Amsterdam, 1959, chap. 5.
74. **Newton, T. W.,** The Kinetics of Oxidation — Reduction Reactions of Uranium, Neptunium, Plutonium, and Americium in Aqueous Solutions, TID 26506, ERDA Critical Review Series, U.S. Energy Research and Development Administration, 1975.
75. *Fed. Regist.,* 31(250), 16584, 1966.

76. **Gourisse, D.**, Laboratory studies of nitrous acid and neptunium behavior in TBP extraction processes, in *Proc. Int. Solvent Extraction Conf.*, Vol. 1, Gregory, J. G., Evans, B., and Weston, P. C., Eds., Society of Chemical Industry, London, 1971, 781.
77. **Poe, W. L., Joyce, A. W., and Martens, R. I.**, $Np^{237}$ and $Pu^{238}$ separation at the Savannah River Plant, *Ind. Eng. Chem. Process Des. Dev.*, 3, 315, 1964.
78. **Tsujino, T. and Aochi, T.**, Reduction stripping of plutonium loaded in TBP with addition of nitrous acid, *J. Nucl. Sci. Technol.*, 13, 321, 1976.
79. **Bathellier, A.**, The use of nitrous acid as a backwashing agent for actinide bearing solvents, in *Solvent Extraction Chemistry of Metals,* McKay, H. A. C., Ed., Macmillan, London, 1965, 295.
80. **Dukes, E. K.**, Kinetics and mechanism for the oxidation of trivalent plutonium by nitrous acid, *J. Am. Chem. Soc.*, 82, 9, 1960.
81. **Koltunov, V. S. and Marchenko, V. I.**, Reaction between Pu(III) and nitrous acid. I. Equilibrium, *Sov. Radiochem.*, 15, 754, 1973.
82. **Koltunov, V. S. and Marchenko, V. I.**, Reaction between plutonium (III) and nitrous acid. II. Kinetics of forward reaction, *Sov. Radiochem.*, 15, 787, 1973.
83. **Koltunov, V. S. and Marchenko, V. I.**, Reaction between plutonium (III) and nitrous acid. III. Kinetics of reverse reaction, *Sov. Radiochem.*, 15, 791, 1973.
84. **Schmieder, H., Baumgartner, F., Goldacker, H., Hausberger, H., and Warmecke, E.**, Electrolytic methods for application in the Purex process, in *Proc. Int. Solvent Extraction Conf.*, Vol. 3, Jeffreys, G. V., Ed., Society of Chemical Industry, London, 1974, 1997.
85. **Koch, G., Baumgartner, F., Goldacker, H., Ochsenfeld, W., and Schmieder, H.**, A solvent extraction flowsheet for a large-scale LWR fuel reprocessing plant, Rep. KFK-2557, *Kernforschungszentrum Karlsruhe Ber.*, 1977.
86. **Schmieder, H., Baumgartner, F., Goldacker, H., and Hausberger, H.**, Electrolytic techniques in the Purex process, Rep. KFK- 2082, *Kernforschungszentrum Karlsruhe Ber.*, 1974.
87. **Barney, G. S.**, A kinetic study of the reaction of plutonium (IV) with hydroxylamine, *J. Inorg. Nucl. Chem.*, 38, 1677, 1976.
88. **Koltunov, V. S. and Zhuravleva, G. I.**, Kinetics of reduction of actinides by hydroxylamine. III. Reduction of plutonium (IV) in nitric acid solution, *Sov. Radiochem.*, 20, 73, 1978.
89. **Talmont, X. L. R.**, Utilization of uranium (IV) nitrate in a second cycle purification of plutonium by TBP, in *Solvent Extraction Chemistry of Metals,* McKay, H. A. C., Ed., Macmillan, London, 1965, 103.
90. **Horner, D. E.**, The Use of Ferrous Nitrate as a Plutonium Reductant for Partitioning Plutonium and Uranium in Purex Processes, Rep. ORNL-4383, Oak Ridge National Laboratory, Oak Ridge, Tenn., 1969.
91. **Bond, W. D., Jao, Y., and Peterson, J. R.**, The oxidation-reduction kinetics involved in the Np(V) + Fe(II) $\rightleftharpoons$ Np(IV) + Fe(III) system in nitric acid solutions, *J. Inorg. Nucl. Chem.*, 39, 1395, 1977.
92. **Biddle, P. and Miles, J. H.**, The rate of reaction of nitrous acid with hydrazine and with sulfamic acid, *J. Inorg. Nucl. Chem.*, 30, 1291, 1968.
93. **Koltunov, V. S. and Marchenko, V. I.**, The kinetics of oxidation of hydrazine by nitrous acid, *Kinet. Cata. USSR,* 7, 202, 1966.
94. **Perrott, J. R. and Stedman, G.**, The kinetics of nitrite scavenging by hydrazine and hydrazoic acids at high acidities, *J. Inorg. Nucl. Chem.*, 39, 325, 1977.
95. **Kelmers, A. D. and Valentine, D. Y.**, Search for Alternate Holding Reductants to Stabilize Plutonium (III) Solutions, Rep. ORNL/TM-6521, Oak Ridge National Laboratory, Oak Ridge, Tenn., 1978.
96. **Dukes, E. R. and Wallace, R. M.**, Formation of Hydrazoic Acid from Hydrazine in Nitric Acid Solutions, Rep. DP-728, Savannah River Laboratory, Aiken, S.C., 1962.
97. **Schlea, C. S., Caverly, M. R., Henry, H. E., and Jenkins, W. J.**, Uranium (IV) Nitrate as a Reducing Agent for Plutonium (V) in the Purex Process, Rep. DP-808, Savannah River Laboratory, Aiken, S.C., 1963.
98. **Clark, C. C.**, Hydrazine, Matheson Chemical Company, Baltimore, Md., 1953, 2.
99. **Koltunov, V. S. and Zhuravleva, G. I.**, Kinetics of reduction of plutonium by hydrazine. II. Reduction of Pu(IV), *Sov. Radiochem.*, 16, 80, 1974.
100. **Pushlenkov, M. F., Tikhonov, M. F., and Shchepetil'nikov, N. N.**, Reductive re-extraction of plutonium in a heterophase system with the aid of various reagents, *Sov. Radiochem.*, 17, 815, 1975.
101. **Koltunov, V. S., Zhuravleva, G. I., Marchenko, V. I., and Tilhonov, M. F.**, Peculiarities of the kinetics and mechanism of certain redox reactions of neptunium, plutonium and uranium, *Sov. Radiochem.*, 17, 305, 1975.
102. **Koltunov, V. S. and Tikhonov, M. F.**, Kinetics of reduction of neptunium by hydrazine. II. Reduction of Np(VI) to Np(V) in nitric acid, *Sov. Radiochem.*, 15, 190, 1973.

103. **Gourisse, D. and Gautier, A.,** Distribution of nitrous acid in extractions with tributylphosphate, *J. Inorg. Nucl. Chem.*, 31, 839, 1969.
104. **Biddle, P., McKay, H. A. C., and Miles, J. H.,** The role of nitrous acid in the reduction of plutonium (IV) by U(IV) in TBP systems, in *Solvent Extraction Chemistry of Metals,* McKay, H. A. C., Ed., Macmillan, London, 1965, 133.
105. **Newman, R. I.,** The Barnwell Nuclear Fuel Plant, *Nucl. Eng. Int.*, 17, 938, 1972.
106. **Cermak, A. F., Gray, J. H., Murbach, W. E., Neace J. C., and Spaunburgh, R. G.,** Development of the electropulse column for BNFP, in *Proc. American Nuclear Society Meeting: Backend of the LWR Fuel Cycle,* CONF-780304, American Nuclear Society, LaGrange Park, Ill., 1974, V-II.
107. **Burger, L. L.,** The decomposition reactions of tributyl phosphate and its diluents and their effect on uranium recovery processes, *Prog. Nucl. Energy Ser. 3,* 2, 307, 1958.
108. **Davis, W. and Kibbey, A. H.,** Aqueous Phase Decomposition of Tributyl Phosphate at 35°C to 75°C in Nitric Acid Solutions, Rep. ORNL/TM-3062, Oak Ridge National Laboratory, Oak Ridge, Tenn., 1970.
109. **Moffat, A. J. and Thompson, R. D.,** The chemical stability of tributyl phosphate in some nitrate and chloride systems, *J. Inorg. Nucl. Chem.*, 16, 356, 1961.
110. **Rigg, T. and Wild, W.,** Radiation effects in solvent extraction processes, *Prog. Nucl. Energy Ser.3,* 2, 320, 1958.
111. **Rofer-DePoorter, C. K. and DePoorter, G. L.,** Photochemical reduction of uranyl nitrate by tri-n-butyl phosphate, *J. Inorg. Nucl. Chem.*, 39, 631, 1977.
112. **Clay, P. G. and Witort, M.,** Radiolysis of tri-n-butyl phosphate in aqueous solution, *Radiochem. Radioanal. Lett.*, 19, 101, 1974.
113. **Nowak, Z.,** Radiolytic degradation of extractant-diluent systems used in the Purex process, *Nukleonkika,* 22, 155, 1977.
114. **Barelko, E. V. and Solyanina, I. P.,** Radiolysis of TBP in contact with nitric acid-formation of radiolysis products of the extraction reagent, *Sov. At. Energy,* 35, 898, 1973.
115. **Barelko, E. V., Solyanina, I. P., and Babakina, G. S.,** Influence of radiation on the extraction of radiozirconium and radioiodine by solutions of tri-n-butyl phosphate (TBP), *Sov. Radiochem.*, 18, 573, 1976.
116. **Blake, C. A.,** Solvent Stability in Nuclear Fuel Processing: Evaluation of the Literature, Calculation of Radiation Dose, Effects of Iodine, and Plutonium, Rep. ORNL 4212, Oak Ridge National Laboratory, Oak Ridge, Tenn., 1968.
117. **Huggard, A. J. and Warner, B. F.,** Investigations to determine the extent of degradation of TBP/odorless kerosine solvent in the new separation plant Windscale, *Nucl. Sci. Eng.*, 17, 638, 1963.
118. **Lane, E. S.,** Performance and degradation of diluents for TBP and the clean-up of degraded solvents, *Nucl. Sci. Eng.*, 17, 620, 1963.
119. **Blake, C. A., Davis, W., and Schmitt, J. M.,** Properties of degraded TBP-Amsco solutions and alternative extractant — diluent systems, *Nucl. Sci. Eng.*, 17, 626, 1963.
120. **Martson, A. L., West, D. L., and Wilhite, R. N.,** Selection, cost and performance of n-paraffin diluents in solvent extraction chemistry of metals, in *Solvent Extraction Chemistry of Metals,* McKay, H. A. C., Ed., Macmillan, London, 1965, 213.
121. **Blake, C. A. and Schmitt, J. M.,** Stability of aromatic diluents and solvent extraction reagents in radiochemical processing, in *Solvent Extraction Chemistry of Metals,* McKay, H. A. C., Ed., Macmillan, London, 1965, 161.
122. **Savel'ev, Yu. I., Ershova, Z. V., and Vladimorova, M. V.,** Alpha-radiolysis of aqueous solutions of nitric acid, *Sov. Radiochem.*, 9, 221, 1967.
123. **Zazanjian, A. R., Miner, F. J., Brown, A. K., Hagan, P. G., and Berry, J. W.,** Radiolysis of nitric acid solutions: L.E.T. effects, *Trans. Faraday Soc.*, 66, 2192, 1970.
124. **Healy, T. V. and Pilbeam, A.,** Metal retention by hydroxamic acids in irradiated TBP solutions, in *Proc. Int. Solvent Extraction Conf.,* Vol. 1, Jeffreys, G. V., Ed., Society of Chemical Industry, London, 1974, 459.
125. **Stieglitz, L.,** Investigation on the nature of degradation products in the system 20% TBP-dodecane-nitric acid, in *Proc. Int. Solvent Extraction Conf.,* Vol. 1, Gregory, J. G., Evans, B., and Weston, P. C., Eds, Society of Chemical Industry, London, 1971, 155.
126. **Becker, R., Baumgartner, F., and Stieglitz, L.,** The identification of complexing radiolytic products in Purex systems, Rep. KFK-2304, *Kernforschungszentrum, Karlsruhe Ber.*, 1976.
127. **Becker, R. and Stieglitz, L.,** Identification of decomposition products in the system 20% TBP-dodecane-$HNO_3$, Rep. KFK-1373, *Kernforschungszentrum, Karlsruhe Ber.*, 1973.
128. **Goldacker, H., Schmieder, H., Steinbrunn, F., and Stieglitz, L.,** A newly developed wash process in nuclear fuel reprocessing decreasing the waste volume, *Kerntechnik,* 18, 426, 1976.
129. **Duyckaerts, G. and Nedden, P.,** Extractions by organophosphorous acids, in *Gmelin Handbuch der Anorganischen Chemie,* Vol. 21, Koch, G., Ed., Springer-Verlag, Berlin, 1975, 249.

130. **Peppard, D. F.,** Applications of liquid-liquid extraction in inorganic separations, *Prog. Nucl. Energy, Ser. 9,* 2, 201, 1961.
131. **Hardy, C. J.,** Alkyl phosphoric acids and their complexes with metals, *Nucl. Sci. Eng.,* 16, 401, 1963.
132. **Hardy, C. J. and Scargill, D.,** Studies on mono- and di-n-butylphosphoric acid. II. The solubility and distribution of mono- and di-n-butylphosphoric acids in aqueous-organic solvent systems, *J. Inorg. Nucl. Chem.,* 11, 128, 1959.
133. **Maya, L. and Bopp, C. D.,** Extraction of zirconium from nitric acid by butyl lauryl phosphoric acid in dodecane, *J. Inorg. Nucl. Chem.,* 40, 1147, 1978.
134. **Hardy, C. J. and Scargill, D.,** Studies on mono- and di-n-butylphosphoric acids. III. The extraction of zirconium from nitrate solution by di-n-butylphosphoric acid, *J. Inorg. Nucl. Chem.,* 17, 337, 1961.
135. **Sokhina, L. P., Solovkin, A. S., Teterin, E. G., Bogdanov, F. A., and Shesterikov, N. N.,** Dibutyl phosphates of tetravalent plutonium, zirconium, and thorium formed in solutions of tri-n-butyl phosphate (TBP), *Sov. Radiochem.,* 20, 19, 1978.
136. **Sokhina, L. P., Bagdanov, F. A., Solovkin, A. S., Teterin, E. G., and Shesterikov, N. N.,** Reaction of plutonium IV with hydrogen di-n-butyl-phosphate in an organic phase, *Russ. J. Inorg. Chem.,* 21, 2471, 1976.
137. **Krutikov, P. G. and Solovkin, A. S.,** (Di-n-butyl phosphate) — compounds of uranyl, *Russ. J. Inorg. Chem.,* 15, 825, 1976.
138. **Wallace, R. M.,** Nitrosylruthenium-organophosphorous complexes, *Nucl. Sci. Eng.,* 19, 296, 1964.
139. **Van Geel, J., Joseph, C., Detilleux, E., Heinz, W., Centeno, J., and Gustafsson, B.,** Chemical aspects of solvent extraction on a plant scale at Eurochemic, in *Proc. Int. Solvent Extraction Conf.,* Vol. 1, Gregory, J. G., Evans, B., and Weston, P. C., Eds., Society of Chemical Industry, London, 1971, 577.
140. **Davis, W.,** The Solubilities of Uranyl and Iron (III) Dibutyl and Monobutyl Phosphates in TBP Solvent Extraction Solutions, Rep. ORNL-3084, Oak Ridge National Laboratory, Oak Ridge, Tenn., 1961.
141. **Ochsenfeld, W., Baumgartner, F., Bauder, U., Bleyl, H. J., Ertel, D., and Koch, G.,** Experience with the reprocessing of LWR, Pu recycle, and FBR fuel in the millifacility, Rep. KFK- 2558, *Kernforschungszentrum Karlsruhe Ber.,* 1977.
142. **Hahn, H. T. and Vander Wall, E. M.,** TBP decomposition product behavior in post-extractive operations, *Nucl. Sci. Eng.,* 17, 613, 1963.
143. **Moffat, A. J. and Thompson, R. D.,** Basic Studies of the Chemical Stability in Extraction Systems. I. The Effect of Zirconium Nitrate and Nitric Acid upon the Stability of Tributyl Phosphate, Rep. IDO-14543, National Reactor Testing Station, Idaho Falls, Idaho, 1961.
144. **Faugeras, P. and Talmont, X.,** Radiolysis and hydrolysis of TBP and their effects, in *Solvent Extraction Research,* Kertes, A. S. and Marcus, Y., Eds.,Interscience, New York, 1969, 411.
145. **Cleveland, J. M.,** Separation from irradiated uranium, in *Plutonium Handbook — A Guide to the Technology,* Vol. 2, Wick, O. J., Ed., Gordon & Breach, New York, 1967, chap. 15.
146. **Boudry, J. C. and Miquell, P.,** Application of the Purex process to the reprocessing of fast reactor fuel, in *Proc. Int. Solvent Extraction Conf.,* Vol. 2, Jeffreys, G. V., Ed., Society of Chemical Industry, London, 1974, 1551.
147. **Breschet, C. and Miquell, P.,** Improvement of procedure used to treat highly irradiated fuels, in *Proc. Int. Solvent Extraction Conf.,* Vol. 1, Gregory, J. G., Evans, B., and Weston, P. C., Eds., Society of Chemical Industry, London, 1971, 565.
148. **Ortega, J.,** Separation of uranium — plutonium with tartaric acid in the Purex process, in *Solvent Extraction Research,* Kertes, A. S. and Marcus, Y., Eds.,Interscience, New York, 1969, 391.
149. **Gourisse, D., Bathellier, A., and Blanchard, J. C.,** Use of sulfate in solvent extraction separation processes, in *Proc. Int. Solvent Extraction Conf.,* Vol. 1, Gregory, J. G., Evans, B., and Weston, P. C., Eds., Society of Chemical Industry, London, 1971, 147.
150. **Germain, M., Bathellier, A., and Berand, P.,** The use of formic acid for the stripping of plutonium, in *Proc. Int. Solvent Extraction Conf.,* Vol. 3, Jeffreys, G. V., Ed., Society of Chemical Industry, London, 1974, 2075.
151. **Davis, M. W. and Jennings, A. S.,** Equipment for processing by solvent extraction, in *Chemical Processing of Reactor Fuels,* Flagg, J. F., Ed., Academic Press, New York, 1961, 271.
152. **Bernard, C., Michel, P., and Tarnero, M.,** Experience with centrifugal contactors: comparison with other types of contactors, in *Proc. Int. Solvent Extraction Conf.,* Vol. 2, Gregory, J. G., Evans, B. and Weston, P. C., Eds., Society of Chemical Industry, London, 1971, 1282.
153. **Warner, B. F., Naylor, A., Duncan, A., and Wilson, P. D.,** A review of the suitability of solvent extraction for the reprocessing of fast reactor fuels, in *Proc. Int. Solvent Extraction Conf.,* Vol. 2, Jeffreys, G. V., Ed., Society of Chemical Industry, London, 1974, 1481.

154. **Karraker, D. G.,** Temperature effects on TBP solvent extraction processes, *Prog. Nucl. Energy Ser. 3*, 3, 188, 1961.
155. **Nichols, G. S.,** Decomposition of Tributyl Phosphate-Nitrate Complexes, Rep. DP-526, Savannah River Laboratory, Aiken, S. C., 1960.
156. **Harmon, H. D., Hyder, M. L., Tiffany, B., Gray, L. W., and Soltys, P. A.,** Behavior of Tributyl Phosphate in A-Line Process, Rep. DP-1418, Savannah River Laboratory, Aiken, S. C., 1976.
157. **Orth, D. A. and Olcott, T. W.,** Purex process performance versus solvent exposure and treatment, *Nucl. Sci. Eng.*, 17, 539, 1963.
158. **Novak, Z. and Novak, M.,** Regeneration of extraction systems used in the Purex process, *Sov. Radiochem.*, 17, 361, 1975.
159. **Schulz, W. W.,** Macroreticular ion-exchange resin clean-up of Purex process tributyl phosphate solvent, in *Proc. Int. Solvent Extraction Conf.*, Vol. 1, Gregory, J. G., Evans, B., and Weston, P. C., Eds., Society of Chemical Industry, London, 1971, 174.
160. **Kennedy, J., Peckett, J. W. A., and Fletcher, J. M.,** The influence of high radiation levels on the retention of ruthenium in processing solvents, in *Solvent Extraction Chemistry of Metals*, McKay, H. A. C., Ed., Macmillan, London, 1965, 187.

Chapter 5

# CHEMISTRY OF VOLATILE FISSION PRODUCT ELEMENTS

**J. C. Mailen and L. M. Toth**

## TABLE OF CONTENTS

## I. INTRODUCTION

The volatile fission products iodine, tritium, and krypton (xenon) have received considerable attention in recent years because of their significant potential contributions to radiation exposure of the population and the relative difficulty of their adequate containment in the fuel reprocessing system. In the past, no attempt has been made (except in the case of iodine) to control emissions of these radionuclides; however, regulations for future facilities will require special provisions for containment of some or all of them.

Many of the candidate processes for control of the volatile fission products are in an early stage of development; in addition, several possible control systems are being considered for each element. This chapter will discuss their chemical and physical properties as they relate to their behavior in a fuel reprocessing plant and in the proposed control systems.

## II. IODINE FORMATION AND CHEMISTRY IN LIGHT WATER REACTOR (LWR) FUEL REPROCESSING

Iodine is a major fission product with a yield, at discharge, of about 275 g/t of heavy metal at a burnup of 33,000 MWd/t and a power level of 30 MW.[1] The isotopes of concern for containment are $^{131}I$ (half-life, 8.05 days) and $^{129}I$ (half-life, $1.6 \times 10^7$ years). Although the $^{131}I$ will have decayed to very low levels under most reprocessing conditions, it may still contribute significantly to containment requirements because of its high specific activity.

Iodine present in the reactor fuel exists in combination with other elements, particularly as CsI. The major condensed phase is CsI with gaseous species consisting of CsI and $Cs_2I_2$.[2] The gaseous species $ZrI_2$, $ZrI_3$, and $ZrI_4$ can be formed[2] in contact with the Zircaloy cladding.

During voloxidation (oxidation with air at about 450°C), the cesium would be expected to react with uranium oxides to form cesium uranates and to free iodine in elemental form.[3] The fraction of iodine released ranges up to 84%, depending on the fuel and its treatment;[4] the remainder is apparently trapped within fuel particles. Dissolution of either oxidized or unoxidized fuel in nitric acid releases the iodine, at least initially, as elemental iodine; the oxidation of any iodide is very rapid in the hot nitric acid-nitrous acid dissolver solution.

Iodine undergoes hydrolysis in aqueous solutions to form hypoiodous acid, (HOI), according to

$$I_2 + H_2O \rightleftarrows HOI + HI \quad (1)$$

However, the quantity of HOI which can be formed at the high temperatures and low pH levels of dissolver solutions is insignificant. It is shown in Figure 1 that, even after the iodine concentration is reduced to $10^{-8}$ *M* (originally about $6 \times 10^{-4}$ *M* if all the iodine is in solution),[1] less than 10% is present as HOI under dissolver conditions.[5,6]

Under dissolver conditions, a portion of the iodine is oxidized to iodate.[7] At first glance, this seems surprising since a large excess of nitrous acid is present to act as a reductant for iodate. A reasonable explanation is that iodine is oxidized by pernitrous acid, which is formed by the reaction of nitrous acid and hydrogen peroxide[8] generated by radiolysis. Pernitrous acid decomposes into free radicals:[9]

$$HO_2NO \rightarrow HO\cdot + \cdot NO_2 \quad (2)$$

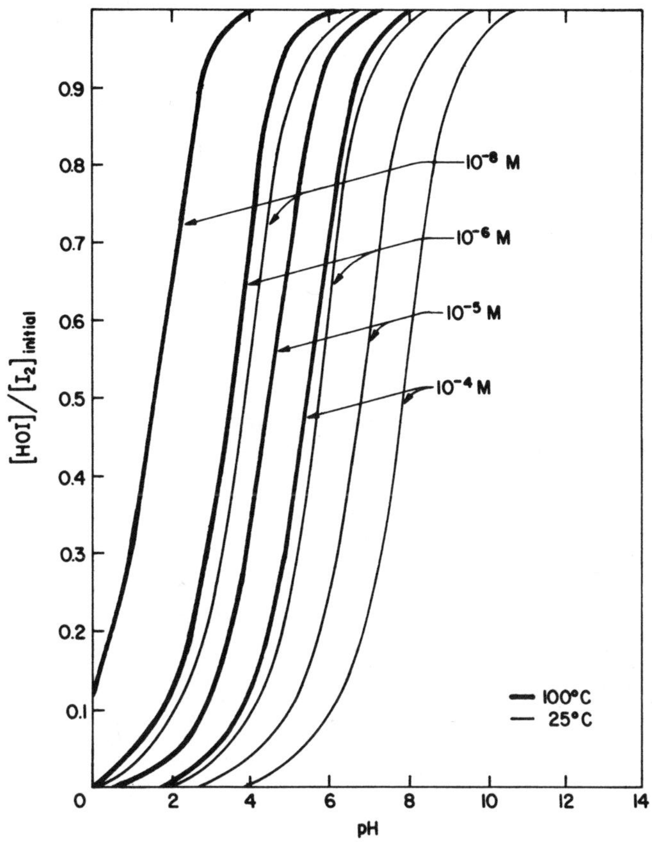


FIGURE 1. Extent of HOI formation under dissolver conditions at 25 and 100°C.

These radicals are highly oxidizing and readily liberate bromine from hydrobromic acid.[8] Since the potentials for oxidation of bromide to bromine (−1.07 V) and for the oxidation of iodine to iodate (−1.2 V) are similar,[10] it is expected that this system will generate some iodate during dissolution. Presence of iodate may retard volatilization of iodine to some extent because of the necessity of reducing iodate to elemental iodine. This retardation must be slight at elevated temperature in the presence of excess nitrous acid (conditions in the dissolver) since iodine is readily removed under process conditions.[4]

A volatile iodine species analogous to the know NOCl is not thought to be likely; NOI has been identified only as an unstable intermediate.[11] Volatile organic iodides are observed in all studies of iodine behavior in fuel reprocessing. Recent Japanese work[12-14] has determined that (1) the formation of organic iodide requires the presence of nitric acid; (2) the formation of the organic iodides occurs in the solution, and not in the gaseous phase or on the wall of the reaction vessel; (3) the $I^-$ to $I^\circ$ oxidation step is not involved in the formation of organic iodides; (4) the formation reaction involves iodine with an oxidation state between 0 and +5; (5) the source of organic material to form the iodides is probably high molecular weight organic impurities in the nitric acid; and (6) the most common volatile organic iodides formed are methyl iodide ($CH_3I$) (predominant), $C_2H_5I$, $n$-$C_3H_7I$, and $n$-$C_4H_9I$. Although these authors state that the major quantities of organic iodides are formed in the nitric acid solution, it is also feasible to form $CH_3I$ in the gas phase:[15]

$$CH_4 + I_2 \rightleftarrows CH_3I + HI \tag{3}$$

Nonvolatile organic iodide compounds may also be present in dissolver solutions. Iodoacetic acid forms when nitric acid containing acetic acid and inorganic iodine is irradiated. However, compounds of this type do not cause retention of iodine in the dissolver solution,[4] probably because they are destroyed by pernitrous acid formed by irradiation. Pernitrous acid has been shown experimentally to remove iodine from iodoacetic acid and convert it to iodic acid,[16] which was subsequently reduced to elemental iodine by excess peroxide or nitrous acid.

Under common dissolver conditions, iodine is swept out of the dissolver solution by the combination of air sparging and nitrogen oxide reactions. However, in a fumeless dissolution[17,18] where a slight overpressure of oxygen continuously reconverts the nitrogen oxides to nitric acid, nearly 100% of the iodine will remain in the solution, principally as the element. Under such conditions, insoluble $PdI_2$ (palladium is present as a fission product) can precipitate; this precipitation can be enhanced by the addition of a mild reducing agent such as hydroxylamine nitrate. The solubility of $PdI_2$ in nitrate and perchlorate solutions[19] is shown in Figure 2. Formation of $PdI_2$ as a method of isolating iodine directly from dissolver solutions was examined in hot-cell tests, with approximately 99% removal being achieved.[20]

From the preceding discussion, it is seen that the off-gas from the dissolver will contain iodine species which are predominantly elemental iodine and $CH_3I$ plus small amounts of higher molecular weight organic iodides. Nitrogen, oxygen, water vapor, nitric acid, and oxides of nitrogen will be major components of the gas stream. The off-gas treatment system must successfully remove all the iodine species and prepare them for delivery to the waste handling system.

## III. IODINE CONTROL METHODS

Good reviews[6,21] of off-gas treatment methods are available. The methods include both scrubbing in caustic, concentrated nitric acid, and mercuric nitrate solutions, and trapping on solid sorbents (principally silver-containing materials). This section will examine the chemical aspects of the various trapping systems.

In any aqueous scrubbing system, the species to be removed from the air stream must first distribute to the aqueous solution. The distribution coefficients ($D_G^L$ = molar concentration in liquid divided by the molar concentration in gas) for elemental iodine and $CH_3I$ between air and nitric acid are given in Figures 3 and 4. As shown by these figures, the distribution of $CH_3I$ to aqueous solutions is lower by factors of 10 to 15 than that for iodine. This behavior is one of the major reasons for the difficulty observed in aqueous scrubbing of $CH_3I$. The unexpectedly high distribution shown for $CH_3I$ to 11 *M* $HNO_3$ probably results from significant nitrolysis of methyl iodide under these conditions.

### A. Iodox System

The Iodox process for removing iodine species from air streams uses hyperazeotropic (about 20 to 22 *M*) nitric acid to absorb the iodine and ultimately to convert it to $HI_3O_8$, a dehydrated form of iodic acid from which $Ba(IO_3)_2$ can be produced for disposal. The initial absorption of iodine, especially elemental iodine, is favored by the high distribution of iodine to concentrated nitric acid from air (Figure 3).

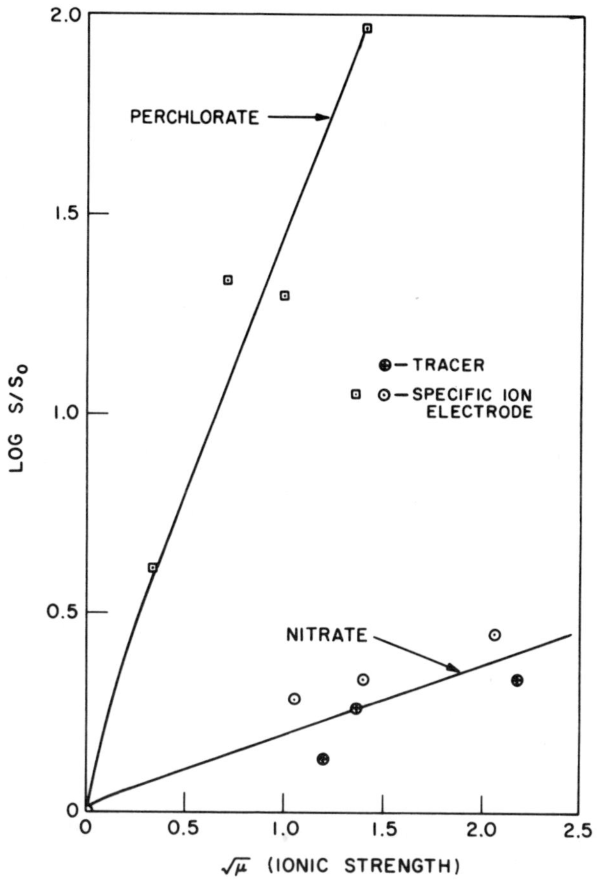


FIGURE 2. Solubilities of $PdI_2$ in nitrate and perchlorate solutions at 25°C. $S/S_o$ is the solubility at a given ionic strength divided by the solubility at zero ionic strength.

The reaction of elemental iodine with hyperazeotropic nitric acid proceeds in at least two steps.[26] The equilibria for two reactions have been determined as follows:[23]

$$I_2 + 4HNO_3 \rightleftarrows 2I^+ + 2NO_3^- + 2H_2O + N_2O_4 \quad (4)$$

$$I^+ + 3HNO_3 + NO_3^- \rightleftarrows IO_3^- + 2N_2O_4 + H_2O + H^+ \quad (5)$$

The oxidation of $I_2$ to $I^+$ proceeds rapidly, but the oxidation of $I^+$ is rather slow.[28]

In cases where the $N_2O_4$ content of the liquid is low, another equilibrium has been found to be controlling. Although the reaction involved in this equilibrium is not known, the empirical equilibrium constant has been determined as a function of nitric acid concentration.[22] Its form is

$$K_3 = \frac{C_{I^+}}{C_{I_2}} \quad (6)$$

At higher $N_2O_4$ concentrations, where the ratio of $I^+$ to $I_2$ calculated from Equation 6

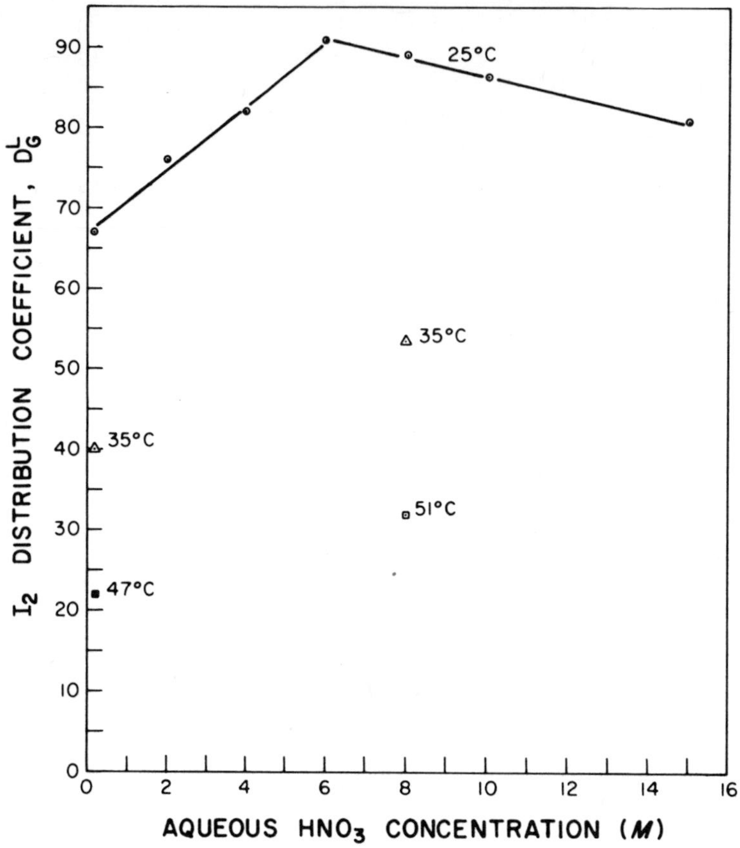


FIGURE 3. Distribution coefficients for $I_2$ from air to nitric acid at several temperatures.

is greater than that calculated for equilibrium Reaction 4, the equilibrium for Reaction 4 is obeyed:

$$K_1 = \frac{C_{I^+}^2 C_{N_2O_4}}{C_{I_2}} \tag{7}$$

The existence of the monovalent iodine species was supported both by the 1:1 ratio of $N_2O_4$ formed per $I_2$ oxidized[24] and by the effect of the $N_2O_4$ concentration in the acid on the equilibrium of $I_2$ with the +1 species[23] (Figure 5). Here, we see that the equilibrium attained at short equilibration times (where the further oxidation of the +1 species to $IO_3^-$ can be neglected) is consistent with 1 mol of $N_2O_4$ being formed per mole of $I_2$ oxidized.

For $CH_3I$, the initial reaction in concentrated nitric acid is a very rapid nitrolysis:

$$CH_3I + 2HNO_3 \rightarrow CH_3NO_3 + 1/2\ I_2 + 1/2\ N_2O_4 + H_2O \tag{8}$$

followed by reaction of the $I_2$ with nitric acid as in Equations 4 and 5. The rate constant defined by the first-order differential equation

$$\frac{d\,(CH_3I)}{dt} = -\,k\,(CH_3I) \tag{9}$$

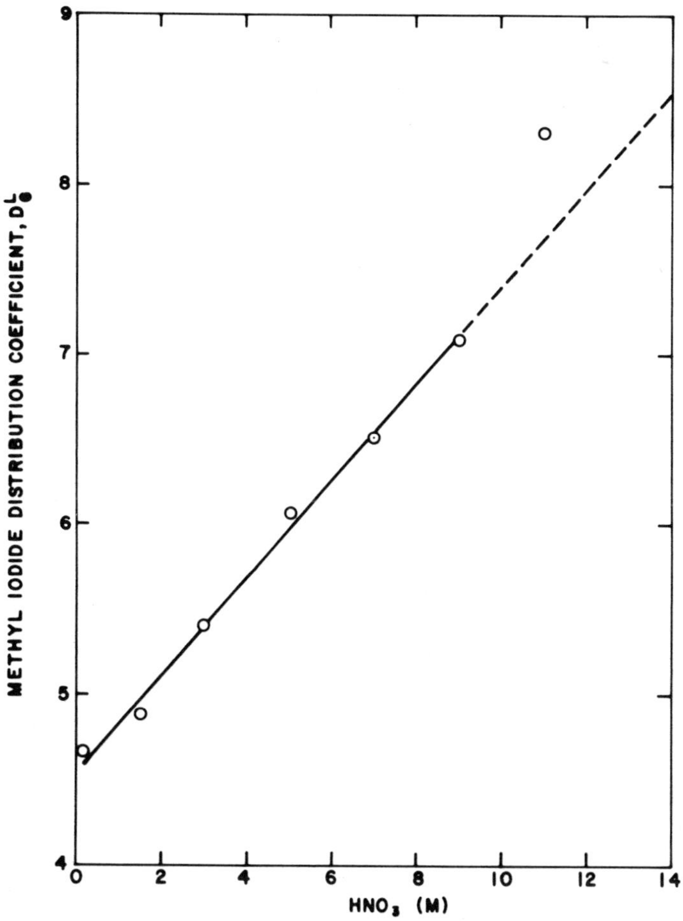


FIGURE 4. Distribution coefficients for $CH_3I$ from air to nitric acid at 25°C.

was found to be

$$\ln k = -93.73 + 32.51 \ln C_{HNO_3} \tag{10}$$

where concentrations are in molarity and time is in seconds. The iodine formed in this reaction would then participate in the reactions and equilibria defined for elemental iodine. The fact that $CH_3I$ is more difficult than iodine to scrub from air is attributed to the lower distribution factors for $CH_3I$.

Since $N_2O_4$ is formed as a reaction product in the oxidation of iodine, the distribution of $NO_2$ between air and nitric acid must be known to define completely the equilibrium constant. Approximate values of the distribution have been measured to yield the equilibrium constant

$$K_4 = \frac{C_{N_2O_4}^{1/2}}{C_{NO_2}} \text{ (molar units)} \tag{11}$$

and are given by

$$\ln K_4 = 0.2\ C_{HNO_3} + 1.56 \tag{12}$$

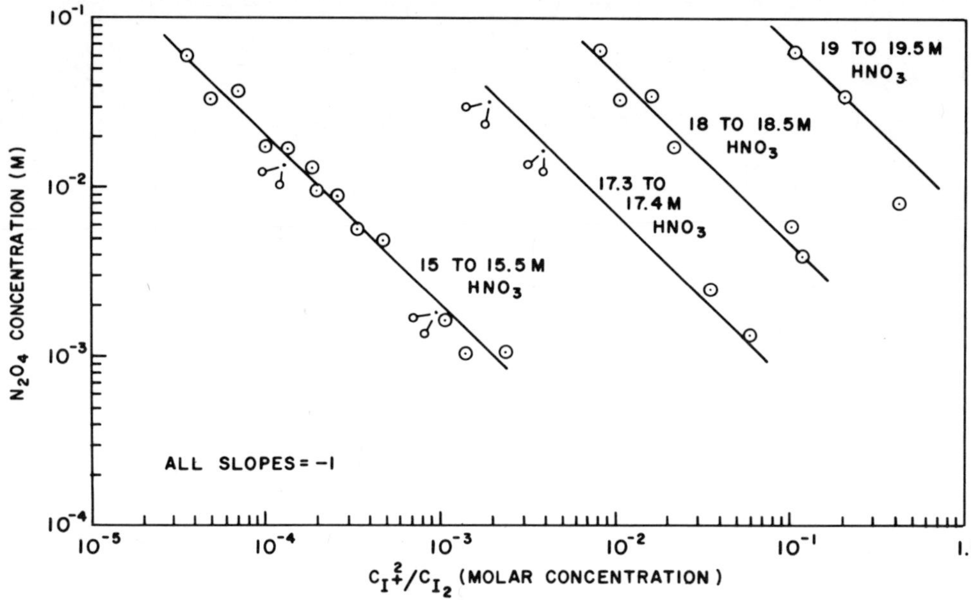


FIGURE 5. First power dependence of $K_1$ (Equation 13) on $N_2O_4$ concentration.

The ultimate capacity of the acid used in the Iodox process is limited by the solubility of iodic acid in hyperazeotropic nitric acid. The solubility is given by[25]

$$\ln S = 13.28 - 0.4515\ C_{HNO_3} - 1170/T \quad (13)$$

where S is the solubility of $HI_3O_8$ in g/ℓ, $C_{HNO_3}$ is the molar concentration of $HNO_3$ (12 to 20 *M*), and T is the absolute temperature in K (298 to 373 K).

## B. Mercuric Nitrate/Nitric Acid Scrubbers

The mercuric nitrate [$Hg(NO_3)_2$]/nitric acid scrubber system for removing iodine species from air streams uses a solution of $Hg(NO_3)_2$ in subazeotropic nitric acid in a gas-liquid contactor to trap the iodine as a mercuric iodide complex.

The reaction of $I_2$ with $Hg(NO_3)_2$-nitric acid solutions has been shown to be

$$6I_2 + 6Hg(NO_3)_2 + 6H_2O \rightarrow 5HgI_2 + Hg(IO_3)_2 + 12HNO_3 \quad (14)$$

in the absence of nitrous acid. Since the $Hg(IO_3)_2$ has a low solubility in nitric acid, this reaction should be avoided. Fortunately, the presence of nitrous acid in the sorbent solution ($NO_x$ in the gas phase) prevents formation of the iodate species.

The reaction of $CH_3I$ with the system has not been studied; it may be either of the following:

$$CH_3I + 1/2\ Hg(NO_3)_2 \rightarrow 1/2\ HgI_2 + CH_3NO_3 \quad (15)$$

$$CH_3I + 1/2\ Hg(NO_3)_2 + H_2O \rightarrow 1/2\ HgI_2 + CH_3OH + HNO_3 \quad (16)$$

The reaction rate of $I_2$ with the mercuric nitrate solution, which is very fast, has not

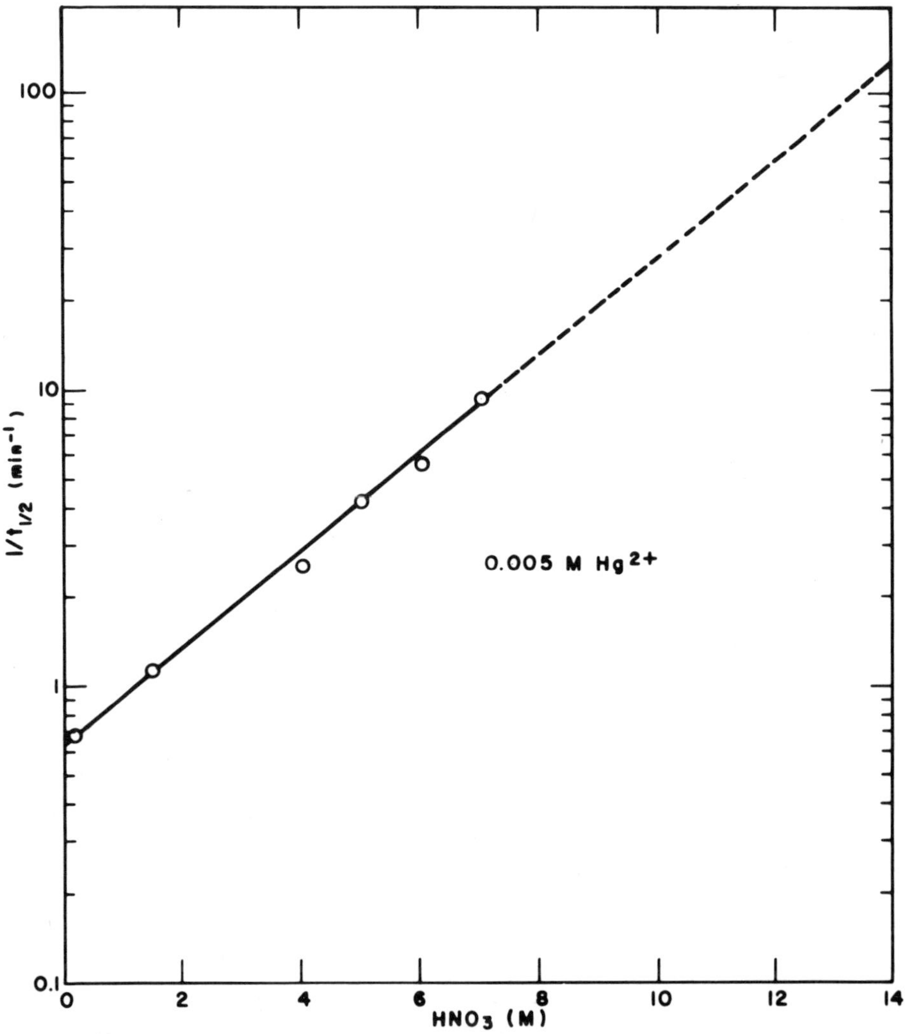


FIGURE 6. Effect of nitric acid concentration on the rate of reaction of $CH_3I$ with $Hg^{2+}/HNO_3$ solutions.

been measured. The kinetics of the reaction of $CH_3I$ have been determined as functions of mercuric ion and nitric acid concentrations. The effects of nitric acid at constant mercuric ion concentration and of mercuric ion at constant nitric acid concentration are illustrated in Figures 6 and 7, respectively.

The controlling mechanism for $CH_3I$ trapping can be either chemical kinetics or diffusion. The controlling mechanism for elemental iodine trapping, on the other hand, is not known. Since the chemical reaction rate is so rapid, diffusion, chemical equilibrium, or generation of a volatile species by unknown reactions could control the trapping efficiency.

A mercury/iodine mole ratio of about 4 must be maintained to prevent precipitation of $HgI_2$.[26] In 0.1 *M* $HNO_3$, the formation of a permanent precipitate occurred at a mercury/iodine ratio of 3.9; in 5 *M* $HNO_3$, it occurred at a ratio of 3.1. If a 0.4 *M* $Hg(NO_3)_2$ solution is used, the safe iodine concentration is around 12.7 g/$\ell$. Another consideration is increased iodine bleed-off at high iodine/mercury ratios,[27] which can,

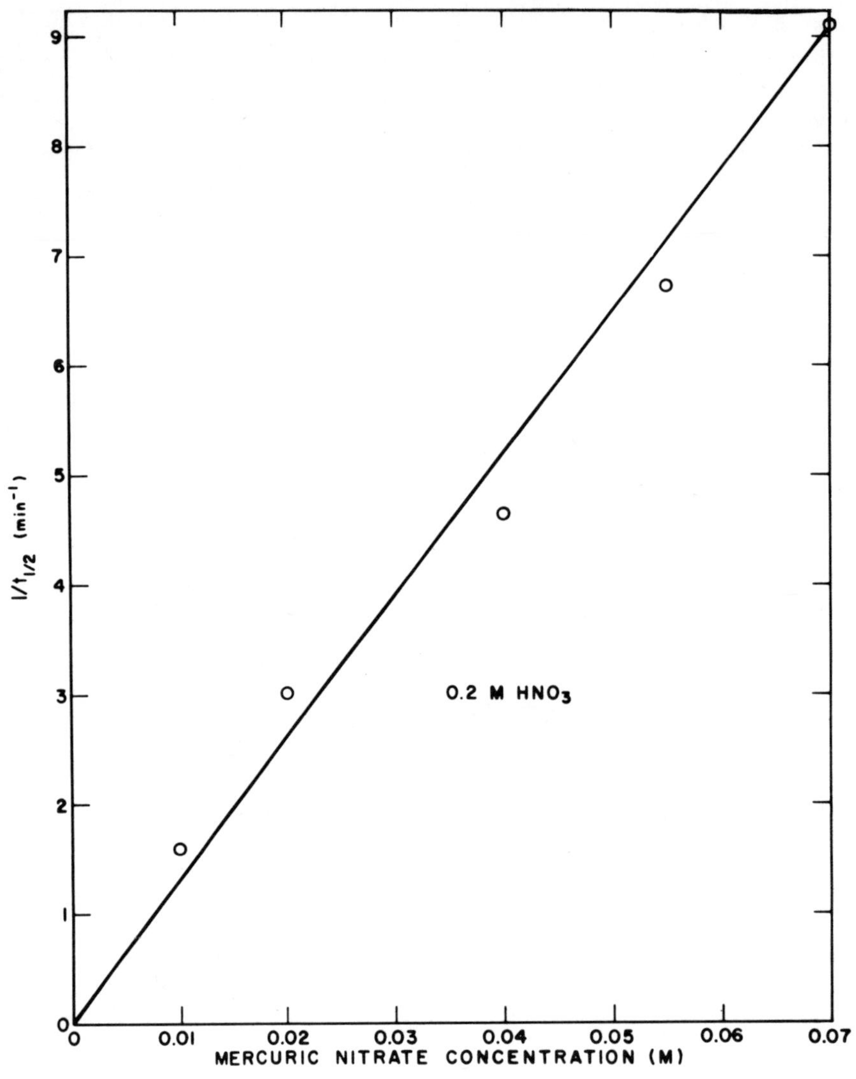


FIGURE 7. Effect of mercury concentration on the rate of reaction of $CH_3I$ with $Hg^{2+}$/$HNO_3$ solutions.

depending on the arrangement of the scrubbing system, lead to greatly decreased decontamination efficiency.

## C. Caustic Scrubbers

Caustic scrubbers should perform satisfactorily for removing elemental iodine from air and converting it to iodide and iodate by the well-known disproportionation reaction:

$$3I_2 + 6KOH \rightarrow 5KI + KIO_3 + 3H_2O \tag{17}$$

However, in practice, iodine removals[28] are usually no higher than 90%. A major factor in this low removal is the failure of caustic to trap organic iodides.[29] In addition, it has been shown that the presence of organic impurities in the caustic leads to the generation of volatile organic iodides. The mechanism for this iodination and loss of trapping ability is based on a compound intermediate in valence between iodide and iodate, as shown by electrolytic trapping tests.[29]

The addition of oxidizing or reducing agents has been examined as a method for improving the operation of caustic scrubbers. Results of tests made under highly radioactive process conditions at the TRU facility at Oak Ridge National Laboratory showed poor trapping performance when 0.01 *M* NaOCl was added to 0.1 *M* NaOH.[30] The addition of 0.1 *M* thiosulfate to 1 *M* NaOH has been reported to give approximately fivefold improvement in iodine removals;[31] however, the presence of nitrogen oxides and nitric acid vapors in the air stream will lead to losses of thiosulfate by oxidation.

Since caustic scrubbers also trap $CO_2$ from the air stream, the low solubility of sodium carbonate dictates the use of potassium hydroxide in the scrubber solution.

Any $NO_x$ present in the dissolver off-gas will consume large quantities of caustic and, therefore, must be removed before the gas enters the scrubber.

## D. Solid Sorbents

Impregnated activated charcoal and metal-ion-impregnated inorganic solids are the most widely studied solid sorbents for removing iodine compounds from fuel reprocessing off-gas. Although charcoal sorbents perform satisfactorily, they have major drawbacks. Charcoal has a relatively low ignition temperature,[32] which is further decreased by absorption of nitrogen oxides;[33] the presence of nitrogen oxides may also lead to the formation of explosive compounds.[33-34] Further, iodine retention is poor at high temperatures,[35] and ignition can lead to catastrophic release of the accumulated iodine. For these reasons, the chemistry of iodine sorption on charcoal will not be reviewed here; the reader is directed to References 6 and 21 if additional information is desired.

The solid sorbents of major interest are silver-exchanged zeolites and silver nitrate-impregnated amorphous silicic acid.

The synthetic zeolites under consideration are sodium aluminosilicates with various ratios of alumina to silica; at the higher silica to alumina ratios, the acid resistance of the material is increased. The sodium in the zeolite is readily exchanged with silver by contacting the zeolite with a silver nitrate solution. A portion of the iodine sorbed by the zeolite is physically adsorbed, while another portion is reacted to form silver iodide.[36] It is assumed that the metal ions exist as oxides in the zeolite structure and that the reaction with iodine may be written as:

$$Ag_2O(s) + I_2(g) \rightarrow 2AgI(s) + 1/2\ O_2(g) \qquad (18)$$

The standard free energy of the reaction is −16.1 kcal/mol. The proposed mechanism[36] for this reaction is

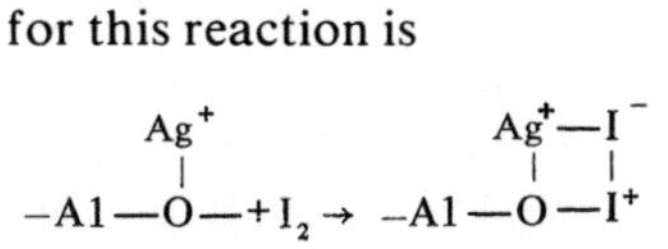

(19)

The reaction with $CH_3I$ may be

$$Ag_2O(s) + 2CH_3I(g) + H_2O(g) \rightarrow 2AgI(s) + 2CH_3OH(g) \qquad (20)$$

although this has not been established.

Studies have shown that silver zeolite can be regenerated by treatment with hydrogen

**Table 1**
**TRITIUM FISSION YIELD DATA**

| Reaction | Tritium atoms[a] per $10^4$ fissions |
|---|---|
| $^{233}U + n_{th}$ | 0.9 |
| $^{235}U + n_{th}$ | 0.9 |
| $^{239}Pu + n_{th}$ | 1.5 |

[a] Average of values given in Reference 39.

at 400 to 500°C. The reaction liberates HI; the equilibrium vapor pressure[36] of HI conforms to the chemical reaction

$$2AgI(s) + H_2(g) \rightleftarrows 2Ag(s) + 2HI(g) \tag{21}$$

When the iodine sorbent consists of an inert solid such as amorphous silicic acid[35] impregnated with silver nitrate, the reactions that occur with elemental iodine have been shown to be[37]

$$AgNO_3 + I_2 \rightarrow AgI + INO_3 \tag{22}$$

$$2INO_3 + AgNO_3 \rightarrow AgIO_3 + 3NO_2 + 1/2\ I_2 \tag{23}$$

$$INO_3 \rightarrow NO_2 + 1/2\,O_2 + 1/2\,I_2 \tag{24}$$

By analogy, the reactions with methyl iodide are probably

$$AgNO_3 + CH_3I \rightarrow AgI + CH_3NO_3 \tag{25}$$

possibly followed by hydrolysis of the methyl nitrate:

$$CH_3NO_3 + H_2O \rightarrow CH_3OH + HNO_3 \tag{26}$$

Regeneration of the silicic acid sorbent with hydrogen should also be possible, but this has not been verified by experimental studies.

## IV. TRITIUM FORMATION AND CHEMISTRY IN LWR FUEL REPROCESSING

Tritium (half-life, 12.26 years) is a rare isotope in the environment; its natural occurrence is estimated[38] as only 6 g for the entire biosphere. Man-made production results principally from weapons tests and nuclear power operations. In a reactor, tritium can be produced either by ternary fission or by a variety of activation routes. The fission yields for thermal neutrons[39,40] are listed in Table 1 for the common fissile isotopes.

Tritium is produced by numerous reactions between neutrons and low-atomic-number elements, including the following:[41]

$$^6Li\ (n, \alpha, \gamma)\ T$$

$$^{10}B\ (n,\gamma)^{8}Be + T$$

$$^{10}B\ (n, 2\alpha)\ T$$

$$^{10}B\ (n,\alpha)^{7}Li\ (n, n\alpha)\ T$$

$$^{14}N\ (n,\gamma)^{12}C + T$$

$$^{14}N\ (n, 3\alpha,\gamma)\ T$$

$$^{9}Be\ (n,\alpha)^{6}Li\ (n,\alpha)\ T$$

$$^{3}He\ (n,p)\ T$$

$$^{2}H\ (n,\gamma)\ T$$

The major source of tritium produced in LWR fuels is ternary fission.

After the tritium has been produced in the $UO_2$ fuel, it can diffuse out of the $UO_2$ matrix. However, studies have shown that such diffusion is slow[42] relative to the bulk diffusion of hydrogen through $UO_2$. This slow diffusion may be due to the chemical form of the tritium or to the fact that the tritium can be formed inside the $UO_2$ crystallites, far from grain boundaries, and may thus exhibit slower rates of diffusion than a bulk gas which has access to the grain boundaries. The slow diffusion of tritium in $UO_2$ is similar to that observed in studies[43] of the diffusion of recoil tritium from $Al_2O_3$ and BeO.[43] The Fick's law diffusion coefficients for tritium in BeO and $Al_2O_3$ are found to be generally lower than those for $UO_2$ (Figure 8). Tritium has been shown to diffuse as atoms[44] in oxides such as $TiO_2$, $Al_2O_3$, and BeO and apparently passes through the solid by alternately making and breaking O–H bonds. However, when the cover gas is nonoxidizing, the tritium that leaves the solid appears largely as HT gas.[43]

The chemical form of gas-phase tritium in contact with the fuel can be influenced both by chemical effects and by radiation effects. Initially, the oxygen-to-uranium (O/U) ratio in LWR fuel is 2.0; however, as the irradiation progresses, the O/U ratio increases[45] due to a net excess of oxygen over that which can be bound by the fission products. The oxygen chemical potential ($\mu_{o_2}$) is a complex function of the fuel temperature (i.e., the axial and radial position).[46] Therefore, the $H_2O/H_2$ ratio, which depends on $\mu_{o_2}$,[47] could vary from extremely large to extremely small values. Moreover, the intense radiation field and the presence of the tritium isolated within the crystal lattice of the $UO_2$ prevent any completely unambiguous statement as to the chemical species of the tritium in the fuel.

Any tritium that diffuses out of the fuel, whether as HT or HTO, may contact the Zircaloy cladding and react with the zirconium to form a Zr-HT solution and zirconium tritide, $ZrT_2$. This reaction has been confirmed in several types of studies. Gas-phase samples[48] from LWR fuel pins show that only about 0.001% of the total tritium is present in the gas, while up to about 50% is present in the Zircaloy.[48] Operating reactors also exhibit retention of tritium in Zircaloy-clad fuel.[49] High-temperature vacuum pumping can remove the tritium from the Zircaloy, as expected from tritide.[48] Metallographic examinations of the Zircaloy reveal zirconium hydride platelets near

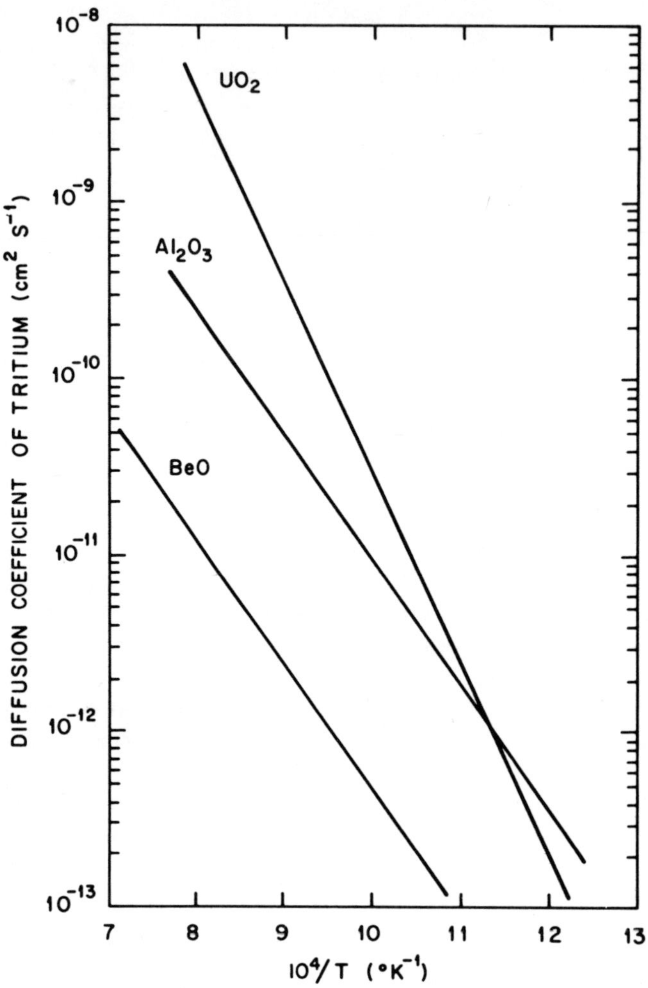


FIGURE 8. Comparison of tritium diffusion coefficients in $UO_2$, $Al_2O_3$, and BeO.

the cooler outer wall.[50] Thus, in the LWR fuel element, the bulk of the tritium will be contained within the fuel or as $ZrT_2$ in the cladding.

During any processing of LWR fuel, the tritium content of the $UO_2$ pellets will be exposed to water, either as vapor in the air or as the aqueous dissolver solution. Any HT which has not been oxidized by the excess oxygen in the LWR fuel[45] can then exchange with water according to the reaction

$$HT + H_2O \rightleftarrows H_2 + HTO \tag{27}$$

thus producing very dilute tritiated water. Experimental values of the equilibrium constant for this reaction are given by:[41]

$$\log K = 0.297 \log T + (336.5/T) - 1.055 \tag{28}$$

over the temperature range 16 to 303°C. This exchange probably requires a catalyst; however, the high radiation level in a fuel solution and the presence of noble-metal fission products should lead to rapid exchange.[41] HT is not observed[51] when LWR fuel is dissolved in the absence of cladding, indicating either that it is not present in the fuel per se or that rapid exchange with water converts it to HTO. When Zircaloy cladding is present, a small amount of HT is found in the off-gas;[52] this HT undoubtedly is produced as a result of slight dissolution of the metal.

## V. TRITIUM CONTROL METHODS

Only two methods have been proposed for control of tritium during fuel reprocessing: voloxidation and retention of tritiated water within the first solvent extraction cycle.

In the voloxidation process[51,53-55] the chopped fuel is air-oxidized at about 450°C; this treatment disintegrates the $UO_2$ fuel matrix as it is oxidized to $U_3O_8$. The much larger unit cell of $U_3O_8$ allows tritium that is located along grain boundaries or in voids to escape immediately, and the disruption of the fuel pellets gives tritium within fuel grains a much shorter diffusion path. The high temperature, oxidizing atmosphere, and many potential catalysts (fission products) in the fuel ensure that most of the tritium (if any is present as HT) is oxidized.[52] Exchange with normal water also readily occurs under these conditions. The water vapor, containing HTO, is then trapped by means of a desiccating agent. The further treatment of these materials is covered in Chapter 6, "Radioactive Waste Management".

Retention of water in the first solvent extraction cycle is being considered as a means of tritium control. In order to examine this method for isolating tritium, the distribution of hydrogen in several forms between the aqueous and organic phases must be known. This is not a simple problem since it involves the dissolution of both water and nitric acid in the organic extractant and physical entrainment of the aqueous phase in the organic phase. The tritium exchanged for active hydrogen would ordinarily be limited to hydrogen present in such impurities as alcohols,[41] but some enhancement of exchange due to the high radiation field may be expected. Aqueous-phase entrainment occurs to the extent of 0.1 to 1% by volume, depending on operating parameters. Another major source of tritium entering the organic phase is extraction of $H_2O$ and $HNO_3$. Both water and nitric acid are extracted by tributylphosphate (TBP); the quantities of each are complex functions of many factors, including nitric acid concentration, uranium concentration, TBP concentration, and temperature. These factors have been studied[56-62] and are being correlated,[63] based on the following assumptions. Water is assumed to be complexed by at least two mechanisms: the straightforward complex formation,

$$H_2O^a + 2TBP^o \rightleftarrows H_2O\cdot(TBP)_2^o \tag{29}$$

and the formation of a double complex including nitric acid,

$$HNO_3\cdot H_2O^a + 2TBP^o \rightleftarrows HNO_3\cdot H_2O\cdot(TBP)_2^o \tag{30}$$

where the superscripts o and a refer to the organic and aqueous phases, respectively. The assumed stoichiometric relationships between water-TBP and acid-TBP complexes result from an assumed conceptual mechanism — namely that the phosphoryl oxygen in the TBP is hydrogen-bonded to the water or acid. Thus, two TBPs coordinate with the two water hydrogens and only one TBP coordinates with the acid hydrogen. It is further assumed that in the double complex of Equation 30, the TBP is bonded to the water hydrogens and the nitric acid is hydrogen-bonded to the water oxygen. The experimental evidence that indicates the existence of the double complex of Equation 30

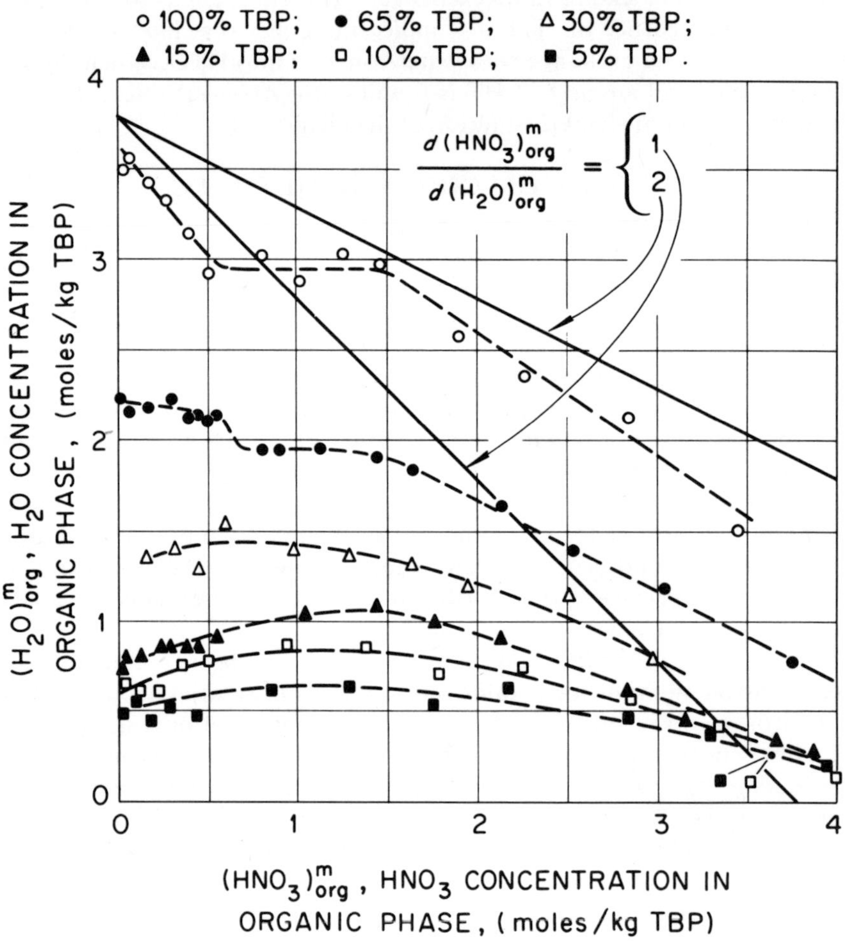


FIGURE 9. Effects of $HNO_3$ and diluent on the distribution of $H_2O$ into the organic phase.

is illustrated in Figure 9.[56] For TBP concentrations less than 30%, a significant increase in the water content of the organic phase accompanies the initial increases in the nitric acid content of the organic phase. At higher acid concentration, nitric acid replaces water:

$$HNO_3^a + HNO_3 \cdot H_2O \cdot (TBP)_2^o \rightleftarrows 2HNO_3 \cdot TBP^o + H_2O^a \qquad (31)$$

and

$$2HNO_3^a + H_2O \cdot (TBP)_2^o \rightleftarrows 2HNO_3 \cdot TBP^o + H_2O^a \qquad (32)$$

The quantity of water extracted is underestimated by the assumed mechanism; water must also be extracted by other means. One possibility is that water extraction can occur by hydrogen bonding to the etheryl oxygens of TBP; in this case, the water that is extracted would not be in competition with the nitric acid or uranyl nitrate if they coordinate only with the phosphoryl oxygen in TBP. Uranyl ion tends to replace both water and nitric acid in the TBP complexes. As a first approximation, the quantities of water and nitric acid extracted can be estimated from the data of Davis,[56] using the "free TBP" concept. As an illustration, when a 3 *M* $HNO_3$ solution containing uranium at a concentration of 300 g/ℓ is equilibrated with 3.75 vol of organic containing 30 vol % TBP, about 0.28 mol of nitric acid and 0.13 mol of water will be extracted. This amounts to slightly more than 0.5% of the total hydrogen atoms originally present in the aqueous phase. Thus, the transfer of tritium under these assumed conditions due to extraction and entrainment would range from 0.6 to 1.5% of that initially in the dissolver solution. Additional transfer of tritium by atomic exchange cannot be assessed from existing data.

## VI. KRYPTON AND XENON CHEMISTRY IN LWR FUEL REPROCESSING[64-66]

Of all the noble (or rare) gases, only krypton and xenon occur as significant fission products in nuclear-fueled reactors. Their isotopic compositions under these conditions are quite different from those occurring naturally; while some of these isotopes are responsible for the immediate need to be concerned with their separation and isolation, the isotopic differences do not have any major influence of their physical-chemical properties. Table 2 shows the isotopes of krypton and xenon produced during fission of $^{235}U$, their fission yields, and their half-lives. Xenon is important only insofar as it is present in amounts which are large compared to the amounts of krypton. The radioactive xenon isotopes are short-lived and so pose no radiation or radioactive waste disposal problems. Nonetheless, separations processes designed for krypton isolation must allow for the stable xenon, which accompanies the krypton.

At the end of a typical fuel cycle, a 1000-MWe reactor would contain[65] about 300 million Ci of radioactive krypton and about 400 million Ci of xenon. Nevertheless, each gas contains several short-lived components so that only about 1 million Ci of krypton, chiefly $^{85}Kr$ ($T_{1/2}$ = 10.73 years), and 100 million Ci of xenon, chiefly $^{133}Xe$ ($t_{1/2}$ = 5 days), would remain after a 24-hr decay period. After about a month, the principal remaining activity from the noble gases would be due to $^{85}Kr$, which is about 6% of all krypton isotopes present. The major activity of $^{85}Kr$ is from beta decay with an energy of 0.65 MeV and the accompanying production of stable rubidium.

The general properties of importance with respect to krypton and xenon separation processes are presented in Table 3. The noble gases are characterized by a high ionization potential, and the small van der Waals' attractive forces associated with them make low temperatures and/or high pressures necessary for liquefaction. Their solubilities are typically very low in water, but considerably higher in certain organic solvents such as fluorocarbons. Under pressure, krypton and xenon can form addition compounds; in particular, with water, the krypton hydrate has a dissociation pressure of 14.2 bars at 0°C and that of xenon, 1.42 bars. Clathrates,[68] formed from certain organic compounds by molecular entrapment, are related to the gas hydrates and are formed by imprisonment of compounds in the vacancies in the lattice structure during crystallization [e.g., hydroquinone, $C_6H_4(OH)_2$].

Because the noble gases are not ordinarily chemically active, the most practical separation and isolation processes are based on the physical properties as outlined above. In fact, the liquefied air industry has made extensive use of cryogenic methods to obtain commercial quantities of naturally occurring krypton and xenon from the atmosphere, which contains about 1 and 0.05 ppm, respectively, of each. Thus, for the purposes of this discussion the "chemistry" of the noble gases will be specifically concerned with these physical-chemical properties, and no further attention will be given

Table 2
THERMAL-NEUTRON FISSION YIELDS FROM $^{235}U$[a]

| Fission product | Fission yield (%) | Half-life[b] |
|---|---|---|
| $^{83}Kr$ | 0.54 | |
| $^{84}Kr$ | 1.00 | |
| $^{85m}Kr$ | 1.5 | 4.4 hr |
| $^{85}Kr$ | 0.29 | 10.73 years |
| $^{86}Kr$ | 2.02 | |
| $^{87}Kr$ | 2.7 | 78 min |
| $^{88}Kr$ | 3.7 | 2.8 hr |
| $^{131}Xe$ | 2.93 | |
| $^{132}Xe$ | 4.38 | |
| $^{133}Xe$ | 6.62 | 5.3 days |
| $^{134}Xe$ | 8.06 | |
| $^{135}Xe$ | 6.30 | 9.2 hr |
| $^{136}Xe$ | 6.46 | |

[a] Data taken from Reference 67.
[b] No value indicates isotope is stable.

Table 3
PHYSICAL PROPERTIES OF KRYPTON AND XENON COMPARED WITH THOSE OF MAJOR ATMOSPHERIC GASES[78,79]

| | Kr | Xe | $N_2$ | $O_2$ | Ar |
|---|---|---|---|---|---|
| Atomic weight | 83.8 | 137.3 | 28 | 32 | 40 |
| Density of liquid at bp (g/cc) | 2.41 | 3.06 | 0.81 | 1.14 | 1.39 |
| Melting point (K) | 116.55 | 161.35 | 63.1 | 54.6 | 83.8 |
| Boiling point (K) | 120.3 | 166.1 | 77.2 | 90.0 | 87.1 |
| Critical pressure (atm) | 54.2 | 58.2 | 33.5 | 50.1 | 48.0 |
| Critical temperature (K) | 209.4 | 289.75 | 125.8 | 154.6 | 150.5 |
| First ionization potential (eV) | 14 | 12.1 | 15.5 | 12.5 | 15.7 |
| Heat of vaporization at bp (cal/mol) | 2310 | 3110 | 1336 | 1629 | 1504 |

to the extraordinary chemical reactivity[68] that has been noted for them, even though these properties have been discussed briefly with respect to possible separation schemes.[69-70]

Since these fission-product noble gases ordinarily do not form chemical compounds and exhibit very low solubility in aqueous solvents, they are emitted entirely into the off-gas system of fuel reprocessing plants. In the case of LWR fuel elements, they are mainly released during $HNO_3$ dissolution of the oxide fuel, but smaller quantities also enter the off-gas when the fuel elements are sheared or are otherwise disassembled. Fuel pin failure can cause some minor gas release during reactor operations, but this can be handled, if necessary, in a reactor off-gas cleanup system.

Previous Nuclear Regulatory Commission standards permitted the controlled release of $^{85}Kr$ from tall stacks into the atmosphere; now, however, revisions in these standards require[71] that the release be "as low as reasonably achievable". Following these revisions, extensive effort has been directed to various methods of noble gas recovery and storage.

At first it may seem strange that anything other than established commercial procedures (cryogenic distillation) for krypton and xenon separation need to be considered. An attempt will be made to explain the needs for other separations, and the various physical-chemical properties of these elements which are important in achieving such separations will be described. In addition, secondary reactions associated with particular separation schemes will be described because they play a dominant role in the selection of a suitable process.

## VII. PROCESSES FOR SEPARATING KRYPTON AND XENON

From a physical-chemical viewpoint, four general methods of separation may be considered: (1) cryogenic distillation, (2) selective adsorption, (3) selective absorption, and (4) diffusion.

### A. Cryogenic Distillation

As stated earlier, cryogenic distillation has been utilized for decades by industry to achieve separation of xenon and krypton from the atmosphere. Separation of krypton and xenon from the more abundant atmospheric gases is achieved by fractional distillation of liquefied air. The boiling points of the major components at atmospheric pressure are included with those of krypton and xenon in Table 3. Although the distillation process is ideal for commercial gas production from atmospheric resources, the exposure of oxygen to radiation from the decaying noble gas isotopes causes the formation of oxygen radicals and, ultimately, of ozone. With the potential for hydrocarbons also to be present in the liquefied gases, ozone presents an intolerable explosion hazard. Several incidents of ozone explosions in cryogenic systems exposed to radiation fields have been reviewed by Cottrell.[72] Thus, current designs specify the complete removal of oxygen prior to cryogenic distillation by combining it with hydrogen over a palladium-platinum catalyst at about 550°C.

Although advantageous use of the physical properties of the noble gases is made in the cryogenic process, problems are incurred with these properties. Freezing points of impurity gases (e.g., nitrogen oxides, carbon dioxide, water, and hydrocarbons) which overlap boiling points of the atmospheric gases cause system plugging.

### B. Selective Adsorption

Active carbons and molecular sieves (i.e., zeolites) have been used to remove krypton and xenon from LWR waste gases.[73] Since zeolites recently have been considered as an alternate means of $^{85}Kr$ storage, the particular discussion concerning them will be deferred until the section on $^{85}Kr$ storage. Although there is considerable opposition to the use of adsorption methods in general,[66] a brief presentation of the fundamental aspects of these methods will be given. A more complete description can be found in the reviews by Kovach.[74,75]

The capacities for krypton and xenon adsorption on carbon are, in general, related to the carbon surface areas. However, both the total surface area and the distribution of this surface in the various-diameter micropores become important in the very low concentration ranges. In the case of activated* carbons, the adsorptive capacity of the material increases with increasing krypton or xenon concentration. The effect of temperature on adsorption is very significant, and since the adsorption process is exothermic, it always takes place to a greater extent at lower temperatures.

* A term used in activated-carbon manufacturing for the heat treatment of the carbon at 800 to 1000°C in a controlled and slightly oxidizing atmosphere.

Typical adsorption isotherm data for krypton, shown in Figure 10, emphasize the effect of temperature on the adsorption process.

### C. Selective Absorption

Selective absorption processes for separating krypton and xenon exploit the solubility differences between various gaseous components in the liquid system. The extent of solubility of a gas in a liquid when no interaction takes place can be estimated from Hildebrand's solution theory for nonelectrolytes.[76] The solubility tends to be maximum when the solubility parameter,* $\delta$, approaches that of the solvent.[76] The solubility parameters for krypton and xenon at their boiling points are 7.9 and 8.4, respectively.[77]

Steinberg and Manowitz[77] studied a large number of solvents and found that the solubilities of krypton and xenon increase as the solvent becomes more nonpolar. Their solubility data tend to pass through a maximum in the vicinity of a solvent solubility parameter of 8.0, which is in agreement with Hildebrand's theory. On the basis of these considerations, various solvents have been proposed for stripping noble gases from contaminated air streams; these include carbon tetrachloride,[78-80] kerosene-base liquids,[77,81] liquid nitrogen,[82] nitrous oxide,[83] and fluorocarbons.[84] Considering solvent capacities, separation factors, and thermal- and radiation-stabilities as well as overall process safety and economic features, Steinberg[83] suggested an absorption process employing dichlorodifluoromethane, i.e., refrigerant-12 (R-12), and significant development experience has been gained using this solvent.[85]

Figure 11 gives the Henry's law constants as a function of temperature for krypton, xenon, and the common atmospheric gases dissolved in R-12. Since Henry's law constant, $K = p_2/x_2$ (where $p_2$ is the partial pressure in atmospheres of the solute over $x_2$, its mole fraction in the liquid phase), is inversely proportional to the solubility, the figure shows that krypton and xenon are more soluble in R-12 than are the other gases. This difference in the relative solubilities makes separation possible. Furthermore, since the differences in the noble gas solubilities relative to the atmospheric components are inversely related to temperature, operation of a fluorocarbon separation process at low temperatures enhances the separation.

Various off-gas contaminants such as $I_2$, $CH_3I$, $CO_2$, $NO_2$, $H_2O$, and $NH_3$ could reach the fluorocarbon system and affect the efficiency of the noble gas separation. Their physical-chemical behavior in the fluorocarbon system has been the subject of investigation by Toth et al.[86] Distribution coefficients, solubility relationships, phase behavior, and some aspects of chemical reactivity (especially corrosion of type 304L stainless steel) have been described. Particularly significant from a chemistry standpoint was the observation that carbon dioxide deviated unusually from regular solution behavior, showing a $-3.5\%$ excess volume of mixing, along with the liberation of about 0.32 kcal/mol on mixing equimolar amounts of solute and solvent. Significant interaction between $CO_2$ and R-12 was suggested by these phenomena, in agreement with previous observations[76] regarding $CO_2$ in unsaturated hydrocarbon solvents.

Radiation effects on fluorocarbon solutions of various fission products are yet to be experimentally tested, but current indications are that the production of chlorine atoms and fluorocarbon radicals, e.g., $\cdot CF_2Cl$, are the most probable consequences. Chlorine atoms may cause some complications in the presence of off-gas components, especially water vapor, where hydrogen abstraction and HCl formation are most probable. Other fluorocarbons generated by recombination of fluorocarbon radicals present few, if any, serious consequences since they are compatible with R-12. Ozone could also be formed by radiolysis of oxygen, but it is expected to be stabilized by the fluorocarbons and thus be much less reactive than in other systems.

* Solubility parameter is defined as the square root of the cohesive energy density and is equal to the square root of the energy of vaporization per cubic centimeter of the ideal gas.

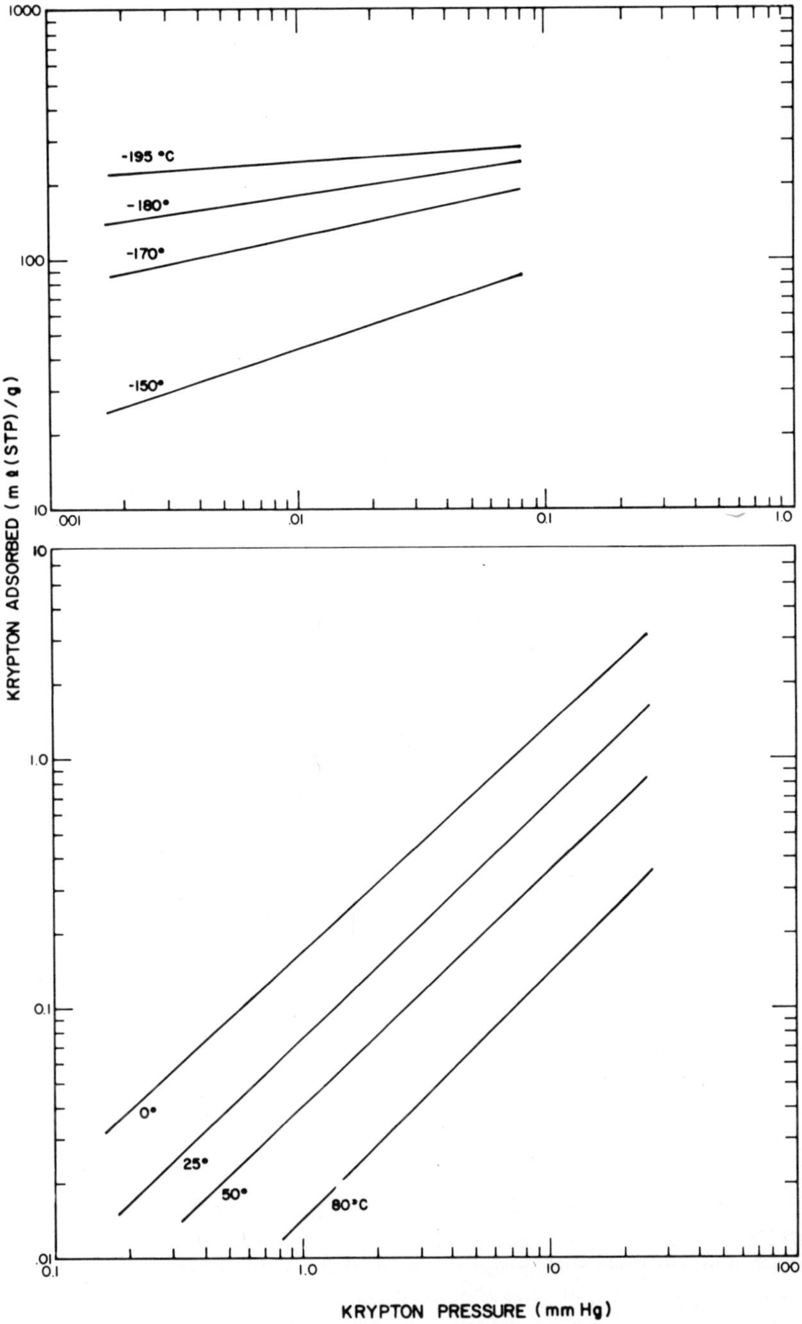


FIGURE 10. Adsorption isotherm data for krypton on activated carbon.[74]

## D. Diffusion

Various diffusion processes, all of which are based on differences in the migration velocities of the gaseous atoms, have been suggested for separating krypton and xenon from off-gas streams. These have been outlined previously[66,87] and discussed in some detail. However, a description of the permselective membrane process will be presented here in order to identify some of the general phenomena which must be considered.

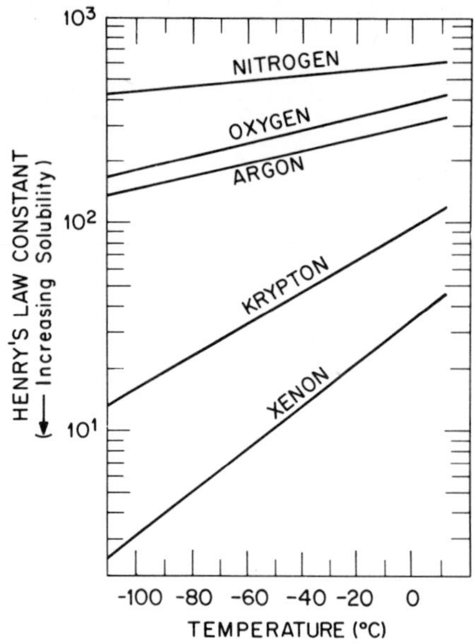


FIGURE 11. Henry's law constants for krypton, xenon, and the major atmospheric gases in refrigerant-12, illustrating the differences in the solubilities of these gases.[84]

Differences in permeability of gases through a special membrane form is the basis of the permselective membrane separation process. Under similar conditions, one gas will permeate faster than another across a thin membrane and thus effect a partial separation. Such a membrane is said to be permselective. The membrane is not porous, and gases do not effuse through minute openings in the material, as in the "gaseous diffusion process" for isotope separation. Instead, the gas dissolves in the membrane on the high-pressure side, diffuses through the membrane, and comes out of solution on the low-pressure side. Figure 12 depicts the differences in the transport mechanisms through porous and permselective materials.

Mitchell[88] first observed that various gases penetrated natural membranes at different rates, but industrial separations of gases by this technique did not become practical until methods were developed for preparing very thin, hole-free sheets of modern polymers. Although any polymer may be used for the separation of gases, silicone rubbers have the highest permeabilities measured thus far; they also yield high gas separation factors.

With a permselective membrane, the separation is proportional to the ratio of the products of the solubilities in, and the diffusivities through, the membrane. Most gases in polymers have been found[165] to obey Henry's law of solubility and Fick's law for diffusion, $F_i = -D_i(dC_i/dt^*)$, where $F_i$ is the gas flux, $D_i$ is the diffusivity of the gas in the membrane, $C_i$ is the concentration of a particular gas in the membrane, and $t^*$ is the distance into the membrane. From these equations, the following expression can be derived for a "permeability factor", $P_i$

$$P_i = F_i\, t/\Delta p_i \tag{33}$$

The units are the same as those defined above except that, in this case, t is the membrane thickness and $\Delta p_i$ is the pressure difference across the membrane. Figure 13

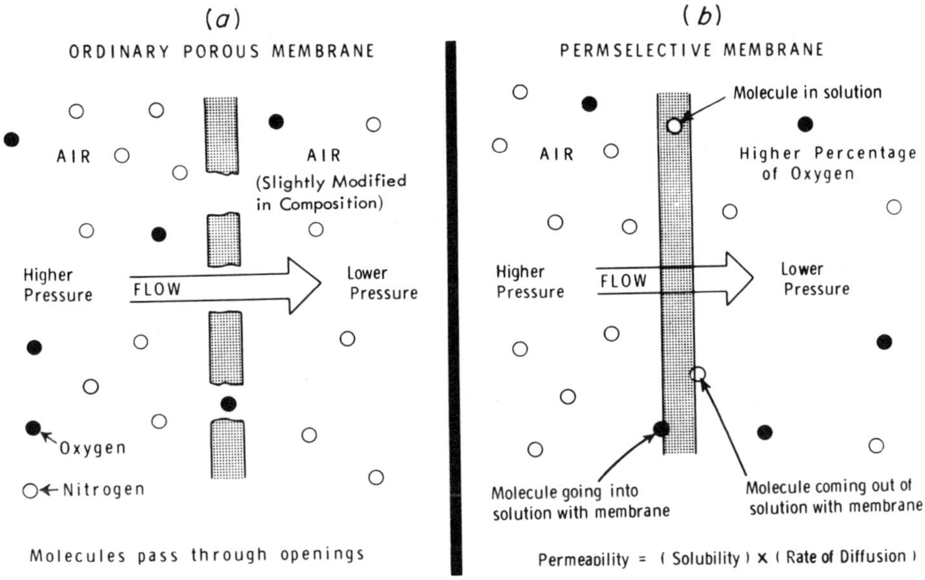


FIGURE 12. Schematic representation of the differences between effusion through a porous membrane (a) and diffusion through a permselective membrane (b).

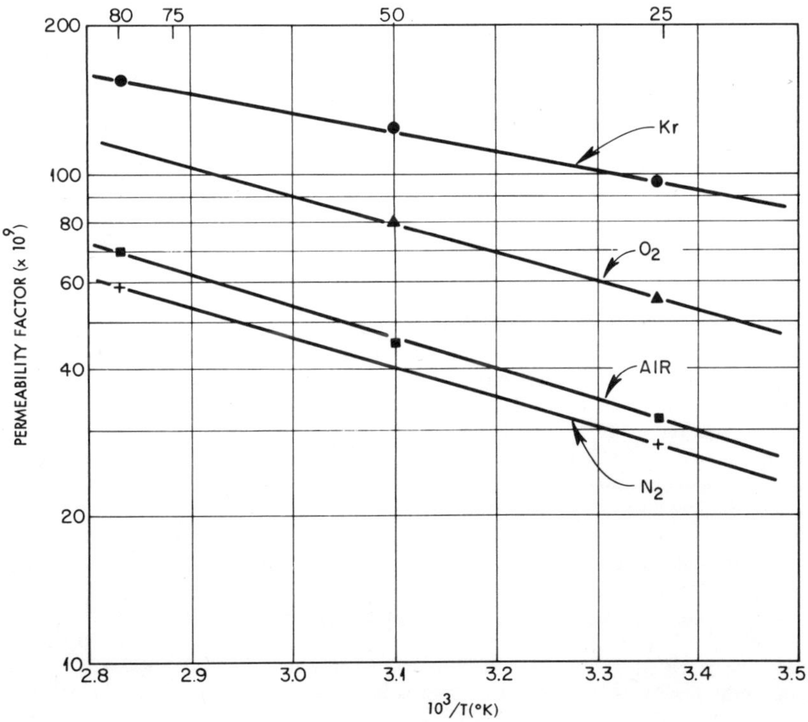


FIGURE 13. Permeability factors for Kr, $O_2$, air, and $N_2$ through silicone rubber membranes.

shows the permeability of krypton and the atmospheric gases as a function of temperature.

## VIII. STORAGE OF KRYPTON AND XENON

The most direct method of storing $^{85}Kr$ after separation is in high-pressure gas cylinders. Obviously, little chemistry is associated with this mehod; however, peculiar characteristics of the isotope, which include both decay heat and the krypton decay product, rubidium, have resulted in the necessity of using stainless steel to ensure structural integrity of the containers.

Another possible alternative for storage is zeolite encapsulation.[89] Zeolites are crystalline materials into which molecules or atoms (noble gases) may diffuse under a given set of conditions and then become entrapped in lattice pockets or "cages" when the conditions conducive to their entry and egress from the cages (e.g., high temperature or pressure) are reduced.[90] Typical zeolites are aluminosilicates,[102] e.g., Zeolite A, $Na_{12}[(AlO_2)_{12}(SiO_2)_{12}]\cdot 27\ H_2O$, which contain cages of 6.6 Å free diameter connected by channels of 2.2-Å diameter. The krypton atom has a minimum (kinetic) diameter of 3.6 Å; thus, any atoms entrapped in the cavity cannot escape. This situation is much like the situation with the clathrates.

The process by which krypton diffuses through such a pore structure is known as activated diffusion,[89] since an activation energy is required for the krypton atoms to pass through the 2.2-Å apertures. High temperatures and pressures (up to 350°C and 4300 atm) are required to overcome this activation energy and move atoms into the cages. At lower temperatures the noble gas becomes effectively trapped. A krypton concentration of 90.5 cc/g has been reported[90] in Zeolite A for the conditions of temperature and pressure specified above. The molecules or atoms can thus be present in the zeolite at a density about half that of the liquefied gas at its normal boiling point.

## REFERENCES


1. **Kee, C. W.,** ORIGEN Code Calculation, Oak Ridge National Laboratory, Oak Ridge, Tenn., 1976; Bell, M.J., ORIGEN - The ORNL Isotope Generation and Depletion Code, ORNL-4628, Oak Ridge National Laboratory, Oak Ridge, Tenn., 1973.
2. **Besmann, T. M. and Lindemer, T. B.,** Chemical thermodynamics of the system Cs-U-Zr-H-I-O in the Light Water reactor fuel-cladding gap, *Nucl. Technol.,* 40, 297, 1978.
3. **Besmann, T. M.,** personal communication, 1978.
4. **Stone, J. A. and Johnson, D. R.,** Measurement of radioactive gaseous effluents from voloxidation and dissolution of spent nuclear fuel, paper presented at the 15th DOE Nuclear Air Cleaning Conf., Boston, August 7 to 10, 1978.
5. **Keller, J. H., Duce, F. A., Pence, D. T., and Maeck, W. J.,** Hypoiodous acid: an airborne inorganic iodine species in steam-air mixtures, in Proc. 11th AEC Air Cleaning Conf., Vol. 2, Rep. CONF-700816, U.S. Atomic Energy Commission, Washington, D.C., 1970, 467.
6. **Jubin, R. T.,** A Literature Survey of Methods to Remove Iodine from Off-Gas Streams Using Solid Sorbents, Rep. ORNL/TM-6607, Oak Ridge National Laboratory, Oak Ridge, Tenn., March 1979.
7. **Bryant, P. M. and Warner, B. F.,** Control of radioiodine release from reprocessing plants, in Control of Iodine in the Nuclear Industry, Tech. Rep. Ser. No. 148, International Atomic Energy Agency, Vienna, 1973, 29.
8. **Halfpenny, E. and Robinson, P. L.,** Pernitrous acid. The reaction between hydrogen peroxide and nitrous acid, and the properties of an intermediate product, *J. Chem. Soc.,* p. 928, 1952.
9. **Halfpenny, E. and Robinson, P. L.,** The nitration and hydroxylation of aromatic compounds by pernitrous acid, *J. Chem. Soc.,* p. 939, 1952.

10. **Latimer, W. M.,** *Oxidation Potentials,* 2nd ed., Prentice-Hall, Englewood Cliffs, N.J., 1952, 62.
11. **Porter, G., Szabo, Z. G., and Townsend, M. G.,** The recombination of atoms. V. Iodine atom recombination in nitric oxides, *Proc. R. Soc. London Ser. A,* 270, 493, 1962.
12. **Tachikawa, E., Tamura, N., Furukawa, K., and Motojima, K.,** Formation of organic iodides on dissolution of irradiated uranium metal in nitric acid, *Int. J. Appl. Radiat. Isot.,* 22, 793, 1971.
13. **Nakamura, K., Saeki, M., and Tachikawa, E.,** Formation of organic iodides upon heating nitric acid solutions containing carrier-free iodine-131, *Nucl. Sci. Technol.,* 10, 367, 1973.
14. **Saeki, M., Numahura, K., and Tachikawa, E.,** Carbon sources and formation mechanisms of organic iodides in nitric acid solutions, *Int. J. Appl. Radiat. Isot.,* 25, 407, 1974.
15. **Parsly, L. F.,** Chemical and Physical Properties of Methyl Iodide and its Occurrence Under Reactor Accident Conditions, A Summary and Annotated Bibliography, Rep. ORNL-NSIC-82, Oak Ridge National Laboratory, Oak Ridge, Tenn., December 1971.
16. **Horner, D. E. and Mailen, J. C.,** unpublished data, 1976.
17. **Miles, G. L.,** The Fumeless Dissolving of Uranium, Rep. AERE-CR-1804, U.K. Atomic Energy Research Establishment, Harwell, Berks., England, 1970.
18. **Rainey, R. H. and Lovelace, R. C.,** personal communication, 1976.
19. **Horner, D. E., Mailen, J. C., and Bigelow, H. R.,** Solubility of $PdI_2$ in nitrate and perchlorate solutions, *J. Inorg. Nucl. Chem.,* 39, 1945, 1977.
20. **Mailen, J. C. and Horner, D. E.,** Removal of iodine from reactor fuel solutions as insoluble $PdI_2$, *Nucl. Technol.,* 33, 260, 1977.
21. **Holladay, D. W.,** A Literature Survey: Methods for Removal of Iodine Species from Nuclear Power Plant and Nuclear Fuel Reprocessing Off-Gases and Liquid Waste Streams, with Emphasis on Solid Sorbents, Rep. ORNL/TM-6350, Oak Ridge National Laboratory, Oak Ridge, Tenn., 1978.
22. **Mailen, J. C.,** Use of chemical equilibrium data to calculate Iodox column operations, *Nucl. Technol.,* 30, 325, 1976.
23. **Mailen, J. C.,** Equilibrium in the oxidation of iodine by hyperazeotropic nitric acid, *J. Inorg. Nucl. Chem.,* 37, 1019, 1975.
24. **Mailen, J. C. and Tiffany, T. O.,** The reaction of iodine with concentrated nitric acid, *J. Inorg. Nucl. Chem.,* 37, 127, 1975.
25. **Rainey, R. H. and Mailen, J. C.,** Solubility of $HI_3O_8$ in concentrated nitric acid, *J. Chem. Eng. Data,* 19, 262, 1974.
26. **Ferguson, D. E. and staff,** Chemical Technology Division Annual Progress Report for Period Ending May 31, 1969, Rep. ORNL-4422, Oak Ridge National Laboratory, Oak Ridge, Tenn., 1969, 84.
27. **Crouse, D. J.,** personal communication, July 1976.
28. Alternatives for Managing Wastes from Reactors and Post-Fission Operations in the LWR Fuel Cycle, Rep. ERDA-76-43, U.S. Energy Research and Development Administration, Washington, D.C., 1976, 13.12.
29. **Mailen, J. C. and Horner, D. E.,** Removal of radioiodine from gas streams by electrolytic scrubbing, *Nucl. Technol.,* 30, 317, 1976.
30. **Collins, E. D.,** personal communication, July 1976.
31. **Cederberg, G. K. and MacQueen, D. K.,** Containment of Iodine-131 Released by the RaLa Process, Rep. IDO-14566, Atomic Energy Division, Phillips Petroleum Company, Idaho Falls, Idaho, 1961, 24.
32. **Evans, A. G.,** Confinement of Airborne Radioactivity, Progress Report, July to December 1972, Rep. DP-1329, Savannah River Laboratory, Aiken, S.C., 1973.
33. **Kovach, J. L.,** The evaluation of the ignition temperature of activated charcoals in air, steam, oxygen, and oxides of nitrogen, in Proc. Symp. Operating and Developmental Experience in the Treatment of Airborne Radioactive Wastes, International Atomic Energy Agency, Vienna, 1968, 438.
34. **Pence, D. T., Duce, F. A., and Maeck, W. I.,** Iodine adsorbents program, in Idaho Chemical Programs Annual Technical Report Fiscal Year 1971, Rep. ICP-1006, Allied Chemical Corporation, Idaho Falls, Idaho, 1972, 69.
35. **Wilhelm, J. G.,** Iodine Filters in Nuclear Power Stations, Rep. KFK-2449, Gesellschaft fur Kernforschung mbH, Karlsruhe, West Germany, 1977.
36. **Staples, B. A., Murphy, L. P., and Thomas, T. R.,** Airborne elemental iodine loading capacities of metal zeolites and a dry method for recycling silver zeolite, in Proc. 14th ERDA Air Cleaning Conf., Rep. CONF-760822, U.S. Energy Research and Development Agency, Washington, D.C., 1976, 363.
37. **Patil, K. C., Rao, C. N. R., Lacksonen, J. W., and Dryden, C. E.,** The silver nitrate-iodine reaction: iodine nitrate as the reaction intermediate, *J. Inorg. Nucl. Chem.,* 29, 407, 1967.
38. **Schnez, H., Laser, M., and Merz, E.,** Tritium in Reprocessing Plants. A Study of the Inventory, Behavior and the Possibilities of Separation of the Tritium Isotope, Rep. JUL-10999-CT, Kernforschungsanlage Julich Gesellschaft mit beschrankter Haftung, Julich, West Germany, 1974, 2. Translation No. GERHTR-139.

39. **Horrocks, D. L.**, Yields of tritium during the fissioning of nuclides of interest in the operation of nuclear power reactors, in *Tritium*, Moghissi, A. A. and Carter, M. W., Eds., Messenger Graphics, Phoenix, 1973, 30.
40. **Fluss, M. J., Dudey, N. D., and Malewicki, R. L.**, Tritium and alpha-particle yields in fast and thermal neutron fission of $^{235}U$, *Phys. Rev. C*, 6 (6), 2252, 1972.
41. **Jacobs, D. G.**, Sources of Tritium and its Behavior upon Release to the Environment, Rep. TID-24635, Division of Technical Information, U.S. Atomic Energy Agency, Oak Ridge, Tenn., December 1968, 4.
42. **Scargill, D.**, Diffusion of Tritium in Irradiated $UO_2$, Rep. AERE-8474, U.K. Atomic Energy Research Establishment, Harwell, Berks., England, 1976.
43. **Fowler, J. D., Chandra, D., Elleman, T. S., Payne, A. W., and Verghese, K.**, Tritium diffusion in $Al_2O_3$ and BeO, *J. Am. Ceram. Soc.*, 60, 155, 1977.
44. **Bates, J. B., Wang, J. C., and Perkins, R. A.**, Mechanisms for hydrogen diffusion in $TiO_2$, *Phys. Rev. B*, 19, 4130, 1979.
45. **Lindemer, T. B. and de Nordwall, J. H.**, An Analysis of Chemical Failure of Coated $UO_2$ and Other Oxide Fuels in the High-Temperature Gas-Cooled Reactor, Rep. ORNL-4926, Oak Ridge National Laboratory, Oak Ridge, Tenn., 1974.
46. **Johnson, C. E., Johnson, P. E., Blackburn, P. E., and Crouthamel, C. E.**, Effects of oxygen concentration on properties of fast reactor mixed-oxide fuel, *React. Technol.*, 15, 303, 1972.
47. **Darken, L. S. and Gurry, R. W.**, *Physical Chemistry of Metals*, McGraw-Hill, New York, 1953, 349.
48. **Goode, J. H. and Cox, C. M.**, The Distribution of Fission Product Tritium in a Zircaloy-clad $UO_2$ Blanket Rod from PWR-1, Rep. ORNL-TM-2994, Oak Ridge National Laboratory, Oak Ridge, Tenn., June 1970.
49. **Ray, J. W., Wooton, R. O., and Barnes, R. H.**, Investigation of Tritium Generation and Release in PM Nuclear Power Plants, Rep. BMI-1787, Battelle Memorial Institute, Columbus, Ohio, October 1966.
50. **Fiorelli, A. J. and Santore, C., Eds.**, PWR Core 1 Component Examination Program Summary Report, Rep. WAPD-320, Westinghouse Atomic Power Development, Pittsburgh, October 1969.
51. **Goode, J. H. and Stacy, R. G.**, Comparative Studies of Head-End Processing Using Irradiated Mechanically Blended and Coprecipitated $(U,Pu)O_2$ Reactor Fuels, Rep. ORNL/TM-6370, Oak Ridge National Laboratory, Oak Ridge, Tenn., April 1978.
52. **Stone, J. A. and Johnson, D. R.**, Measurement of radioactive gaseous effluents from voloxidation and dissolution of spent nuclear fuel, paper presented at the 15th DOE Nuclear Air Cleaning Conf., Boston, August 7 to 10, 1978.
53. **Goode, J. H. and Vaughen, V. C. A.**, ORNL Experiments on the Behavior of Tritium During Head-End Processing of Irradiated Reactor Fuels, Rep. ORNL-TM-2793, Oak Ridge National Laboratory, Oak Ridge, Tenn., February 1970.
54. **North, E. D. and Booth, R. L.**, Fission Product Gas Retention Study, Final Report, Rep. ORNL-TM-4409, Oak Ridge National Laboratory, Oak Ridge, Tenn., August 1973.
55. **Goode, J. H., Ed.**, Voloxidation — Removal of Volatile Fission Products from Spent LMFBR Fuels, Rep. ORNL-TM-3723, Oak Ridge National Laboratory, Oak Ridge, Tenn., January 1973.
56. **Davis, W., Jr.**, Thermodynamics of extraction of nitric acid by tri-*n*-butyl phosphate-hydrocarbon diluent solutions. I. Distribution studies with TBP in Amsco 125-82 at intermediate and low acidities, *Nucl. Sci. Eng.*, 14, 159, 1962.
57. **Davis, W., Jr.**, Thermodynamics of extraction of nitric acid by tri-*n*-butyl phosphate-hydrocarbon diluent solutions. II. Densities, molar volumes, and water solubilities of TBP-Amsco 125-82-$HNO_3$-$H_2O$ solutions, *Nucl. Sci. Eng.*, 14, 169, 1962.
58. **Davis, W., Jr.**, Thermodynamics of extraction of nitric acid by tri-*n*-butyl phosphate-hydrocarbon diluent solutions. III. Comparison of literature data, *Nucl. Sci. Eng.*, 14, 174, 1962.
59. **Davis, W., Jr., Mrochek, J., and Hardy, C. L.**, The system: tri-*n*-butyl phosphate (TBP)-nitric acid-water. I. Activities of TBP in equilibrium with aqueous nitric acid and partial molar volumes of the three components in the TBP phase, *J. Inorg. Nucl. Chem.*, 28, 2001, 1966.
60. **Davis, W., Jr. and Mrochek, J.**, Activities of tributyl phosphate in tributyl phosphate-uranyl nitrate-water solutions, in *Solvent Extraction Chemistry*, Dryssen, D., Liljenzin, O. U., and Rydberg, J., Eds., North-Holland, Amsterdam, 1967.
61. **Davis, W., Jr., Mrochek, J., and Judkins, R. R.**, Thermodynamics of the two-phase system: water-uranyl nitrate-tributyl phosphate-Amsco 125-82, *J. Inorg. Nucl. Chem.*, 32, 1689, 1970.
62. **Leroy, P.**, Study of the solvent tributyl phosphate in dodecane, Rep. CEA-3207, *Commis. Energ. At. Fr. Rapp.*, 43, 1967; ORNL-tr-4344, Oak Ridge National Laboratory, Oak Ridge, Tenn., 1967.
63. **Tedder, D. W.**, personal communication, July 1978.
64. **Keilholtz, G. W.**, Krypton-xenon removal system, *Nucl. Saf.*, 12, 591, 1971.

65. **Rainey, R. H., Carter, W. L., and Blumkin, S.,** Completion Report — Evaluation of the Use of Permselective Membranes in the Nuclear Industry for Removing Radioactive Xenon and Krypton from Various Off-Gas Streams, Rep. ORNL-4522, Oak Ridge National Laboratory, Oak Ridge, Tenn., April 1971.
66. **Schnez, H.,** Treatment and Separation of Radioactive Fission Products, Tritium, Rare Gases, and Iodine in Nuclear Fuel Reprocessing Plants, Rep. GERHTR-158, Kernforschungsanlage Jülich Gesellschaft mit beschänkter Haftong, Jülich, West Germany, July 1975 (English translation of Jul-1223).
67. **Kuller, B. J., Trevorrow, L. E., and Steindler, M. J.,** Tritium and Noble-Gas Fission Products in the Nuclear Fuel Cycle. II. Fuel Reprocessing Plants, Rep. ANL-8135, Argonne National Laboratory, Argonne, Ill., March 1975.
68. **Cotton, F. A. and Wilkerson, G., Eds.,** *Advanced Inorganic Chemistry,* 3rd ed., John Wiley & Sons, New York, 1972.
69. **Stein, L.,** Removal of xenon and radon from contaminated atmospheres with dioxygenyl hexafluoroantimonate, $O_2SbF_6$, *Nature (London),* 243, 30, 1973.
70. **Liebmann, J. F.,** Oxidant for trapping radiokrypton, *Nature (London),* 244, 84, 1973.
71. Code of Federal Regulations, Title 10, Chap. 1, Part 50, Appendix I.
72. **Cottrell, W. B.,** unpublished data, Oak Ridge National Laboratory, Oak Ridge, Tenn., July 24, 1958.
73. **Mahn, J. A. and Perry, C. A.,** General Atomic's radioactive gas recovery system, in Proc. 13th AEC Air Cleaning Conf., U.S. Atomic Energy Commission, San Francisco, 1974, 228.
74. **Kovach, J. L.,** Review of Kr-Xe Adsorber Design Part I, Rep. NACAR-1005, North American Carbon, Columbus, Ohio, July 1, 1970.
75. **Kovach, J. L.,** Krypton-Xenon Adsorption on NACAR Carbons, Rep. NACAR-10004, North American Carbon, Columbus, Ohio, December 28, 1969.
76. **Hildebrand, J. H., Prausnitz, J. M., and Scott, R. L.,** *Regular and Related Solutions,* Van Nostrand Reinhold, New York, 1970.
77. **Steinberg, M. and Manowitz, B.,** Recovery of fission product noble gases, *Ind. Eng. Chem.,* 51, 47, 1959.
78. **McIlroy, R. W., Gleuckauf, E., deNordwall, H. J., and Pummery, F. C. W.,** The recovery of radiokrypton from dissolver waste gases by fractional extraction into solvents, in Proc. 2nd United Nations Int. Conf. Peaceful Uses Atomic Energy, Vol. 13, United Nations, New York, 1958, 309.
79. **Taylor, R. F.,** The recovery of fission product krypton from nuclear fuels, *Ind. Chem.,* p. 358, 1963.
80. **Taylor, R. F. and Wall, G. P.,** Development of a production process for radiokrypton recovery by fractional absorption, *Prog. Nucl. Energy Ser. 4,* 5, 307, 1963.
81. **Steinberg, M. and Manowitz, B.,** An Absorption Process for Recovery of Fission Product Noble Gases, Rep. BNL-493, Brookhaven National Laboratory, Brookhaven, N.Y., February 1958.
82. **Keilholtz, G. W.,** Radioactive noble gases from off-gas streams, *Nucl. Saf.,* 8 (2), 155, 1966-67.
83. **Steinberg, M.,** The Recovery of Fission Product Xenon and Krypton by Absorption Processes, Rep. BNL-542, Brookhaven National Laboratory, Brookhaven, N.Y., January 1959.
84. **Merriman, J. R., Pashley, J. H., Habiger, K. E., Stephenson, M. J., and Anderson, L. W.,** Concentration and collection of krypton and xenon by selective absorption in fluorocarbon systems, Paper SM-110/25, in IAEA Symp. Operating and Development Experience in the Treatment of Airborne Radioactive Wastes, International Atomic Energy Agency, Vienna, 1968.
85. **Merriman, J. R., Stephenson, M. J., Pashley, J. H., and Dunthorn, D. I.,** Removal of radioactive krypton and xenon from contaminated off-gas streams, in Proc. 11th AEC Air Cleaning Conf., Vol. 1, Rep. CONF-700816, U.S. Atomic Energy Commission, Washington, D.C., 1970, 175.
86. **Toth, L. M., Bell, J. T., and Fuller, D. W.,** Chemical and Physical Behavior of Some Contaminants in the Fluorocarbon Off-Gas Process, Rep. ORNL-6484, Oak Ridge National Laboratory, Oak Ridge, Tenn., November 1978.
87. **Slansky, C. M.,** Separation processes for noble gas fission products from the off-gas of fuel reprocessing plants, *At. Energy Rev.,* 9, 432, 1971.
88. **Mitchell, J. K.,** On the penetrativeness of fluids, *J. R. Inst. G.B.,* 2, 101, 1831; Mitchell, J. K., On the penetration of gases, *Am. J. Med. Sci.,* 13, 100, 1833.
89. **Bartlett, J. W., Carrell, J. R., Kreiter, M. R., Platt, A. M., and Powell, J. A.,** Report Coordinators, Alternatives for managing wastes from reactor and post-fission operations in the LWR fuel cycle, in Alternatives for Waste Treatment, Vol. 2, Rep. ERDA-76-43, U.S. Energy Research and Development Administration, Washington, D.C., May 1976, 13.0, 14.0.
90. **Breck, D. W.,** Crystalline molecular sieves, *J. Chem. Educ.,* 41, 678, 1964.

Chapter 6

# RADIOACTIVE WASTE MANAGEMENT

A. G. Croff*

## Table of Contents



* Chemical Technology Division, Oak Ridge National Laboratory, operated by Union Carbide Corporation under Contract W-7405-eng-26 with the U.S. Department of Energy.

# I. INTRODUCTION

The overall objective of waste management is to implement processes which will effectively preclude the radioactive waste constituents from returning to the biosphere in hazardous concentrations during the time required for the waste to decay to innocuous levels. Thus, the time that a particular radioactive waste must remain isolated is heavily dependent on its initial concentration and half-life. In general, a nuclide can be considered to have decayed to innocuous levels when a time equal to 10 half-lives of the major isotope(s) present has elapsed, leaving about 0.1% of the amount of the isotope that was present originally.

A major problem associated with nuclear energy is the "permanent" disposal of the radioactive wastes produced by fission and transmutation reactions. The technical problems arise from the effectively infinite radioactive lifetime of some of the wastes, and from the fact that an appreciable fraction of the earth's air, crust, and water would be required to dilute these wastes to concentrations considered to be acceptable. Together these factors impose a requirement either of disposing of wastes in a manner permitting and ensuring indefinitely long surveillance and attention to ensure their confinement or of disposing of them in a place so free from natural disturbance or remote from man that such attention is not required. The second course of action, if it is possible, would relieve mankind of long-term surveillance. Another alternative method of permanent waste disposal is simply to destroy the very long-lived radioactive isotopes. While all attempts to accelerate significantly the process of radioactive decay have failed so far, there is a way which may be practical to destroy a very significant part of the radioactive wastes.

This is to further process the wastes to recover the potentially troublesome isotopes and subject these to special treatment and/or disposal techniques. The further processing of the wastes to recover these isotopes is called "partitioning". The special treatment and disposal options that have been considered for the isotope fractions recovered via partitioning are transmutation of the isotopes to less toxic species and extraterrestrial disposal of the isotopes. This alternative waste management concept is described further in Section V.

It is generally accepted that the primary hazard presented to people by radioactive wastes devolves from their radioactivity. Thus, it is worthwhile to consider for a moment what the general nature of that hazard is. Most radioactive materials decay with the emission of $\alpha$ or $\beta$ particles or of $\gamma$ rays. Some have multiple decay modes and emit two or all three types of radiation. In general, a given number of curies (1 Ci = $3.7 \times 10^{10}$ disintegrations per second) of an $\alpha$ particle emitter in the body is more hazardous than the same number of curies of a gamma ray emitter inside the body. This results from the fact that all the energy of the $\alpha$ particle is deposited in the body in the immediate vicinity of its emitting isotope; the $\gamma$ ray, on the other hand, penetrates tissue easily, and some of the rays may leave the body unattenuated. Conversely, an $\alpha$ particle emitter outside the body is less hazardous than a $\gamma$ ray emitter outside the body because shielding from $\alpha$ particles, which hardly penetrate the skin, is much easier than shielding from the much more penetrating $\gamma$ rays. Beta particle emitters are, in general, intermediate between $\alpha$ and $\gamma$ emitters in their biological effects.

Because many radioisotopes emit more than one type of radiation during decay, the above generalizations are an oversimplification. Further, some plants and animals eaten by man preferentially take up certain elements, and for those radioactive materials which get inside a person, either by ingestion or respiration, there is the important consideration of preferential uptake or concentration in specific organs such as the thyroid gland, the bones, or the liver. Some radioelements or their compounds are taken up in a highly specific manner by the body. This biological concentration can lead to serious damage because of the resultant high local dose, when that amount of radioactive material dispersed uniformly over the body might be relatively harmless.

It is apparent that the hazard posed by radioactive wastes must be considered from a number of points of view: type of radiation emitted; ecological pathways followed to arrive in food and ultimately in man; biological concentration; sensitivity of organs to the radiation; amounts of various types of radioactive materials and their sources; and the location of the wastes. In addition to the above may be added lifetime (more correctly, half-life) of the radioisotope and the method of its disposal.

The radioactive wastes produced by a commercial nuclear fuel cycle vary substantially in their radioactive constituent make-up and in their physical characteristics. In light of the above, this means that each different waste might be conditioned and isolated in a different manner, depending on the degree and longevity of the hazard posed by the waste. However, in practice, the multitude of radioactive waste types can be reduced to a relatively small number by aggregating the individual wastes into categories. The definitions of these sources of the wastes in the commercial nuclear fuel cycle will be described in Section II.

In general, the transformation of raw radioactive wastes into conditioned and isolated waste forms consists of three steps:

1. Recovery of the raw radioactive waste (if required) (Section II)
2. Treatment of the waste to yield a waste form suitable for disposal (Section III)
3. Disposal of the waste form in a manner appropriate to the waste's characteristics (Section IV)

A useful parameter when considering the disposal of radioactive wastes is the volume of conditioned waste that results from the entire nuclear fuel cycle and which must be isolated. Projections of the amounts of wastes that will be generated during the next few decades have been made based on an assumed nuclear power growth rate and typical waste volumes produced from a "unit-sized" fuel cycle. These radioactive waste projections are summarized in Section VI.

Finally, a brief bibliography of waste-management-related documents is given for the reader who wishes to pursue one or more subjects in more detail.

## II. TYPES, SOURCES, AND CHARACTERISTICS OF RADIOACTIVE WASTES

The light water reactor (LWR) fuel cycle will produce a wide variety of radioactive wastes. Unfortunately, previous attempts to define a universally accepted categorization scheme for these wastes have not been entirely successful. The waste types discussed in this section are those that are generally used and accepted (or at least recognized) within the technical community. These waste types are

1. High-level TRU waste
2. Intermediate-level TRU waste
3. Intermediate-level, non-TRU waste
4. Low-level TRU waste
5. Low-level, non-TRU waste
6. Fuel assembly structural material waste
7. Iodine waste
8. Tritium waste

9. Carbon waste
10. Noble gas wastes
11. Uranium mill tailings
12. Spent fuel
13. Decontamination and decommissioning waste

It is likely that future actions by regulatory agencies such as the Nuclear Regulatory Commission will result in the promulgation of different, and probably more general, waste categories.

It should be noted that many of the waste types that will be defined in subsequent sections will be designated "transuranic-contaminated" or "TRU" wastes. This term simply means that the wastes are expected to contain more than 10 nCi of isotopes that decay via the emission of an alpha particle per gram of treated waste. These wastes are considered to be "long-lived" and must be disposed of in a manner different from that of wastes containing only isotopes that decay to innocuous levels within a relatively short time. The "TRU" appellation originated because materials that are contaminated only with uranium will not fall within this category because of the very low radioactivity (i.e., low radiotoxicity per gram) of the uranium isotopes as compared to the transuranic actinides.

### A. High-Level TRU Waste

High-level TRU waste (HLW) has been defined in federal regulations as "... those aqueous wastes resulting from the operation of the first cycle solvent extraction system, or equivalent, and the concentrated wastes from subsequent extraction cycles, or equivalent, in a facility for reprocessing irradiated reactor fuels." In practice, this complex definition reduces to the following: liquid HLW is an aqueous nitric acid solution containing virtually all of the nonvolatile fission products and actinides, except uranium and plutonium, that were originally in the spent fuel. Approximately 0.5% of the uranium and plutonium in the spent fuel may also be lost to the HLW stream.

As-produced liquid HLW has a volume of approximately 5000 $\ell$/t of heavy metal (uranium plus plutonium) initially in the reactor fuel (i.e., metric tonnes initial heavy metal or MTIHM). However, this waste is immediately sent to an evaporator that concentrates the liquid HLW to about 400 $\ell$/MTIHM and produces concentrated nitric acid for recycle within the reprocessing plant. A typical, concentrated liquid HLW composition produced by reprocessing pressurized water reactor (PWR) fuel is given in Table 1.

The presence of most of the fission products in the liquid HLW results in its being intensely radioactive. The liquid HLW characterized in Table 1 would have a thermal power of about 20 W/$\ell$ 1 year after it was generated and about 3.5 W/$\ell$ 10 years after it was produced. The radioactivity of the liquid HLW at these decay intervals is about 4000 and 700 Ci/$\ell$, respectively.

Other wastes that might be combined with the HLW include some of the liquid, intermediate-level TRU wastes (see Section II. A), and the "bottoms" resulting from the concentration of liquid, low-level TRU wastes (see Section II.B). The addition of these wastes to the liquid HLW can have a substantial impact on the composition of the HLW, principally by greatly increasing the nonradioactive salt concentrations.

### B. Intermediate-Level and Low-Level Wastes

Intermediate-level and low-level wastes are comprised of diverse materials. They can be solid or liquid, combustible or noncombustible, highly radioactive or essentially

**Table 1**
**TYPICAL CHARACTERISTICS OF LIQUID HIGH-LEVEL WASTE[1]**

| | Parameter[a] | | | | |
|---|---|---|---|---|---|
| | Mass (g/ℓ) | Radioactivity (Ci/ℓ) | | Thermal power (W/ℓ) | |
| | 5 years[b] | 6 years[a] | 10 years[b] | 1 year[b] | 10 years[b] |
| Fission products | | | | | |
| $^3H$ | 8 E-6 | 0.10 | 0.06 | — | — |
| Rb | 0.8 | — | — | — | — |
| Sr | 1.5 | 140 | 110 | 0.16 | 0.13 |
| Y | 0.86 | 150 | 110 | 0.80 | 0.62 |
| Zr | 7.0 | 17 | — | 0.08 | — |
| Nb | — | 38 | — | 0.18 | — |
| Mo | 6.9 | — | — | — | — |
| Tc | 1.7 | 0.03 | 0.03 | — | — |
| Ru | 4.5 | 540 | 1.1 | 0.03 | — |
| Rh | 1.1 | 540 | 1.1 | 5.30 | 0.01 |
| Pd | 3.1 | — | — | — | — |
| Ag | 0.16 | 52 | — | 0.38 | — |
| Cd | 0.18 | 0.07 | 0.04 | — | — |
| Sn | 0.11 | 0.83 | — | — | — |
| Sb | 0.03 | 16.8 | 1.7 | 0.05 | — |
| Te | 1 | 7.2 | 0.42 | — | — |
| Cs | 5.4 | 470 | 190 | 2.80 | 0.31 |
| Ba | 3.3 | 200 | 160 | 0.80 | 0.65 |
| La | 2.5 | — | — | — | — |
| Ce | 4.8 | 820 | 0.3 | 0.54 | — |
| Pr | 2.3 | 830 | 0.3 | 6.29 | — |
| Nd | 8.0 | — | — | — | — |
| Pm | 0.08 | 210 | 19.5 | 0.08 | — |
| Sm | 1.8 | 2.9 | 2.7 | — | — |
| Eu | 0.3 | 21.7 | 9.7 | 0.16 | 0.08 |
| Gd | 0.2 | — | — | — | — |
| Total F.P.[c] | 57.75 | 4074 | 614.6 | 17.7 | 1.81 |
| Actinides | | | | | |
| U | 12.65 | — | — | — | — |
| Np | 0.674 | 0.36 | 0.36 | — | — |
| Pu | 0.4 | 3.96 | 3.03 | 0.03 | 0.03 |
| Am | 2.3 | 2.07 | 2.07 | 0.06 | 0.06 |
| Cm | 0.65 | 99.7 | 45.8 | 3.56 | 1.60 |
| Total Act.[c] | 16.79 | 106.2 | 51.3 | 3.65 | 1.69 |
| Total (F.P. + Act.) | 74.5 | 4180 | 664 | 21.35 | 3.50 |
| Chemicals | | | | | |
| $HNO_3$ | 126 | — | — | — | — |
| Gd | 23.8 | — | — | — | — |
| $PO_4^-$ | 4 | — | — | — | — |
| Fe | 3 | — | — | — | — |
| Cr | 0.5 | — | — | — | — |
| Ni | 0.2 | — | — | — | — |
| Na | 0.1 | — | — | — | — |
| Total Chem. | 157.6 | — | — | — | — |

### Table 1 (continued)
### TYPICAL CHARACTERISTICS OF LIQUID HIGH-LEVEL WASTE[1]

[a] Based on 378 ℓ/t initial uranium as metal.
[b] Years of decay following reprocessing 160 days after discharge.
[c] Total for all elements whether listed or not.

nonradioactive, and TRU or non-TRU, depending on their source. Thus, there are a large number of attribute combinations that can be used to describe these wastes. Of these many combinations, the attributes of radiation level and the TRU contamination have been used in this section, resulting in the following four waste types:

1. Intermediate-level TRU waste
2. Intermediate-level, non-TRU waste
3. Low-level TRU waste
4. Low-level, non-TRU waste

The other attributes will be discussed within this framework.

*1. Intermediate-Level TRU Waste*

Intermediate-level TRU wastes (ILW) are defined as those materials, other than high-level and fuel assembly structural material wastes, which contain long-lived alpha activities greater than 10 nCi/g of treated waste and have gamma radiation levels sufficient to require biological shielding and remote handling techniques. The TRU contamination in the ILW results from the presence of transuranic elements, which can only occur if the fuel material has been dissolved. The gamma radiation from the ILW, which typically yields surface dose rates ranging from 10 to 1000 mrem/hr, results from the presence of fission products and/or activated structural material corrosion products in the wastes. Thus, only wastes from fuel reprocessing plants can fall into this category.

As-produced ILW are comprised of both liquid and solid materials. The sources of liquid ILW in the reprocessing plant are

1. Fuel dissolver off-gas scrub liquors
2. Incinerator off-gas scrubber solutions
3. Decontamination solutions
4. Solvent (i.e., TBP) scrub solutions
5. HLW demister scrubber solutions
6. Laundry and laboratory wastes
7. Process cell sump wastes

The reprocessing plant liquid ILW is usually acid (nitric) in the as-produced state. It is expected that this acid waste will be neutralized with sodium hydroxide and then concentrated in an evaporator (if it is not combined with the HLW) resulting in the liquid ILW becoming a concentrated salt waste. The composition and characteristics of the neutralized ILW are given in Table 2.

The solid ILW from reprocessing plants consist of a diverse assortment of materials such as spent ion-exchange resins, silica gel, silver zeolite (containing iodine scrubbed from off-gases), combustible trash (gloves, paper, etc.), noncombustible trash (tools, glassware), failed equipment, and High Efficiency Particulate Air (HEPA) filters.

**Table 2**
**TYPICAL PROPERTIES OF INTERMEDIATE LEVEL WASTE SOLUTION AFTER NEUTRALIZATION[1]**

| | Volume[a] (g/ℓ) | Radioactivity (Ci/year) | Mass (kg/year) |
|---|---|---|---|
| Chemical composition | | | |
| $NO_3^-$ | 400 | | 66,000 |
| Na | 150 | | 24,750 |
| Hg | 15 | | 2,475 |
| Ca, Mn, $SO_4^-$, $Cl^-$ | 1 each | | 165 each |
| K, Fe, Ag, F | 0.5 each | | 82 each |
| Misc. chem.[b] | Minor | | |
| Radioactivity | | | |
| U | 20 | | 3,300 |
| Pu | 0.05 | | 8.25 |
| Fission products | Trace | 3,300 | |
| Physical properties | | | |
| Volume (ℓ/year) | 165,000 | | |
| Solids (kg/year) | 97,000 | | |
| Density (estimate) (g/ℓ) | 1.400 | | |
| Heat generation (W/ℓ) | 0.01—0.1 | | |

[a] Based on 110 ℓ/MTHM, 1500 MTHM/year.
[b] Oxalate, citrate, phosphate, acetate, ammonium, chromate, surfactants, detergents, etc.

*2. Intermediate-Level, Non-TRU Wastes*

Intermediate-level, non-TRU wastes are those materials which contain less than 10 nCi of long-lived alpha activity per gram of treated waste but still have gamma radiation levels sufficient to require biological shielding. The presence of the high radiation levels and the absence of TRU contamination means that the fuel is still within the cladding, but that some fission products and activated corrosion products are present. The fuel cycle facilities whose wastes fit this description are the nuclear reactors and the spent fuel storage facilities.

The ILW from a nuclear reactor is principally comprised of wastes resulting from the cleanup of the coolant water streams in the plant. The major source of liquid waste is the bottoms from evaporators used in cleaning up the coolant water. The major sources of solid wastes are filter cartridges, ion exchange resins, HEPA and charcoal filters, and assorted trash.

The ILW generated at spent fuel storage facilities, whether they be at the reactor, the reprocessing plant, or independently located, will be similar in type to those described for the nuclear reactor. This similarity results from the types of operations being performed in the two cases being nearly the same, viz., cleaning up water coolant that has been in contact with irradiated fuel assemblies.

*3. Low-Level TRU Wastes*

Low-level TRU wastes are defined as those materials that contain long-lived alpha activities greater than 10 nCi/g of treated waste but have external radiation levels low enough to permit handling without supplementary shielding (i.e., dose rates less than 10 mrem/hr). The TRU contamination means that the transuranic elements must be present outside of the fuel cladding. The low external radiation level means that the transuranics have been purified and the fission and activation product concentrations are virtually nil. These conditions will prevail in the portions of the fuel reprocessing plant where the purified plutonium nitrate is being converted to plutonium oxide in preparation for shipment and in the mixed-oxide (MOX) fuel fabrication facility where the plutonium oxide is fabricated into fuel elements.

The only source of low-level liquid TRU wastes is in the MOX fuel fabrication plant, where $HNO_3$-HF etch solutions, $Al(NO_3)_3$ solutions, and water rinses are contaminated with plutonium. In the likely event that a scrap recovery system is incorporated in the MOX fuel fabrication plant, there would also be an acidic low-level TRU waste resulting from the solvent-extraction operation in this system.

Low-level, solid TRU wastes are produced by both the fuel reprocessing plant and the MOX fuel fabrication plant. These materials are both combustible and noncombustible and are diverse in nature. Included in this category are paper, rubber, plastics, laboratory equipment, HEPA filters, rejected fuel cladding (from a MOX fuel fabrication plant), and ashes from waste combustion and uranium fluorination.

#### *4. Low-Level, Non-TRU Wastes*

Low-level, non-TRU wastes are those materials that contain less than 10 nCi of long-lived alpha contamination per gram of treated waste and have radiation levels low enough to permit handling without supplementary shielding. This means that these wastes have not been in contact with materials containing significant amounts of transuranic elements or fission products or that they have been treated to remove these species. The principal source of low-level, non-TRU wastes that do not result from the treatment of other wastes is the facility for conversion of relatively pure uranium to $UF_6$ in the reprocessing plant.

The principal source of liquid low-level, non-TRU wastes is the "bottoms" from the evaporator used in the evaporation of dilute aqueous wastes from the uranium and plutonium conversion operation in the fuel reprocessing plant. These bottoms contain small amounts of fission products and actinides along with a major fraction of the tritium that was initially present in the fuel. This waste will be considered further in Section II. D. where tritiated wastes are discussed.

A wide variety of solid, low-level, non-TRU wastes are generated by the various fuel cycle operations processing relatively pure uranium. These operations are uranium oxide-to-hexafluoride conversion, uranium isotope enrichment, fuel fabrication, and product purification and conversion in the reprocessing plant. The types of materials comprising these wastes are generally the same as those for solid, low-level, TRU wastes plus a $CaF_2$ sludge that results from the cleanup of fluorinator off-gases.

### C. Fuel Assembly Structural Material Wastes

Fuel assembly structural material wastes (FASM) consist principally of short segments (a few centimeters long) of fuel cladding with all but a trace of the spent oxide fuel dissolved from the interior with nitric acid. A photograph of these segments before the fuel has been dissolved is given in Figure 1. Also included in this waste type are the fuel assembly and pieces and plenum springs (usually stainless steel), fuel assembly grid spacers (Inconel® or Zircaloy), as well as the fuel channel in the case of boiling water reactor (BWR) fuels. All of these materials are radioactive due to the neutron activation of the metals during irradiation. This waste is generated by the head-end operations of spent fuel chopping and dissolution in the fuel reprocessing plant (except for the BWR fuel channel, which may be removed at the reactor or before chopping the fuel elements). The cladding is also expected to contain from 0.05 to 0.1% of the original spent oxide fuel which was not dissolved and may contain up to 50% of the

FIGURE 1. Chopped fuel element segments before dissolution.

tritium produced by the fuel, part in the form of $ZrT_2$ and part as a Zr-HT solution.

A small, but important, component of the fuel assembly structural material wastes is the Zircaloy "fines". These are very fine particles of Zircaloy that result from the mechanical shearing operation. Although small in mass, they are important because of their tendency to be pyrophoric.

### D. Wastes Containing Volatile Elements

There are four principal volatile elements that are produced in significant quantities in nuclear reactors and which have isotopes having half-lives long enough to require the recovery and isolation of the elements. These elements and the isotope of interest are iodine ($^{129}I$), tritium ($^{3}H$), carbon ($^{14}C$), and krypton ($^{85}Kr$).

#### *1. Iodine*

Iodine is a semivolatile fission product element. The isotope of interest for waste management purposes is $^{129}I$, which is a low-energy beta-emitter with a half-life of 15.9 million years. For most isotopes, this extraordinarily long half-life would result in such low specific radioactivity that the isotope would not be a problem. However, iodine is a special case since it tends to concentrate in the human thyroid gland. Thus, iodine must be recovered from the dissolver off-gas in the fuel reprocessing plant and treated as a waste while other materials with comparable half-lives need not be recovered because they do not concentrate in a specific organ.

Five different processes have been or are being developed for the removal of iodine from reprocessing plant off-gas streams. These are based on either (1) scrubbing with aqueous solutions or (2) trapping on silver-containing sorbents.

1. Scrubbing with aqueous solutions
   a. Iodox process — Scrubbing with 20 to 22 *M* $HNO_3$. The product is solid $HI_3O_8$, which can be converted to $Ba(IO_3)_2$ for disposal.
   b. Mercuric nitrate/nitric acid — Scrubbing with an aqueous solution of mercuric nitrate in nitric acid. The product is solid $Hg(IO_3)_2$, which can be converted to $Na(IO_3)_2$ for disposal.
   c. Caustic — Scrubbing with KOH solution. This process also removes $CO_2$ from the off-gas. The product is solid $K_2CO_3$-KI-$KIO_3$-KOH.
2. Trapping on silver-containing sorbents
   a. Silver-exchanged zeolites — Sorption on zeolites in which silver has replaced the sodium normally present in the crystal. The product is the aluminosilicate zeolite plus the sorbed iodine and silver.
   b. Silver nitrate-impregnated silicic acid — Sorption on silicic acid loaded with silver nitrate, which is the reactive material. The product is the silicic acid plus the sorbed iodine and silver compounds.

The iodine in the reprocessing plant off-gases is in two distinctly different forms: elemental iodine (i.e., $I_2$) and organic iodides (e.g., methyl iodide). All of the processes listed above are effective with elemental iodine. However, only Iodox and trapping on silver-containing sorbents are totally effective on organic iodides. Mercuric nitrate/nitric acid scrubbing is somewhat effective on organic iodides if a high enough nitric acid concentration (about 8 to 12 molar) is employed.

#### *2. Tritium*

Tritium is a volatile (when elemental) or semivolatile (as tritiated water) hydrogen isotope produced in several ways: as a fission product, as a low-energy neutron activation product of any lithium impurities in the fuel, and as a high-energy neutron activation product of many structural materials. Tritium is also produced in the coolant water of nuclear reactors by the neutron activation of naturally occurring deuterium ($^2H$). Tritium is a low-energy beta emitter with a half-life of 12.3 years and is of concern because, if released, it would be present in the biosphere as tritiated water.

The recovery of tritium in the spent fuel in a form other than very dilute tritiated water involves either removing the tritium from the fuel before any contact with water or using advanced isotope separation techniques. A process being developed for removing the tritium from the fuel before contact with water is called voloxidation. This process involves heating the chopped fuel element segments in air to 400 to 500°C before dissolution. This heating converts the high-density $UO_2$ to lower-density $U_3O_8$ with an attendant phase change, resulting in the disintegration of the fuel material and the release of most of the tritium not combined with the cladding. This process, after catalytic oxidation of the elemental tritium, will result in a relatively concentrated stream of HTO. The concentrated tritiated water vapor can then be removed by refrigeration or by molecular sieves. Advanced isotope separation techniques might be used to recover tritium from very dilute streams. Processes that have been proposed include $HT/H_2O$ catalytic exchange, laser enrichment using selective exitation of HTO, and enrichment by reversible electrolysis. All of these promise to be relatively expensive because of the large volumes of dilute tritiated water that must be processed. If the reprocessing plant tritium is not recovered by voloxidation of the fuel, the most likely alternative is to recover the tritium as a low-level, non-TRU aqueous waste as described in Section II.B.4.

The tritium in the reactor coolant water is produced in an already dilute form. Thus, it can only be recovered using isotopic enrichment techniques or treated as a low-level, non-TRU waste, with the latter alternative being the most probable by far.

### *3. Carbon*

Carbon, since it is generally present as carbon dioxide, may be treated as a volatile species which is produced in the fuel principally by the activation of nitrogen impurities. The isotope of interest is $^{14}C$, which is a low-energy beta emitter with a half-life of 5730 years. This isotope is important because of the importance of carbonaceous compounds in the food cycle.

Processes that have been proposed for recovering the $^{14}C$-contaminated carbon dioxide from the reprocessing plant off-gases include caustic scrubbing with sodium or calcium hydroxide, sorption on molecular sieves, and absorption in liquid fluorocarbon. All of the processes will produce calcium carbonate as the final product because a calcium hydroxide scrub cycle is used to convert the carbon dioxide recovered by other means to the carbonate. It should be noted that virtually all of the carbon recovered in these processes is naturally occurring, nonradioactive $^{12,13}C$ present in the process air used in the reprocessing plant.

### *4. Krypton*

Krypton is a volatile noble gas produced as a fission product in the fuel. The isotope of interest is $^{85}Kr$, which is a beta-gamma emitter with a half-life of 10.7 years. Krypton-85 is significant because of the radiation to the external surface of the body that would result if it were in the atmosphere.

Since krypton is not chemically reactive under most circumstances, its recovery depends on physical processes. Processes that can recover the krypton are cryogenic distillation, selective sorption, and diffusion through permselective membranes. Cryogenic distillation results in a relatively concentrated krypton product that is combined with some stable fission product xenon. The krypton product is about 6% $^{85}Kr$. Selective sorption processes include both adsorption on activated carbon or molecular sieves and absorption in fluorocarbon liquids. The fluorocarbon adsorption process is the same as that used to recover $^{14}C$ (see Section II.D.3). Unless it is purified by cryogenic distillation, the krypton product will also have the stable fission product xenon combined with it, resulting in a $^{85}Kr$ concentration of about 0.3%. Diffusion is not being seriously considered at this time.

## E. Uranium Mill Tailings

Uranium ores being mined today typically contain between 0.15 and 0.2 wt % uranium, the remainder consisting of a variety of other minerals. There are two principal milling processes for the recovery of uranium from its ores: the acid leach processes and the alkaline leach process. These processes are described in more detail in Chapter 2, Section III.D.

Briefly, the acid leach processes involve the dissolution of the uranium and some other constituents of the ore in sulfuric acid followed by the solvent extraction of the uranium from the acid using an amine-kerosene-alcohol mixture or, in some mills, anion exchange of the uranyl sulfate complex. The alkaline leach process involves the selective solubilization of the uranium in the ore using a sodium carbonate solution. However, in either case, the net result is a large volume of waste material because of the low concentration of uranium in the ore. This waste material, called mill tailings, contains the undissolved sand and slimes and the process liquids containing the solubilized minerals along with process chemicals. This waste material is discharged to a retention area called a tailings pond. It is here that the liquids are allowed to evaporate, eventually resulting in a dry waste material very much like the original ore except that (1) most of the uranium is absent, (2) the sulfate or carbonate concentration is increased, (3) the material is much more porous than the original ore, and (4) it is in a more exposed condition. After evaporation, the tailings may be moved to a central area where they are placed in a large tailings pile.

The mill tailings contain most of the radioactivity present in the original ore, which was principally due to the decay daughters of uranium. Since nearly all of the uranium has been removed from the tailings, the long-lived parent controlling the rate of decay becomes $^{230}Th$ with a half-life of 83,000 years. The radioactivity of this material is typically about 3 nCi/g of tailings. This very low level of radioactivity would not ordinarily be of concern, except for the fact that two of the daughters of $^{230}Th$ are the gaseous $^{222}Rn$ and the soluble $^{226}Ra$. The porosity of the tailings piles allows a significant fraction of the $^{222}Rn$ (half-life = 3.8 days) to escape into the atmosphere, where it could be inhaled by the general population. The $^{226}Ra$ could be solubilized by rain and eventually distributed in the biosphere where it may be ingested by the general population. Of these two radioisotopes, the $^{222}Rn$ seems to be of greater concern.

## F. Spent Fuel

The radioactive isotopes that would have been in the other waste types (i.e., high-level waste, intermediate-level and low-level TRU wastes, fuel assembly structural material waste, and the wastes containing volatile elements) all remain in the spent fuel if reprocessing does not occur.

Typical physical characteristics of PWR and BWR fuel elements are given in Table 3. The characteristics most relevant to spent fuel disposal are the total assembly weight (BWR = 276 kg and PWR = 658 kg), and the overall assembly length (BWR = 4.47 m and PWR = 4.06 m). A picture of a typical PWR fuel assembly is shown in Figure 21, Chapter 2.

The radioactivity characteristics of the spent fuel are also significantly different from those of the high-level waste which results from fuel reprocessing. These differences

Table 3
TYPICAL PHYSICAL CHARACTERISTICS OF UNIRRADIATED LWR FUEL ASSEMBLIES

| Characteristics | BWR | PWR |
|---|---|---|
| Overall assembly length (m) | 4.470 | 4.059 |
| Cross section (cm) | 13.9 × 13.9 | 21.4 × 21.4 |
| Fuel element length (m) | 4.064 | 3.851 |
| Active fuel height (m) | 3.759 | 3.658 |
| Fuel element OD (cm) | 1.252 | 0.950 |
| Fuel element array | 8 × 8 | 17 × 17 |
| Fuel rods per assembly | 63 | 264 |
| Assembly total weight (kg) | 275.7 | 657.9 |
| Uranium/assembly (kg) | 183.3 | 461.4 |
| $UO_2$/assembly (kg) | 208.0 | 523.4 |
| Zircaloy/assembly (kg) | 56.9[a] | 108.4[b] |
| Hardware/assembly (kg) | 9.77[c] | 26.1[d] |
| Total structural metal/assembly (kg) | 67.7 | 134.5 |

[a] Includes Zircaloy fuel-element spacers.
[b] Includes Zircaloy control-rod guide thimbles.
[c] Includes SS tie-plates and Inconel® springs.
[d] Includes SS nozzles and Inconel®-718 grids.

result principally from the fact that all of the plutonium is still present in the spent fuel, whereas less than 0.5% is left in the high-level waste. The magnitude of these differences is shown in Figure 2 for three characteristic parameters: the thermal power, the activity, and the ingestion toxicity. The thermal power, typically measured in watts, is the amount of decay heat being generated by the waste. The activity, typically measured in curies, is the rate at which radioactive decay events are occurring in the waste. The ingestion toxicity, typically measured in cubic meters of water, is the total amount of water required to dilute the radioactive isotopes in the waste to the maximum permissible concentrations for unrestricted consumption as given in Title 10, Part 20 of the U.S. Code of Federal Regulations. Thus, a more toxic waste would require a larger volume of water to dilute it to safe levels. This does not imply that these wastes would be disposed of in this manner; it is simply a measure of one of the relative biological effects of the waste.

From Figure 2, it is clear that the spent fuel generates more heat, is more toxic, and is more radioactive than the high-level waste. The thermal power and toxicity are a factor of 15 to 30 higher for the spent fuel for decay times longer than 1000 years while the activity is a factor of 3 to 10 higher for the spent fuel. These increases are almost totally due to the 200-fold greater amount of plutonium in the spent fuel as compared to that in the high-level waste. The effects of the plutonium are nil at times less than 100 years because of the dominance of the fission products (principally $^{90}Sr$ and $^{137}Cs$) at the shorter decay times. The effects of the plutonium on the thermal power and activity ratios decrease at longer times because of the more rapid decay of the plutonium in the spent fuel as compared to the long-lived $^{237}Np$ which is equally present in both spent fuel and HLW.

### G. Decontamination and Decommissioning Wastes

Decontamination and decommissioning (D&D) wastes result when an obsolete nuclear facility ceases operation and some effort is expended to remove the radioactivity from the facility. The types and amounts of wastes resulting from D&D operations

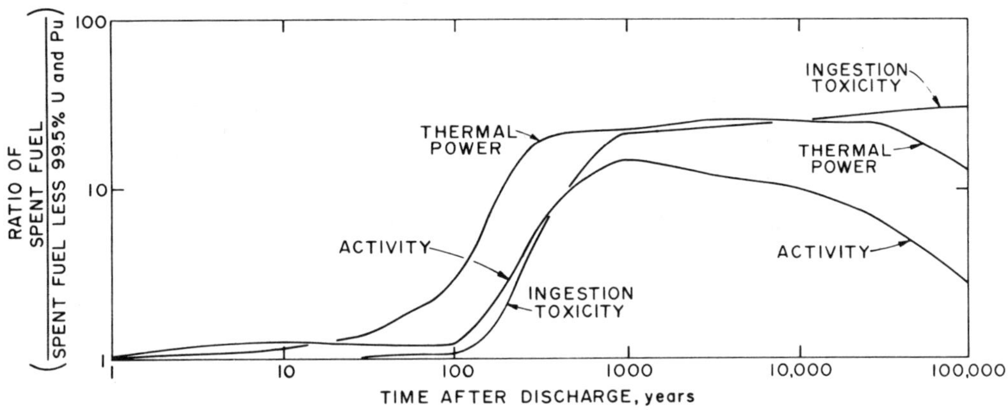


FIGURE 2. Comparison of the characteristic parameters of spent fuel and high-level waste.

are highly variable, depending on the type of facility and the level of D&D. The level of D&D can range from partial cleanup and layaway with restricted access to complete decontamination and dismantling, resulting in the site being returned to unrestricted access. Although the exact nature of D&D wastes is highly variable, they are generally comprised of a wide assortment of intermediate-level and low-level wastes, both TRU and non-TRU. The principal components of the D&D wastes are liquid decontamination solutions and whole or dismantled process equipment with surface contamination and/or components made radioactive by neutron irradiation. In any case, the D&D wastes are simply another source of ILW and LLW and can be treated and disposed of in the same manner. Therefore, the D&D wastes will not be considered separately in the remainder of this section.

## III. TREATMENT OF RADIOACTIVE WASTES

The treatment of radioactive wastes is concerned with the processing of the wastes to (1) make them easier to handle and less dispersible during transportation and disposal and (2) less mobile after disposal. In most cases, the treatment involved is to immobilize the waste in a monolithic waste form, such as glass, cement, etc. This section is concerned with the identification and description of the treatment options available for each of the waste types described in Section II.

It should be noted that the treatment of radioactive wastes usually results in the volume of the final wastes being larger than that of the waste being treated. The treatment processes are used because they give the waste better physical characteristics, as noted above, and because it is often possible to concentrate most of the radioactivity in a relatively small fraction of the treated wastes which is easier and less expensive to handle.

### A. High-Level Waste

The treatment of high-level waste is unquestionably the most heavily investigated topic in the radioactive waste management area. Although the disposal of untreated high-level waste by several methods has been proposed (see Section IV.D), the most likely event is that the high-level waste will be immobilized in a solid form resistant to the relatively high heat generation rate of the waste and suitable for storage, transportation, and disposal. The type of treatment that has been favored for years is calcination of the liquid high-level waste to a stable solid oxide material followed by vitrification of the oxide to form a glass matrix. Other high-level waste treatment alternatives currently being investigated include "supercalcines" and metal matrices.

As was noted above, the purpose of the treatment of the waste is to make it more amenable to handling during disposal operations and to limit the mobility of the radioactive species after disposal. However, it should be recognized that waste treatment cannot reasonably be expected, nor is it intended, to immobilize the radioactive waste for the tens to hundreds of thousands of years that the high-level wastes (and other TRU wastes) must remain isolated from the environment. Isolation of the wastes from the environment for long times will depend mainly on geologic isolation. The tremendous dilution capacity of the biosphere may be relied on to ensure that no harm comes from any small releases from the disposal site after very long decay times. This aspect of waste management will be considered further in Section IV.

In general, each of the proposed waste treatment processes involves four major operations: feed preparation, waste immobilization, immobilized waste packaging, and treatment of secondary effluents. Feed preparation generally involves evaporation of the liquid HLW to a volume of about 400 ℓ/MTIHM (if this had not been done previously). There are a number of different immobilization processes which will be discussed in more detail subsequently. The immobilized waste may then be packaged in metal cylinders 40 to 60 cm in diameter and 3.0 to 4.5 m long to prevent the spread of surface radioactivity, and to facilitate handling. The treatment of the secondary effluents from immobilization of the liquid HLW may include nitric acid concentration and recovery, scrubbing of particulate matter from off-gases, and removal of semivolatile iodine and possibly ruthenium compounds from the off-gas.

A summary of liquid HLW treatment processes and their development status is given in Table 4. The preferred method for treating high-level wastes is drying and calcination of the liquid HLW followed by vitrification (making into a glass-like material) of the HLW calcine. There are, however, alternative immobilization processes listed in Table 4 which have been proposed and are under investigation. The following sections briefly describe each of these processes for treating high-level wastes.

### *1. Drying and Calcination Processes*

As shown in Table 4, there are five different drying and calcination processes. The net result of all of these processes is to convert the concentrated liquid HLW into a dry oxide powder. A typical composition of calcined HLW is given in Table 5.

A schematic diagram of the fluidized-bed calcination system is given in Figure 3. The liquid HLW, along with atomizing air, oxygen, kerosene, and fluidizing air, are injected into a chamber where the liquids are evaporated and the resulting solids suspended in the air moving through the chamber until they grow large enough to sink to the bottom. There they are removed to a canister where they are baked to eliminate residual nitrates.

Pot calcination is a relatively simple process involving injection of the liquid HLW into a metal cylinder where it is heated to 900°C. The liquids are vaporized and the oxides remain in the cylinder. Liquid HLW is continuously fed into the cylinder until it contains the desired amount of calcined product.

The rotary kiln calciner was developed by the French specifically for close-coupling to a continuous melter to convert the calcine to glass (see the next section). The calciner is a slightly sloped, externally rotated cylinder that is heated externally. The cylinder has a slight slope, the waste is injected into the high end and, by the time it has moved to the lower end, the result is a finely divided oxide powder. The product is collected in a canister which is baked at 900°C to remove residual nitrates.

### Table 4
### DEVELOPMENT STATUS OF HLW TREATMENT PROCESSES[1]

| Treatment category | Ready for demonstration[a] | Requires further development | |
|---|---|---|---|
| | | Short-term[b] | Long-term[c] |
| Drying and calcination | Fluidized-bed<br>Pot calcination<br>Rotary kiln<br>Spray calcination<br>British rising level glass | Wiped/scraped film evaporator | |
| Vitrification | In-can melting | German spray calciner-continuous melter | |
| | French rotary kiln-continuous melter[d] | Continuous ceramic melter | |
| Other immobilization processes | | | Supercalcine<br>Metal matrix<br>Sintering<br>Glass ceramic<br>Coated particles<br>Inorganic ion exchange |

[a] Less than 1 year to be ready for engineering detail design.
[b] Less than 5 years to be ready for engineering detail design.
[c] Greater than 5 years to be ready for engineering detail design.
[d] Construction of a plant which will use this process to vitrify the HLW from the Marcoule reprocessing plant is nearing completion.

The spray calcination process is very similar to the fluidized-bed calcination process. The principal difference is that the spray calciner simply atomizes the liquid HLW into a heated chamber where it is calcined and drops into the storage canister and the nitrate bakeout furnace. Fluidizing air is not used in this process.

The wiped/scraped film evaporator process is similar to the pot calcination process, wherein the calcination takes place in the storage canister. The difference between the two is that this process uses an evaporator with a wiper rotating at 1600 rpm to concentrate the incoming HLW to a thick, flowable slurry or a free-flowing solid before the high-temperature calcination.

*2. Vitrification Processes*

Vitrification processes are used to convert HLW into a monolithic glass product. Research has shown that the best type for waste immobilization is borosilicate glass. The composition of a typical borosilicate glass containing the calcined HLW composition given in Table 5 is given in Table 6. A photograph of glass prepared from HLW is shown in Figure 4.

The most heavily investigated vitrification process in the U.S. is the in-can melting process. This process involves flowing a continuous stream of mixed calcined waste and glass frit into a storage canister heated to 1000 to 1100°C. When cooled, the result is a monolithic glass product containing the HLW. This process could be fed with calcined HLW taken from a large storage bin. However, the preferred approach is to directly couple the in-can melting vitrification process with a continuous calcination process such as the fluidized-bed or spray calciners or the wiped/scraped-film evaporator. The direct coupling eliminates the need for the nitrate bakeout step in these

Table 5
TYPICAL COMPOSITION OF CALCINED HLW[2]

| Category | Compound | kg/MTIHM | Category | Compound | kg/MTIHM |
|---|---|---|---|---|---|
| Fission Products | $Rb_2O$ | 0.354 | Inerts | $Na_2O$ | — |
| | $SrO$ | 1.059 | | $Fe_2O_3$ | 1.511 |
| | $Y_2O_3$ | 0.598 | | $Cr_2O_3$ | 0.345 |
| | $ZrO_2$ | 4.944 | | $NiO$ | 0.141 |
| | $MoO_3$ | 5.176 | | $P_2O_5$ | 0.672 |
| | $Tc_2O_7$ | 1.291 | | $MnO_2$ | — |
| | $RuO_2$ | 2.972 | | $Gd_2O_3$ | — |
| | $Rh_2O_3$ | 0.480 | | Zircaloy-2 Fines ($ZrO_2$) | — |
| | $PdO$ | 1.483 | | | |
| | $Ag_2O$ | 0.088 | | | |
| | $CdO$ | 0.097 | Actinides | $U_3O_8$ | 1.169 |
| | $TeO_2$ | 0.725 | | $NpO_2$ | 0.865 |
| | $Cs_2O$ | 2.880 | | $PuO_2$ | 0.010 |
| | $BaO$ | 1.567 | | $Am_2O_3$ | 0.181 |
| | $La_2O_3$ | 1.480 | | $Cm_2O_3$ | 0.040 |
| | $CeO_2$ | 3.323 | | | |
| | $Pr_6O_{11}$ | 1.482 | | | |
| | $Nd_2O_3$ | 4.522 | | | |
| | $Pm_2O_3$ | 0.123 | | | |
| | $Sm_2O_3$ | 0.924 | | | |
| | $Eu_2O_3$ | 0.200 | | | |
| | $Gd_2O_3$ | 0.137 | Total | | 40.84 |

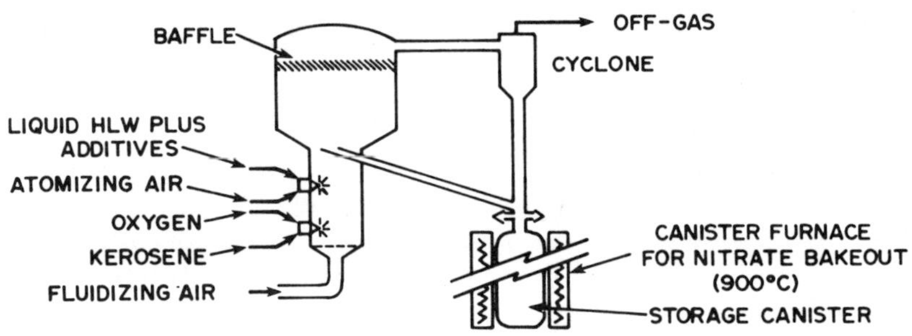


FIGURE 3. Schematic diagram of fluidized-bed calcination system.[1]

calcination processes. The vitrification portion of this operation is semicontinuous, since it is necessary to stop vitrification in order to remove filled storage canisters and insert empty canisters.

The French have developed a continuous vitrification process which will be attached to the continuous rotary kiln calciner described previously. The continuous glass making process involves the use of an induction-heated continuous melter in which the glass and calcined HLW are mixed and melted. The storage canisters are filled from the melter through an induction-heated freeze valve in the bottom of the melter.

The British have developed a vitrification process, called rising-level glass, whch is analogous to the pot calcination process. In the British process, liquid HLW containing glass frit is semicontinuously fed to a storage canister heated to about 1000°C. The HLW is dried, calcined, and vitrified all in a single operation. This process eliminates the need for preliminary treatment steps such as drying or calcination.

**Table 6**
**TYPICAL BOROSILICATE GLASS COMPOSITION**[2]

| Constituent | Weight % |
|---|---|
| $SiO_2$ | 27.3 |
| $B_2O_3$ | 11.1 |
| $Na_2O$ | 4.0 |
| $K_2O$ | 4.0 |
| ZnO | 21.3 |
| CaO | 1.5 |
| MgO | 1.5 |
| SrO | 1.5 |
| BaO | 1.5 |
| Waste[a] | 26.3 |

[a] See Table 5 for waste composition.

The Germans, like the French, are developing a close-coupled calciner-vitrifier process. This process involves the use of a spray calciner as described in Section III.A.1. with a continuous vitrification process similar to that being developed by the French.

Another process being developed in the U.S. is the continuous ceramic melter. In this process, liquid HLW, preconcentrated HLW (i.e., by a wiped-film evaporator), or calcined HLW is mixed with glass frit and continuously charged to a ceramic-lined melter that is electrically heated. The liquid glass/waste product is withdrawn semicontinuously into a storage canister by changing the level of one end of the melter using a screw jack.

*3. Other HLW Immobilization Processes*

Alternative processes to calcination and vitrification of HLW are being studied. Some of these processes are substitutes for standard calcination, and some for vitrification. However, as noted in Table 4, all of these treatment processes will require substantial development before they are available for demonstration.

***a. Supercalcine***

The ''supercalcine'' process is designed to produce a calcined HLW product wherein all of the long-lived waste constituents are in a thermodynamically stable crystalline phase. A standard calcine is not necessarily thermodynamically stable with respect to the disposal medium since it is formed from the materials in the liquid HLW with essentially no modifications. The formation of a supercalcine is accomplished by the addition of materials such as nitrate solutions of calcium, strontium, and aluminum and a stabilized colloidal suspension of silica in amounts dependent on the composition of the liquid HLW. Cesium, which forms some of the most water-soluble compounds of all the fission product elements, is fixed as $CsAlSi_2O_6$, the mineral in which it is found in nature. The trivalent rare earth elements (RE = Nd, La, Pr, Sm, Pm, Eu, and Gd) can be incorporated into a structure similar to that of the mineral apatite: $Sr_{2+x}\,RE_{8-x}\,(SiO_4)_{6-x}(PO_4)_xO_2$. Some of the actinide elements (e.g., americium and curium) may also be incorporated into this type of structure since their chemistry is very similar to that of several trivalent lanthanides. Other stable crystalline phases that are formed include scheelite type [$(Sr,Ca,Ba)MoO_4$], fluorite type [$(Ce,Zr)O_2$], perovskite type [$(Ba,Sr)ZrO_3$], and spinel type [$Ni(Fe,Cr)_2O_4$]. About 35 to 40 wt % of additives is required to form a supercalcine of the HLW. To prepare the supercalcine the additives are mixed with the liquid HLW and the mixture is calcined using a spray calciner or fluidized-bed calciner as described previously. After this step the calcine is granular and largely amorphous and must be heated to 900 to 1100°C and cooled slowly to form the stable crystalline phases. As with other calcines, this product will most likely be consolidated into a monolithic waste form using glass, some other matrix material, or by coating the supercalcine material. The latter two consolidation processes are described later in this section.

#### *b. Metal Matrix*

The metal matrix process is proposed as a substitute for the vitrification of particulate HLW materials such as calcines and supercalcines. The process can be accomplished (1) by mixing the calcine with molten metal or (2) by mixing the material with metal powder, pressing the mixture, and sintering the compact. In either case, the result is that the calcines are dispersed in the continuous metal matrix. Metals that have been investigated for the molten metal variation include lead, zinc, and aluminum. In the sintering variation of this process, iron, iron alloys, or copper have been studied. The HLW can constitute up to 70 vol % of the final product.

#### *c. Sintering*

The sintering HLW treatment process begins with a calcined product. This product is then mixed with a fluxing material and a binder-lubricant and sintered into a large disk at 970 to 1070°C or mixed with a binder and glass frit, formed into pellets, and sintered at 850 to 1000°C. The disks can be packaged directly or consolidated in a matrix. The pellets must be consolidated in a matrix.

#### *d. Glass Ceramic*

The glass ceramic HLW process also begins with a calcined product. The product is vitrified and loaded into small, thin-wall containers which are then subjected to a carefully controlled thermal treatment at 500 to 800°C. This thermal treatment results in the crystallization of some of the glass, thus reducing its susceptibility to uncontrolled devitrification. The crystallized "glass" would then be consolidated with a metal matrix.

#### *e. Coated Particles*

The coated particle process involves the coating of calcined, supercalcined, or sintered HLW in nonradioactive, leach-resistant material. Proposed coating materials include enameled clay, SiC over pyrolytic carbon, $SiO_2$, and $Al_2O_3$. The coated product could then be formed into a monolithic solid using the metal matrix process.

#### *f. Inorganic Ion Exchange*

In the inorganic ion exchange process the liquid HLW is neutralized to pH 1 and then flowed through the inorganic ion exchanger sodium titanate ($Na_2Ti_3O_7$). This ion exchange material removes nearly all of the chemical constituents of the waste except ruthenium, cesium, and technetium, which are removed either by zeolites (cesium) or by anion exchange (technetium and ruthenium). The loaded titanate and zeolite media are mixed, dried, pressed, and sintered at about 1100°C.

### *4. Comparison of Treated HLW Products*

The treated HLW products are compared in Table 7. Unreprocessed spent fuel is

FIGURE 4. Typical vitrified high-level waste product. Radioactive glass prepared from TBP-25 aluminum waste.

also included in this table since its characteristics are most smilar to that of the HLW (see Section II.F). The treatment of unreprocessed spent fuel will be considered further in Section III.F.

The calcine, the coated particle, and the sintered waste products characterized in Table 7 must be consolidated into a monolithic waste form because as produced they are in a dispersible form and have high specific surface areas, which could result in the radioactive constituents of the waste being leached too quickly.

The leach rates, thermal conductivities, surface areas, densities, specific volumes, waste oxide contents, forms, and specific areas of all of the monolithic HLW product forms appear to be very similar to each other considering the variability of these parameters for a specific waste form. This would seem to indicate that the HLW treatment process should be judged on the basis of additional development required, cost, and operational simplicity.

The leach rates given in Table 7 deserve more discussion because they are a very important characteristic of the waste and their implications may be somewhat startling. The units commonly used for leach rate (i.e., g $cm^{-2}$ $day^{-1}$) are somewhat unfortunate since this parameter depends on the element and waste form being considered as well as the characteristics of the leachant. The leach rate ranges given include all elements being leached by water under either ambient or repository conditions. Individual elements could have leach rates that differ by two or three orders of magnitude and still fall within the range. For example, cesium and strontium generally have relatively high leach rates because they are relatively soluble. On the other hand, the rare earths and actinides tend to have relatively low leach rates. Furthermore, the leach rate varies substantially during the early stages of leaching because of the rapid depletion of the elements in near-surface areas. The initially diffusion-controlled leach rate can decrease by a factor of 10 or more until it reaches a nearly constant value that is thought to be controlled by the rate at which the waste form itself is attacked. Thus, assuming that the waste form is in contact with water moving rapidly enough so that significant element concentrations do not build up in the water, the rate at which elements are leached from the waste form appears to be constant at long times.

It is interesting to use a "typical" leach rate value to estimate the time that would be required for a particular element to be completely leached from a typical glass form prepared from HLW. A typical long-term leach rate for plutonium from borosilicate glass at ambient conditions is $2 \times 10^{-6}$ g $cm^{-2}$ $day^{-1}$. Multiplying this by the surface-to-volume ratio of a typical HLW glass form (assuming some fracturing) of 0.5 $cm^{-1}$ and dividing the glass density (about 3 $g/cm^3$) by this product yields a leach time of about 8200 years. This time could be substantially shortened by either further fracturing (i.e., increasing the surface area) the waste form or by increasing the temperature. It could be lengthened substantially if the water outside the canister were moving slowly enough so that the concentrations of the elements in the waste form built up significantly and retarded deterioration of the waste form. In any case, it is clear that if the waste form is contacted with moving water, all of the elements would be leached out within a time period that is relatively short compared to geologic times. As noted previously, the primary purpose of the waste form is to consolidate the wastes into a form that would be safe during handling and disposal and to immobilize the radioactive elements during the first several hundred years when the decay heat (and the radiotoxicity) would be the highest. Long-term waste isolation is provided by the waste repository medium, the repository's geospheric environment, and dilution in the biosphere.

### B. Intermediate-Level and Low-Level Wastes

In general, the treatment of intermediate-level wastes is exactly the same as that for

# Table 7
## COMPARISON OF HLW TREATMENT PRODUCTS AND SPENT FUEL[3]

| Property | Units | Calcine | Borosilicate glass | Supercalcine ceramic[a] | Metal matrix | Sintered product | Glass ceramic | Coated particles | Inorganic ion exchange | Spent fuel |
|---|---|---|---|---|---|---|---|---|---|---|
| Radioactivity leach rate | | | | | | | | | | |
| Ambient conditions | $g/cm^2$-day | $10^{-1} - 10^{-3}$ | $10^{-4} - 10^{-7}$ | $10^{-7} - 10^{-9}$ | $10^{-4} - 10^{-7}$ | $10^{-4} - 10^{-7}$ | $10^{-5} - 10^{-7}$ | $10^{-4} - 10^{-6}$ | $10^{-4} - 10^{-5}$ | $10^{-4} - 10^{-7}$ |
| Repository conditions[b] | | | $7 \times 10^{-4}$ | $4 \times 10^{-4}$ | $4 \times 10^{-5}$ | $4 \times 10^{-5}$ | | | | $2 \times 10^{-4}$ |
| Specific volume | $m^3$/MTIHM | 0.03 — 0.06 | 0.04 — 0.1 | 0.07 | 0.08 | 0.03 — 0.06 | 0.08 | 0.13 | 0.03 | 0.004 — 0.005 |
| Typical waste oxide content | wt% | ≤100 | 20—35 | 45 | 33 | 45 | 25 | 45 | 25 | 4 |
| Form | | Powder | Monolithic | Monolithic | Monolithic | Pellets | Monolithic | Beads | Monolithic | Pellets in fuel cladding |
| Density | $g/cm^3$ | 1.0 — 2.4 | 3.0 — 3.6 | 3.5 — 4.0 | 2.4 — 5.5 | 24 — 3.3 | 2.9 | 1.7 | ≤4.5 | 10.0 — 10.5[c] |
| Thermal conductivity | W/m K | 0.2 — 0.3 | 0.9 — 1.3 | 0.6 | 5 — 35 | 0.7 | 2.2 | 17 | 1.2 | 2 — 6 |
| Specific area | $m^2$/kg | $10^2 - 10^4$ | 0.005 — 0.05 | 0.005 — 0.05 | 0.005 — 0.05 | 0.1 — 1.0 | 0.005 — 0.05 | $10^2 - 10^4$ | 0.005 — 0.05 | 0.05 — 0.07 |
| Advantages | | Simple | Low leach rate | Low leach rate | Very low leach rate | Low leach rate | Low leach rate | Low leach rate | None | Low leach rate |
| | | | Well developed | High waste content | High thermal conductivity | High waste content | | | | Few secondary wastes |
| Disadvantages | | Very high leach rate | Can devitrify | Complex additives procedure | Not well developed | Not well developed | Not well developed | Not well developed | Not well developed | Contains volatile and semivolatile materials and plutonium |
| | | Very high surface area | | Not well developed | | High surface area | Complex | Very high surface area | Complex | Awkward shape |

[a] After consolidation.
[b] 250°C, 1000 psi.
[c] Pellet density, not bulk density.

low-level wastes and depends on the chemical and physical characteristics of the waste under consideration. There are generally two aspects of ILW/LLW treatment to be considered: volume reduction and immobilization. The relatively large volume of low-density materials, such as cellulosics and laboratory glassware, makes it desirable and cost-effective to reduce the volume of some of the waste by various techniques, depending on the waste characteristics. Then, many of the wastes, particularly those that are readily dispersible, are immobilized to facilitate handling and to retard leaching in the intermediate term after disposal. It should be noted that even though low-level TRU and low-level non-TRU wastes may be treated identically, their differing isotopic compositions could require that the technology used for their disposal be substantially different. This point will be considered further in Section IV.E.

### *1. Volume Reduction of ILW/LLW*

When considering the volume reduction of ILW/LLW it is useful to group these wastes into the following three categories: noncombustible solid wastes, noncombustible liquid wastes, and combustible wastes. Within each of these categories the available volume reduction options are essentially the same for wastes from different sources.

#### ***a. Noncombustible Solid Wastes***

The types of material included in this category are glove boxes (air-tight enclosures for handling radiotoxic, low-activity materials such as plutonium), machinery, piping, glassware, HEPA filters, and miscellaneous trash. The volume reduction options available for these wastes are (1) no reduction, (2) mechanical disassembly, (3) crushing, (4) melting, and (5) dissolution.

Before volume reduction, some of the TRU materials may be pretreated to recover the small amounts of plutonium contained therein. Since the plutonium is normally present as the oxide, which is relatively difficult to dissolve, highly corrosive chemicals are generally required. One common agent for removing plutonium surface contamination is the acid mixture 10 $M\,HNO_3$-0.1 $M\,HF$, the HF being required as a catalyst. Physical techniques such as ultrasonic cleaning, ball-milling, and high-pressure cleaning have also been used. Anodic dissolution of the contaminated surface area, called electropolishing, has also been investigated.

The "no reduction" option is generally used for materials where no significant volume reduction can be expected by further treatment. Massive pieces of metal and floor sweepings might be two examples of such materials.

The mechanical disassembly option involves dismantling failed equipment to reduce its volume or to facilitate the handling of a large piece of equipment. A variety of standard techniques, such as hand tools, cutting torches, saws, and shears, are used, although the disassembly may have to be performed remotely (i.e., behind shielding) if the equipment is too radioactive. Mechanical disassembly may reduce the overall waste volume by about 20%.

Crushing (or compaction) is a very commonly used volume reduction technique for materials such as glassware and HEPA filters. Compaction is also used quite often for combustible solid materials (e.g., paper trash) if these materials are not to be incinerated. Overall, crushing can be expected to reduce the preimmobilization waste volume by about 60%.

The melting option consists of melting the disassembled metallic wastes such as failed equipment and piping and casting the metal into ingots. This option is in the development stage and would result in about a 90% reduction in the volume of these wastes.

The dissolution option involves the chemical reaction of the waste with a molten sodium carbonate solution and air to oxidize the wastes. The resulting oxides will be in solution with the sodium carbonate solvent. This process is still under development and would result in a volume reduction of about 80%.

### *b. Noncombustible Liquid Wastes*

The types of materials included in this waste category are generally aqueous wastes containing varying amounts of chemical salts and radioactive isotopes. The volume reduction options that have been proposed for these wastes are (1) evaporation, (2) calcination, (3) concentration using membranes, (4) filtration and centrifugation, and (5) ion exchange.

Evaporation is widely used in reactor plants to reduce the volume of the aqueous ILW/LLW streams. Although conventional evaporator configurations are generally used, they are designed to leave the maximum amount of radioactivity in the bottoms as opposed to minimizing the amount of heat used per unit evaporative capacity. Volume reductions resulting from evaporation can be as high as 90%, depending on the solute concentration in the initial waste. Evaporation can also be used to concentrate the wastes such that they can be combined with a more concentrated waste stream. For example, it was noted previously that the ILW could be combined with the HLW. Concentrated LLW can be combined with ILW streams. Evaporation can also be used to prepare a lower-volume feed stream for other ILW/LLW volume reduction processes. The vapor from a properly designed LLW evaporator has a sufficiently low radioactivity concentration that it can be released to the environment. The condensate from the ILW evaporator can be reused as process water.

The calcination of the ILW without combination with the HLW has received limited development. The calcination of the ILW combined with the HLW appears to be feasible based on results of limited development work. The calcination processes reduce the volume of the ILW by about 90%. These processes are described in Section III.A.1.

The membrane processes that have been considered are reverse osmosis and electrodialysis and are for LLW only. The reverse osmosis process pressurizes an aqueous waste in contact with a semipermeable membrane above the osmotic pressure for that membrane. The result is the transport of essentially pure water through the membrane accompanied by few of the radioactive constituents contained in the waste. Reverse osmosis has been used on steam generator blowdown in reactor plants. The electrodialysis process uses an electrical potential to cause the migration of ionic species through a membrane, thus resulting in a purified residual aqueous stream. Electrodialysis has not been used on radioactive wastes in other than developmental work, but is routinely used in desalination applications.

The filtration/centrifugation option first involves precipitation of the solutes in the waste using chemical additives. The solid materials are then removed using standard filtration or centrifugation technology, both of which are routinely used in treating reactor wastes. Filtration is also used as a pretreatment before ion exchange processing to prevent plugging of the ion exchange bed.

Ion exchange treatment of LW/LLW removes the ionic radioactive constituents from the waste by exchanging the radioactive ions in the liquid phase for those in the solid ion exchange media. Ion exchange can be used as the primary treatment process or can serve as a "polishing" process. The concentrated waste from this process is either a saturated ion exchange resin or a concentrated aqueous solution formed by the regeneration of the ion exchange resin. Ion exchange resins are in routine use at reactor plants.

### *c. Combustible Wastes*

The treatment processes for combustible wastes involve oxidation of the wastes to ashes, which are then treated as a noncombustible solid waste. Other secondary wastes result from the scrubbing of the off-gases from the oxidation process.

Oxidation of combustible radioactive wastes has been practiced for years, resulting in the development of a large number of generic processes and an extremely large number of specific designs. The generic oxidation systems and their characteristics are summarized in Table 8.

*2. Immobilization of ILW/LLW*

The immobilization techniques used for ILW and LLW are much more conventional than those being used for the HLW because the ILW and LLW generate essentially no heat from radioactive decay. Since ILW and LLW have been produced for years at reactor plants, previously operated reprocessing plants, and by national defense programs, several technologies for immobilizing these wastes are presently being used commercially. In addition, several other processes are being developed which may offer better waste forms and/or economy. The presently identified ILW/LLW treatment options have been grouped according to their development status in Table 9. The characteristics of the immobilization technologies presently in use will be discussed in some detail below. The characteristics of the technologies under development will be summarized briefly.

***a. Absorbents***

Absorbents are useful materials for immobilizing materials that are difficult to solidify using other means. Typical absorbent materials used for liquid immobilization are listed below:

- Vermiculite
- Various clays
- Diatomite (biological deposits of calcium silicate)
- Calcium sulfate dessicants
- Silica gel
- Portland cement
- Plaster of paris

The liquids are fixed by simply mixing them with the absorber until it has been nearly saturated. Since the absorbed liquids are only immobilized physically, they can be displaced by other liquids and are relatively dispersible due to the particulate nature of the absorbents.

***b. Salt Matrix***

Salt matrix immobilization can be implemented by either of two methods. The first is to evaporate aqueous ILW or LLW to the point where the "bottoms" from the evaporator crystallize into a solid salt cake when cooled to ambient temperature. The second method is to cast the solid salt cake from the molten salt. This type of product would be produced by the salt removed from the molten salt oxidation process described in Table 8. The product is a monolithic salt cake which is not readily dispersible, but which is readily soluble upon contact with water.

***c. Urea-Formaldehyde Resins***

Urea-formaldehyde resins are materials used to immobilize liquid and particulate solid wastes. The resin is a cross-linking polymerization product of the reactants urea and formaldehyde and consists of a mixture of monomethylol urea

## Table 8
## CHARACTERISTICS OF GENERIC OXIDATION SYSTEMS

| Oxidation process | Description |
|---|---|
| Excess air incinerator | Older single-chamber incinerators supply excess air in addition to that required for combustion. The use of surplus air results in lower burning temperatures and more fly ash carry-over than are obtained in the more modern controlled air incinerators. |
| Controlled air incinerator | Controlled air systems use the concept of multiple-chamber burning to achieve complete combustion of solid waste. Charges of waste enter the first chamber where they burn in the presence of substoichiometric quantities of air. The products of partial oxidation and volatilization flow into a secondary heated combustion chamber where excess air provides complete combustion. This mode of operation produces a nonturbulent combustion environment in the first chamber and minimizes the entrainment of fly ash. |
| Pyrolysis incinerator | In a pyrolysis incinerator, combustible material is thermally decomposed in an oxygen-deficient atmosphere to yield a volatile gas which is subsequently combined with excess air and burned in a secondary combustion chamber. A pyrolysis incinerator depends on accurate control of the amount of air fed to the burning chambers for efficient operation. |
| Vortex incinerator | The vortex or suspended burning incinerator consists of a stationary, refractory-lined cylinder into which the finely ground waste is pneumatically fed. Excess air ($\sim$300% of the stoichiometric requirement) is introduced tangentially, giving a spiral burning pattern to the suspended waste as it passes through the incinerator. The excess air, coupled with its high velocity through the chamber, causes much fly ash carry-over and may require a correspondingly large gas treatment system. Primarily designed for high burning rates, this incinerator is used in the lumber industry to burn bark and scrap. Its primary success has been in applications using only one type of feed which can be easily shredded or ground to a fine consistency for pneumatic transport. |
| Fluidized-bed incinerator | Shredded wastes are introduced into a vertical, fluidized bed of inert material. The highly agitated fluidized system improves mixing and combustion efficiency. It also provides improved temperature control due to the heat dissipation properties provided by the fluidized inert material. Substitution of pelletized sodium carbonate for the inert bed material provides additional benefits through neutralization of acidic combustion products, such as HCl, thus permitting use of less expensive construction materials in the incinerator and off-gas treatment system. |
| Rotary kiln incinerator | The primary combustion chamber is a tilted horizontal cylinder that rotates on its longitudinal axis to provide a tumbling action for improved combustion efficiency. The unit also provides positive ash removal from the combustion chamber, which is a definite advantage for assuring plutonium inventory control. Off-setting these advantages are the complicated mechanical drives and the need for effective seals. The seals represent a containment problem and cause difficulties in control of the combustion air. |
| Moving grate incinerator | The moving grate or rotating grate incinerator uses the motion of the grate to move waste material and constantly expose new unburned material for combustion. The moving grate also provides positive ash discharge for improved plutonium inventory control. This system, due to its internal moving parts, presents greater maintenance problems than other types of incinerators. |
| Molten salt oxidation | Combustible wastes are burned with air in sodium carbonate and sodium sulfate. Inorganic products (metallic oxides, ash, sodium chloride, and fissile material) are retained in the melt. The salt-ash product can be cast in storage drums or processed to recover salt and transuranics. |
| Acid digestion | Combustible wastes are converted to gaseous products ($CO_2$, $H_2O$, HCl) and sulfate-oxide residues by digestion in $H_2SO_4$ at 230—250°C. |
| Waste solvent burner | Liquid mixtures of n-dodecane and tributylphosphate used in solvent extraction steps during reprocessing are burned. |

**Table 9**
**ILW/LLW IMMOBILIZATION ALTERNATIVES**

| Technologies in use | Technologies under development |
|---|---|
| Absorbents | Aluminosilicate minerals (clay) |
| Salt matrix | Vitrification |
| Urea-formaldehyde resins | Pelletization/compaction |
| Cement | Polyethylene and other organic polymers |
| Bitumen (asphalt) | Metal matrix |

($NH_2 \cdot CO \cdot NH \cdot CH_2 \cdot OH$) and dimethylol urea ($CH_2 \cdot OH \cdot NH \cdot CO \cdot NH \cdot CH_2 \cdot OH$), depending on the ratio of reactants. The reaction of the urea and formaldehyde is initiated using a weak acid such as sodium bisulfate or phosphoric acid at a pH between 1.0 and 2.0. The cross-linked structure of the resin results in the trapping of the aqueous waste within the solid resin matrix. During the curing of the resin to a reasonably hard, free standing solid some water is often released which is usually immobilized using one of the absorbent materials.

***d. Cement***

Hydraulic cements (usually portland cement) are used in one of the most popular ILW/LLW immobilization techniques in the U.S., along with the urea-formaldehyde resin described above. The procedures for using cement as an immobilizing agent are basically the same as those for the urea-formaldehyde resin except that catalysts are not required. The immobilization characteristics of cements can be improved markedly by the inclusion of certain additives such as sodium silicate (improves setting properties and increases allowable waste concentration), fly ash (improves strontium retention), attapulgite clay (aids in suspending the cement), grundite (improves cesium retention), delta gluconolactone (retards setting), and tributyl phosphate (prevents foaming during mixing).

A relatively recent innovation with the cement immobilization process is to impregnate the surface of the monolithic cement/waste product with styrene monomer and polymerize the monomer to polystyrene at 50 to 70°C. This results in an essentially impermeable layer on the outer surface that also improves the waste form's strength, durability, and resistance to chemical attack.

***e. Bitumen***

Bitumen (i.e., asphalt) is a high molecular weight hydrocarbon that is found in natural beds and also produced as a residue in petroleum refining. The chemical inertness, insolubility in water, and good coating properties have made it a popular material for immobilizing ILW and LLW in foreign countries, although its use in the U.S. has been limited to development studies. Bitumen is used by heating it to 150 to 230°C and mixing in the wastes to be immobilized. The high temperatures (compared to those used with cement and urea-formaldehyde) result in evaporation of the water in the wastes, which gives a smaller volume of bitumen, but also results in a residual low-level stream that must still be treated. Bitumens have not been used in this country because of the possibility that they might ignite at temperatures of 275 to 330°C if oxidizing compounds such as $NaNO_3$, $NaNO_2$, or $MnO_2$ were present in the bituminized waste in appreciable quantities. However, foreign experience to date indicates that this is not a problem.

Descriptions of the ILW/LLW processes that require further development are given

Table 10
CHARACTERISTICS OF ILW/LLW IMMOBILIZATION PROCESSES UNDER DEVELOPMENT

| Process | Description |
|---|---|
| Aluminosilicate mineral fixation | Immobilize high-salt wastes in the chemical structure of minerals by reacting the aqueous waste with kaolin or bentonite clay to form a cancrinite mineral; similar to supercalcine for HLW |
| Vitrification | Incorporation of the waste solids in a glass melt; very similar to vitrification process for HLW; most HLW development work should be applicable |
| Pelletization | Waste solids are mixed with a binder, pressed into pellets, and heat treated to unify the pellet; very similar to the sintering process for HLW |
| Polyethylene and other organic polymers | Incorporation of the wastes in a polyethylene melt (similar to bituminization) or in water-extended polyester polymer (similar to urea-formaldehyde process) |
| Metal matrix | Incorporation of the wastes in a molten metal; similar to metal matrix process for HLW |

in Table 10. The most striking aspect of many of these processes is their similarity to processes for immobilizing high-level wastes.

*3. Comparison of Waste Forms*

A comparison of the three principal waste forms for ILW and LLW, viz., cement, bitumen, and urea-formaldehyde, is given in Table 11. One noteworthy aspect of these three materials is their varying tolerances of the various chemicals expected in the ILW/LLW from different sources. The data for the calcine, glass, and ceramic waste forms for ILW and LLW would be very much like the data given in Table 7 for the calcine, borosilicate glass, and supercalcine, respectively. The other waste forms for ILW/LLW cannot be adequately characterized at this time.

**C. Treatment of Fuel Assembly Structural Materials**

The treatments of the fuel assembly structural materials such as the cladding, end pieces, and grid spacers are of two general types: pretreatment and final treatment.

*1. Pretreatment*

There are three pretreatment options available: no pretreatment, sorting, and surface cleaning. The first of these simply involves applying one of the immobilization techniques described below to the as-generated fuel assembly structural materials.

The sorting option involves separating the end pieces and possibly the grid spacers from the fuel elements, before the elements are chopped in the fuel reprocessing plant. This sorting accomplishes two things:

1. It separates the fuel assembly structural materials by metal type (iron-nickel-chromium metals from zirconium alloys).
2. It permits the non-TRU waste materials (which have not been in contact with the spent fuel) to be separated before fuel element chopping and disposed of as non-TRU wastes.

Sorting is a necessary prerequisite to some of the final treatment options described below.

**Table 11**
**COMPARISON OF THE PRINCIPAL ILW/LLW PRODUCT FORMS[1,4]**

| | Waste form | | |
|---|---|---|---|
| **Property** | **Cements** | **Bitumens** | **Urea-formaldehydes** |
| Leachability ($g/cm^2$-day) | | | |
| Alkali metals and alkaline earths | $4 \times 10^{-6}$—$10^{-1}$ | $3 \times 10^{-7}$—$3 \times 10^{-4}$ | $1.5 \times 10^{-2}$—$1.5 \times 10^{-1}$ |
| Rare earths and actinides | $2 \times 10^{-9}$—$1.1 \times 10^{-7}$ | $2 \times 10^{-8}$—$1 \times 10^{-7}$ | [a] |
| Thermal conductivity (W/m-°C) | 0.3—1.0 | 0.2—0.3 | [a] |
| Maximum stable temp (°C) | 300—500 (dehydrate and lose strength) | ~350—450 (ignite spontaneously) | [a] |
| Radiation resistance (rad) | $\geq 10^{10}$ | $\sim 10^{9}$ | $\geq 10^{8}$ |
| Waste solids (wt%) | 5—55 | 40—60 | 8—17 |
| Bulk density (g/cc) | 1.6—2.2 | 1.1—1.9 | 1.1—1.3 |
| Volume of product/volume of waste | 1.5—3.0 | 0.4—1.1 | 1.3—2.0 |
| Chemical tolerances | | | |
| Boric acid soln | Poor | Good | Good |
| $Na_2SO_4$ acid soln | Fair | Good | Reduced efficiency |
| $NaNO_3$ acid soln | Good | Not recommended | Fair |
| Alkaline acid soln | Good | Good | Reduced efficiency |
| Laundry detergent | Poor | Fair | Poor |
| Organic liquids | Poor | Fair | Poor |
| Ion exchange resins | Fair | Fair | Good |
| Sludges | Good | Good; hazardous with $MnO_2$ sludge | Good; may require pH adjustment |
| Mechanical strength | Good | Good | Fair |
| Product stability | Very good | Good | Fair (loses water and strength in open systems) |

[a] Data not yet available.

There are processes under development for decontaminating the surface of the Zircaloy cladding after most of the fuel material has been dissolved with nitric acid. One process initially treats the Zircaloy cladding, which has a nonreactive $ZrO_2$ layer on its surface, with gaseous hydrogen fluoride, which penetrates the oxide and reacts with the underlying metal. Subsequent treatment with an aqueous solution containing 0.1 *M* ammonium fluoride, 0.4 *M* ammonium oxalate, 0.16 *M* ammonium citrate, and 0.3 *M* hydrogen peroxide at 95°C quickly removes both the oxide and fluoride layers on the cladding, leaving clean metal. The resultant aqueous stream can then be treated as an intermediate-level TRU waste, while the cladding may be decontaminated to the point that it can be considered to be an intermediate-level non-TRU waste.

A second cladding decontamination process involves the use of a fused salt such as 37 mol % LiF—37 mol % NaF—26 mol % $ZrF_4$ at 700°C as the cleaning agent. This salt readily attacks the $ZrO_2$ layer on the cladding and results in the production of clean metal, a contaminated salt stream, and a contaminated aqueous stream from washing the salt off the metal after cleaning.

### *2. Treatment*

There are five treatment options available or being developed for fuel assembly structural materials: no treatment, matrix formation, compaction, melting, and conversion to nonmetals. The "no treatment" option simply involves packaging the as-generated cladding and end pieces in a metal container for disposal. This treatment is likely for the non-Zircaloy components of the fuel assembly if they are separated by sorting.

The matrix formation option involves placing the as-generated fuel assembly structural materials in a container and filling the void space with some type of a matrix. Matrix materials that have been proposed include cement, bitumen, glass, and sand. One of the principal purposes of the matrix would be to essentially eliminate the pyrophoricity of the Zircaloy fines. Although this area has not been studied extensively, it appears that sand is the medium of choice because it readily fills the voids in the structural materials, and it is relatively simple and inexpensive to use.

The compaction of the fuel assembly structural materials using mechanical devices can reduce the volume of the as-generated waste by about a factor of three. This type of volume reduction before packaging would substantially reduce the cost of handling, transporting, and disposing of these materials. Although compactors are commercially available, none have been tested in this particular application.

Melting of the fuel assembly structural materials offers a waste volume reduction factor of approximately six. However, this process is substantially more complex than the compaction process described above. Although there are several possible methods for melting the fuel assembly structural material wastes, the most heavily investigated method in the U.S. is "Inductoslag melting". This process involves the induction melting of the Zircaloy-Inconel®-stainless steel mixture from the fuel assembly in an inert atmosphere to prevent ignition of the zirconium. A high-purity slag material such as calcium fluoride is also required. This process produces an ingot of the melted structural materials without further melting or casting. Although the material produced from nonradioactive test runs using chopped Zircaloy cladding was of sufficiently high quality that it could be remanufactured into cladding, this will probably not occur with the Zircaloy from spent fuel because of the expense of remotely fabricating the highly radioactive metal.

The final treatment process is conversion of the fuel assembly structural materials, principally the reactive Zircaloy, to a stable chemical compound. Proposed processes include the following:

1. Dissolution in a mixture of nitric and sulfuric acids, resulting in the production of zirconium sulfate
2. Dissolution in an aqueous mixture of $NH_4F$-$NH_4NO_3$ to form $(NH_4)_2ZrF_6$ which can then readily be converted to $ZrO_2$ for disposal
3. Volatilization of the zirconium using HCl or $ZnCl_2$ to form $ZrCl_4$ which can then readily be converted to $ZrO_2$
4. Oxidation of the Zircaloy and stainless steel directly at 500 to 600°C using a HF catalyst

These processes eliminate the pyrophoric tendencies of zirconium but require substantial development.

### D. Treatment of Wastes Containing Volatile Elements

The volatile or semivolatile elements assumed to be recovered are iodine ($^{129}I$), tritium ($^{3}H$), carbon ($^{14}C$), and noble gases ($^{85}Kr$). The recovery of these elements is discussed in Section II.D. The following sections will discuss the treatment of the recovered wastes to render them less mobile.

*1. Iodine Waste Treatment*

The treatment of the iodine wastes depends primarily on the process used to recover the iodine initially. If silver-loaded or other metal-loaded absorbents are used, it is likely that the iodine will not be recovered from these materials. Thus, the most likely treatments would be simply to package the solid iodine-laden absorbents or to immobilize the absorbents using materials such as cement, urea-formaldehyde, or bitumen.

If the other iodine recovery processes are being used, the iodine will most likely be recovered as a solid material, as described in Section II.D.1. These solids may be packaged either without further treatment, or packaged with low-temperature immobilization as described above, or converted to a different solid material. With respect to the latter option, one of the more promising solid materials is barium iodate, $Ba(IO_3)_2$. An immobilized iodine waste containing about 9% iodine as barium iodate in cement appears to have relatively favorable leach properties. The addition of 0.9% butyl stearate to the cement reduces the leach rate by an additional factor of three. However, because of the relatively high solubility of iodine and its compounds, the best obtainable leach rates at ambient conditions are on the order of $10^{-4}$ g cm$^{-2}$ day$^{-1}$.

*2. Tritium Waste Treatment*

The proposed treatment options for tritium wastes, as with iodine wastes, include a wide variety of processes designed to reduce the mobility of the tritium and thus limit the amount of tritium-hydrogen exchange that would take place if the waste were contacted with water after disposal. The immobilization of tritium effectively for its entire radioactive life should be feasible, since it will have decayed to innocuous levels in about 100 years.

The simplest of the treatment options uses the raw tritium waste (which is usually tritiated water) as the aqueous agent in making hydraulic cement. The cement can either be a standard cement, a high alumina cement (which holds more water per unit volume), or a polymer-impregnated cement (see Section II.B.2).

A more complex tritium waste treatment is to make the tritium into a relatively insoluble hydrogenous compound which may either be packaged directly or immobilized in cement during packaging. Some of the compounds that have been investigated are discussed in the following paragraphs.

**a. Polyacetylene**

The tritiated water is reacted with calcium carbide to form tritiated acetylene and tritiated calcium hydroxide according to the reaction

$$CaC_2 + 2HTO \rightarrow C_2HT + Ca(OH)(OT)$$

The tritiated calcium hydroxide is calcined to recover the tritiated water, which is recycled and reacted again with more calcium carbide. The acetylene monomer is then polymerized to a solid waste form using either gamma rays from $^{60}Co$ or heat and a cupric oxide catalyst.

**b. Bakelite® Analog Polymers**

In this process tritiated acetylene (see above) is reacted in a tritiated water-sulfuric acid solution of mercuric sulfate to form tritiated acetaldehyde according to the reaction

$$C_2HT + H_2O \xrightarrow{Hg^{+2}} CH_2TCHO$$

Resorcinol or phenol and the tritiated acetaldehyde are then mixed in the presence of aqueous sodium hydroxide to form the polymer.

### c. Polyacrylonitrile

Tritiated acetylene is reacted with hydrogen cyanide using a copper catalyst to form acrylonitrile according to the reaction

$$C_2HT + HCN \xrightarrow{Cu^+} CHT{=}CHC{\equiv}N$$

The acrylonitrile is then polymerized using a radical initiator such as potassium persulfate according to the reaction

$$n\,CHTCHCN \xrightarrow{\text{radical initiator}} (\text{—}\underset{\substack{|\\ C\equiv N}}{CHTCH}\text{—})_n$$

### d. Polystyrene

The fixation of tritium in polystyrene first requires that elemental tritium be generated via the electrolysis of water. The fixation is then accomplished by hydrogenating the polystyrene in an organic solvent using nickel-on-Kieselguhr, Raney nickel, or rhodium-on-alumina catalyst.

### e. Hydrides

Metallic hydrides have also been considered for fixing tritium. Although Zr, Ti, Hf, Y, Nb, Ta, and U have all been considered, imposing the requirements of stability to water and air, low dissociation pressure, and reasonable cost and availability reduces this list to Zr and Ti. The hydride is made by heating outgassed metal in a hydrogen atmosphere at 500 to 900°C. The resulting zirconium product, which has been studied the most, is approximately $ZrH_{1.4}$.

Tritiated compounds have lower leach rates than tritiated cement mixtures if they are made into monolithic solid waste forms to reduce the surface area that might be exposed to ground water.

## 3. Carbon Waste Treatment

The treatment of carbon wastes, assumed to be in the form of $^{14}C$-contaminated $CaCO_3$, has not been studied to any great extent. Therefore, no information is currently available concerning alternatives to calcium carbonate for containing the carbon. However, based on the information supplied by the search for better tritium-containing compounds, it would seem that a hydrophobic organic compound containing the $^{14}C$ might prove to be a superior waste form.

In any case, there are only two options that have been considered for treating the calcium carbonate: (1) packaging without further treatment and (2) mixing the calcium carbonate with cement and then packaging. At the present time, both of these options are conceptual and have not been investigated for leach resistance, integrity, or the effects of the carbonate on the cement.

## 4. Noble Gas Waste Treatment

The treatment of the noble gases presents a unique problem since they are volatile under all conditions expected during disposal and they form only a few marginally stable compounds. Thus, the treatment of these gases is based on physical rather than chemical methods. The most obvious and most highly developed treatment option for the noble gases is to store them pressurized in cylinders similar to those used to contain welding and laboratory gases. Since most of the $^{85}Kr$ will decay to the stable daughter rubidium in about 100 years, this type of treatment can be expected to be adequate as long as the cylinders are stored under controlled conditions. The principal aspect of this treatment currently being investigated is the material the cylinder should be made of to withstand the high pressures and somewhat elevated temperatures, and the corrosive effects of the rubidium daughter.

A second, somewhat novel, approach to containing the noble gases is called zeolite encapsulation. Zeolites are crystalline aluminosilicate compounds that contain a regular array of interstitial cavities of a uniform size. The cavity size may be selected within limits. The process by which a noble gas diffuses through a zeolite material is known as activated diffusion; this means that an activation energy is required for the krypton atoms (which have a room-temperature kinetic diameter of 3.5 Å) to diffuse through the 2.3 Å apertures between the interstitial cavities. Therefore, it is possible to introduce the noble gases into the interstitial cavities at temperatures and pressures elevated sufficiently (e.g., 450°C and 1000 atm) to supply the required activation energy. The zeolite is then reduced to roughly ambient temperatures and pressures. Under these conditions insufficient activation energy is available for rare gas movement; thus, each krypton atom is effectively trapped in its interstitial cavity. Calculated net losses occurring over the time it takes for the krypton to decay to innocuous levels are on the order of 0.01% for the better zeolites. This process, which is still under development, is better than storing pressurized gas in cylinders in that (1) it eliminates having to monitor the waste in a controlled environment (since the zeolite can be immobilized and disposed of like other solid wastes) and (2) it eliminates the possibility of a sudden release of noble gases that is present with a pressurized cylinder. However, the zeolite encapsulation process is disadvantageous in that (1) high temperatures and pressures are required, (2) the krypton must be separated from the xenon, and (3) the krypton may be released by moisture.

### E. Treatment of Uranium Mill Tailings

The very large volume of the mill tailings effectively precludes the use of the types of waste treatment processes (i.e., immobilization with cement, bitumen, etc.) used for other waste types. Fortunately, the very dilute nature of these wastes makes this type of treatment unnecessary. However, some type of treatment is required to prevent public access to these materials, to prevent their becoming airborne from erosion, to prevent the leaching of $^{226}Ra$, and to restrict the emanation of radon gas. This can generally be accomplished by covering the tailings pile with a nonradioactive earth overburden to a depth of 10 to 20 ft. This effectively precludes public access and delays release of 3.8 day $^{222}Rn$ sufficiently that virtually all of it decays to its nongaseous daughter before it can be released to the atmosphere. Additional radon retardation could be achieved by covering the tailings pile with a layer of asphalt before the overburden was piled on.

### F. Treatment of Spent Fuel

The treatment of as-generated spent fuel, which has many of the characteristics of high-level waste (see Section II.F), will most likely not be the same as that used for high-level waste. The reasons for this are

1. The spent fuel is a solid material, whereas the as-produced HLW is a liquid.
2. The spent fuel elements contain the volatile fission product gases (krypton, iodine, etc) under relatively high pressure (1000 to 3000 psia).

3. The Zircaloy cladding is not likely to withstand the elevated temperatures that accompany vitrification processes.
4. The spent fuel assembly is relatively fragile and somewhat awkward because of its length.
5. The presence of all of the $^{235}U$ and plutonium makes criticality a potential problem.

Treatment options for spent fuel assemblies are under intensive study. It is not possible to predict at the present time what option will finally be chosen.

## IV. RADIOACTIVE WASTE DISPOSAL

For those wastes considered to be "long-lived" (e.g., all TRU wastes, fuel assembly structural material wastes, iodine wastes), only a properly chosen disposal medium and surrounding environment can provide the required long-term isolation (see Section III.A.1). For "short-lived" wastes (e.g., tritium and noble gases), the waste form itself will also be an effective barrier to the release of radioactivity.

Many waste disposal concepts have been rejected for technical reasons, and the rest can be reduced to a few generic methods. These methods are

1. Shallow land burial
2. Geologic disposal in continental formations
3. Surface storage
4. Other disposal methods
   A. Seabed disposal
   B. Ice sheet disposal
   C. Extraterrestrial disposal

After describing each of these disposal options, it will be useful to examine each of the waste types described in Sections II and III and consider which type of disposal concept is applicable to each.

### A. Shallow Land Burial

The shallow land burial of radioactive wastes involves placement of the waste at relatively shallow depths (as compared to geologic disposal) above the water table with no provisions for retrievability at a later date. Shallow land burial has been used to dispose of radioactive wastes since the inception of the nuclear weapons program during World War II.

The first burial grounds were operated by the U.S. Government for managing the defense wastes. However, the Atomic Energy Commission opened the government burial grounds to commercial wastes in the 1950s to eliminate the then-current practice of disposing of the commercial wastes by dumping at sea. In the early 1960s the first commercial burial ground was opened at Morehead, Ky., and soon afterward the government stopped accepting commercial wastes. There are now six commercial burial grounds licensed in the U.S., although they are not all operating. The identity and characteristics of these facilities are given in Table 12.

The U.S. Nuclear Regulatory Commission has jurisdiction over all of the commercial burial grounds except when it relinquishes this authority to the state in which the burial ground is located. At present, five of the six commercial burial grounds (the Sheffield site is the exception) are under the jurisdiction of these so-called Agreement States. In the past the burial grounds have accepted all waste types except high-level

Table 12
CHARACTERISTICS OF COMMERCIAL BURIAL SITES[5]

| Characteristic | Burial site | | | | | |
|---|---|---|---|---|---|---|
| | West Valley, N.Y. | Morehead, Ky. | Barnwell, S.C. | Sheffield, Ill. | Beatty, Nev. | Richland, Wash. |
| Mean annual precipitation (mm) | 1000 | 1200 | 1100 | 900 | 100 | 200 |
| Surface material | Glacial drift silty clay/ interbedded sand and gravel | Weathered shale; clay and sand | Sand and clay-sand | Glacial drift; sand, silt, and gravel | Alluvial sand and gravel | Clay, sand, and gravel |
| Thickness (m) | 20—30 | 3—5 | 0—10 | 20—30 | Over 200 | Over 150 |
| Interstitial permeability to water (cm/day)[a] | Low (0.5) | Very low (0.02) | Very low (0.02) | Variable (0.04—40) | Variable (0.02—0.1) | Variable |
| Bedrock material | Shale | Shale | Clay, sand, and sandstone | Shale, claystone, and coal | Metamorphic and sedimentary | Volcanic |
| Structure | Flat-lying | Flat-lying | Flat-lying | Flat-lying | Folded | Flat-lying |
| Groundwater depth to shallowest saturated zone (m) | Variable (1—20) | 1—2 | 10—20 | 5—20 | 80—90 | 100 |
| Depth of continuous groundwater zone (m) | 20 | 10—15 | 10—20 | 5—20 | 80—90 | 100 |
| Depth to regional aquifier (m) | None present | None present | 200 | 100 | 80—90 | 100 |
| Surface-water proximity | On site | 500 m | At site boundaries | At site boundaries | 3 km | 10 km |
| Flow characteristics | Small, perennial | Small, perennial | Small, perennial | Lake to north; small, perennial to south | Ephemeral, after storms | Large, perennial (Columbia River) |
| Sorptive or ion-exchange capacity of material surrounding burial[a] | High | High | Moderate | Low | Moderate | Moderate |
| Principal flow paths away from burial | Shale fractures and lenses in drift | Shale fractures | Pore spaces in sand | Pore spaces in till | Unsaturated flow in pores | Unsaturated flow in pores |

[a] Interstitial permeability and sorptive capacity can be bypassed by flow along fractures or other high-permeability zones.

waste. Along with wastes generated by the commercial nuclear fuel cycle, radioactive wastes from hospitals, universities, and industry also contribute substantially to the volume of waste committed to shallow land burial. The sources of wastes at the Nuclear Fuel Services site are given in Table 13.

The technology used in the shallow-land burial of radioactive wastes is relatively simple. Trenches are dug in the ground using conventional earth-moving equipment such as bulldozers and road scrapers. Typical trench dimensions are 20 to 70 ft wide at the land surface, 10 to 50 ft wide at the bottom of the trench, 300 to 600 ft in length, and 17 to 25 ft deep. The bottoms of the trenches have a slope of 1 or 2% to minimize the water retention time should water intrusion occur. The bottoms of the trenches are filled with a porous material such as gravel or sand, also to promote drainage. Sumps are placed in the trench to allow water that has entered the trench to be pumped out, immobilized, and properly disposed of.

The radioactive wastes are placed in the trenches, either randomly (i.e., dumping the waste packages into the trench) or in an orderly manner by stacking. A photograph of a typical trench is shown in Figure 5. The trench is usually filled to within 2 to 4 ft of surface level.

The first layer of backfill, which is typically some type of compacted clay, is immediately placed over each day's deposit of waste material in the partially filled trench to prevent contact by rainwater. This layer typically fills the trench to ground level. After the trench is completely filled, earth is mounded on top of the trench to a height of 4 to 8 ft. The mound is then planted with some type of vegetation and the trench is marked with monuments containing information such as the location and definition of the trench, the date it was closed, the activity and most hazardous isotopes in the trench, and the waste volume in the trench. Based on current experience, an average of about 275,000 $ft^3$ of radioactive waste can be buried in one acre of land using these techniques.

There are relatively few restrictions on the form of the wastes sent to the commercial burial grounds. One of the most common is that liquids must be immobilized, generally by using one of the processes described previously. In general, absorbents are not acceptable for immobilization. There are usually no limits on the radiation levels of the waste packages, although the cost of disposal does rise rapidly above a certain radiation level, typically 200 mrem/hr surface dose. Packaging requirements are generally those required by the Department of Transportation for the shipping of radioactive materials to the site. More recently, restrictions have been placed on the transuranium element content of the wastes.

### B. Geologic Disposal in Continental Formations

Generally, this option involves placing the wastes in a deep continental geologic formation at a depth of about 1000 m. Within this option there are an extremely large number of possibilities; this area of waste management has been and is being intensively studied. A wide variety of disposal concepts have been proposed, including (1) conventional mined cavities, (2) solution-mined cavities, (3) a matrix of drilled holes, (4) "superdeep" holes, (5) hydrofracture emplacement, (6) deep-well injection, and (7) various rock-melting concepts. For each of these concepts one or more types of geologic media have been considered. The principal media being considered are salt, crystalline rocks, agrillaceous rocks, and basalt.

Mined cavities are being considered much more actively than the other disposal concepts listed above. Therefore, the approach that will be taken here is first to discuss the disposal of wastes in a conventionally mined salt repository. Then, the characteristics of other repository media will be compared to those of salt. Finally, each of the other deep continental geologic disposal concepts will be described.

**Table 13**
**HISTORICAL USE OF THE NUCLEAR FUEL SERVICES COMMERCIAL WASTE BURIAL SITE**

| Source of wastes | Waste volume (%) |
|---|---|
| Nuclear power plants | 22.6 |
| Institutional, educational, and hospitals | 1.8 |
| Federal government | 12.0 |
| Industrial and pharmaceutical manufacturing and research and development | 21.5 |
| Waste disposal and decontamination companies | 19.2 |
| Nuclear Fuel Services, Inc. | 23.0 |
| | 100.0 |

From Jump, M. J., *AIChE Symp. Ser.*, 72, 154, 1976. With permission.

The undesirable event that waste disposal seeks to avoid is the release of the waste repository inventory at rates and/or concentrations that would be harmful to the biosphere. Although there are many events that can be postulated that would result in this type of release, the release scenario normally considered most probable is the so-called "slow-leach incident". This incident assumes that, through natural or human-induced events, water enters the repository, contacts the waste, slowly dissolves the waste form, and transports the leached radioactive isotopes to the biosphere. Due to the depth and impermeability of any likely repository environment, the velocity of the water is normally very low, on the order of 5 ft/year or less. The property a repository should possess to minimize the consequences of this event is that it should not be capable of failing in a manner such that the radioactive isotopes in the water can move very quickly from the geosphere of the repository to the biosphere. The two specific repository characteristics that result in this property are

1. An environment (including the host medium and surrounding rocks) that does not fracture, since water can move much more quickly through cracks than through the impermeable medium itself
2. Repository surroundings having a high ion exchange capacity, since this will retard the migration of most of the radioactive isotopes such that their net velocity is much less than that of the water itself

It is, of course, also desirable that the probability of failure be as low as possible. Thus, if the presence of the waste might adversely affect the repository integrity too greatly or if the medium occurs in geologic areas that are potentially unstable, for example from earthquakes, the medium is a poor choice. The first of these effects can generally be accommodated by varying the density of heat-generating waste (i.e., high-level waste or spent fuel) in the repository to reduce the impact of the waste's presence to acceptable levels. It should be noted that the waste form itself is not particularly

FIGURE 5. Typical shallow-land burial trench.

important in the slow leach incident since it can be dissolved in several thousand years (see Section III.A.4), whereas typical isotope migration times in the slow-leach incident are on the order of hundreds of thousands of years. This is not the case, however, if the water is moving relatively quickly, since the leach rate from the waste would then be controlling.

### *1. Waste Repository in Conventionally Mined Salt*

In the use of conventionally mined cavities in salt, a series of parallel corridors would be mined and the waste packages emplaced in the floor at regular intervals (for heat generating wastes) or stacked in rooms (for nonheat-generating wastes). When the repository is filled, the corridors would be backfilled with the salt previously mined, the vertical shaft would be sealed, and the repository site would be monitored for as long as possible. A schematic diagram of a conceptual waste repository is shown in Figure 6.

The disposal of waste in a conventionally mined salt repository was demonstrated in Project Salt Vault during the period 1965-67. This project involved the emplacement of electric heaters and spent fuel assemblies from the Engineering Test Reactor in a bedded salt mine in Lyons, Kan., for 19 months. An array of electric heaters is shown in Figure 7. A vehicle used to emplace HLW is shown in Figure 8. The effects of the

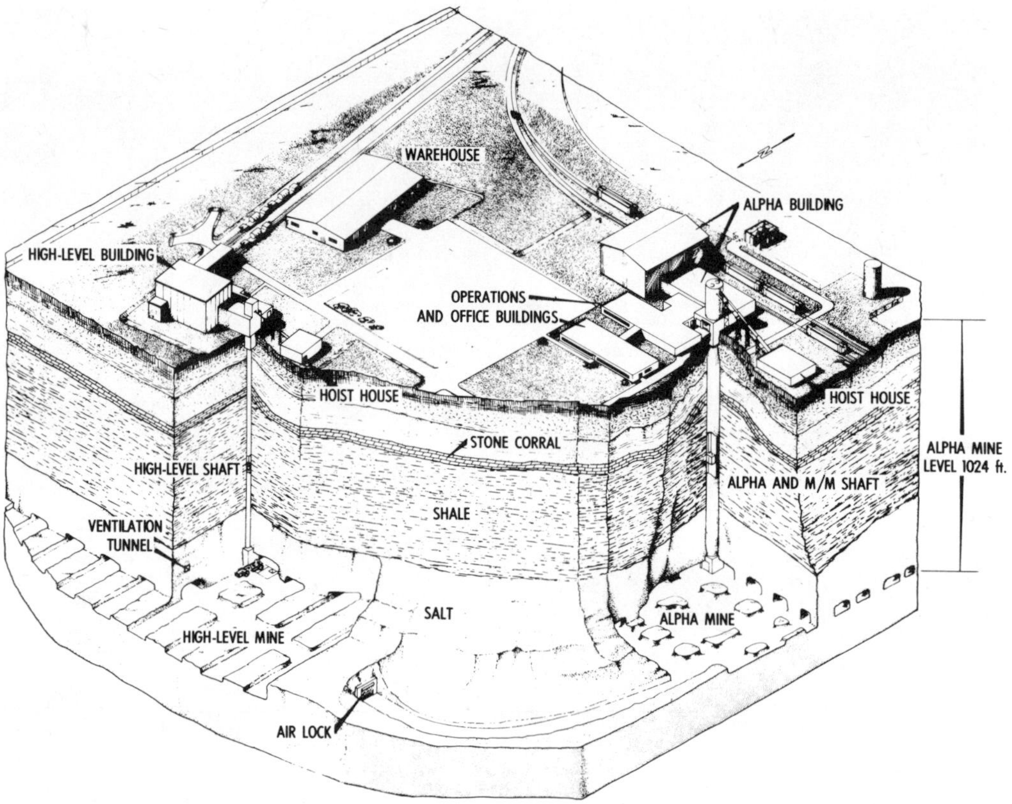


FIGURE 6. Layout of a conceptual geologic repository in salt (federal repository).

heaters and assemblies on the repository were studied; the assemblies were then removed and sent to the Idaho Chemical Processing Plant to be reprocessed.

One of the principal activities currently under way concerning deep geologic waste repositories is the development of analytical tools to predict the behavior of the repository after closure. The emplacement of waste in a repository may result in three different kinds of disturbances: (1) mechanical disturbances caused by the repository cavity itself, (2) chemical disturbances caused by the introduction of materials not in chemical equilibrium with the geologic environment, and (3) thermal disturbances caused by the decay heat of the waste. The mechanical disturbances would result from the settling of the mine since the backfill would be less dense than the original salt. Chemical disturbances in the salt repository would principally be in the form of salt brines interacting with the waste packages. The thermal disturbances are potentially troublesome in that they could cause additional mechanical or chemical disturbances or exacerbate existing disturbances. The results of analytical tool development are predictive models that will allow limits to be placed on various waste parameters, such as the maximum allowable decay heat per package and the package spacing in the repository.

### *2. Comparison of Deep Geologic Media*

Although there are a very large number of geologic media that could be considered for waste disposal purposes, four of these are receiving the most attention: salt, crystalline rocks, agrillaceous rocks, and basalt. The following discussion includes only

FIGURE 7. Salt repository studies using electric heaters.

the generic properties of each disposal medium even though site-specific characteristics of both the disposal medium and surrounding environment must be considered.

*a. Salt*

Salt is being considered in both its bedded and domed forms. Salt offers the following advantages as a disposal medium:

1. It is essentially impermeable because of its plastic properties.
2. It is widely distributed and abundant, underlying about 500,000 $mi^2$ in the U.S. and with known reserves greater than $6 \times 10^{13}$ ton.
3. The cost of developing disposal space is relatively low as compared with other rock types.
4. The heat-transfer properties of salt are good as compared with other rock types.
5. Salt formations in the U.S. are located in areas of low seismicity.

The very existence of the salt bed or dome means that flowing water has not been present for millions of years.

Because salt is soluble in water, if contacted with moving water it would be dissolved and removed rapidly. Salt water is highly corrosive to many materials, making waste packaging design difficult if the option to retrieve the waste while the repository is still

FIGURE 8. Vehicle used to transport and emplace high-level waste in a repository.

open is desired. Also, salt is often found in conjunction with hydrocarbons, thus raising the possibility that someone exploring for petroleum might drill through the formation and allow the entry of water. (However, it is interesting to note that even if the entire contents of a repository were dissolved in water, by the time the salt water was diluted to potable levels, the radioactivity level in the water would be well below the Radionuclide Concentration Guide values for unrestricted release given in the Code of Federal Regulations.)

### *b. Crystalline Rocks*

The crystalline rocks of interest are generally of the siliceous type, typified by granite. The advantages of these rocks for waste disposal purposes are their lack of bedding, wide availability, high strength, low primary (i.e., unfractured) permeability, and probable thermal stability. Known disadvantages are (1) that the rocks can have a high secondary (i.e., after fracturing) permeability as a result of fracturing which does not "heal" in the way that deformable salt does and (2) that the cost of mining hard rock is higher than for mining salt.

### *c. Agrillaceous Rocks*

The agrillaceous rocks of interest are fine-grained sedimentary deposits which can be either soft or hard and are typified by shales. The advantages of these materials from a waste disposal standpoint include wide availability, low permeability, relatively high plasticity, high ion-exchange capacity, and great homogeneity. Possible disadvantages include relatively low strengths, structural instability of the mines (e.g., heaving and roof collapse), presence of gases that can result from carbonaceous shales, the presence of contiguous aquifers, high water contents, and the presence of boreholes in adjacent strata containing hydrocarbons.

### *d. Basalt*

Volcanic rocks, typified by basalt, are also being considered as media for waste disposal. The advantages of basalt are the same as those of crystalline rocks. The disadvantages are also the same, but, in addition, the basalts (1) are located in seismically active areas, since they are volcanic in origin, (2) have some unfavorable hydrological characteristics, and (3) have a relatively low thermal conductivity.

A comparison of some of the principal physical properties of salt, granite, shale, and basalt is given in Table 14. Another characteristic of the disposal media, which results from the application of repository design criteria, is the thermal load limit for the repository (see Table 15). This is important since a lower thermal load limit means the repository must be larger to accommodate a given amount of waste. As is evident from inspection of Table 15, salt and shale are at a disadvantage, principally because of their tendency to plastically creep and thus close the mined cavities during the operational phase. The substantial reduction in allowable areal thermal loading resulting from the extension of the retrieval period to 25 years should also be noted. This factor is of prime importance when considering the retrievable disposal of spent fuel since the energy resource represented by the fissile material in the spent fuel may be needed at some point. However, the price paid for this retrievability in a geologic repository is large.

It should be noted that the choice of disposal medium has been examined by many people and groups over the years. One of the most highly regarded of these groups is the National Academy of Sciences-National Research Council, which examined this question in 1955 and again in 1961. In both instances, the conclusion was that salt was the medium of choice for a waste repository. More recently, recommendations by the U.S. Geological Survey[11] have been somewhat more neutral, indicating that salt is acceptable but that other media should still be carefully considered, as they may offer greater advantages. However, based on the information summarized herein and the wealth of background information from years of analytical and experimental study on salt, both in the U.S. and Germany, this prospect would seem to be rather unlikely.

### *3. Other Deep Geological Waste Disposal Concepts*

The remaining six deep geological waste disposal concepts will be described briefly in this section. Although each of these concepts has been proposed seriously and studied analytically, it is unlikely that they will ever be implemented on a large scale due to the widespread acceptance of geologic disposal in conventionally mined cavities.

**Solution-mined cavities** — This concept is only applicable to media that are soluble in water, such as salt. A waste shaft is drilled into the soluble medium and then the cavity is formed by injecting water down the shaft and pumping the resulting solution to surface level. Once the cavity is formed, packaged wastes are lowered into the cavity and allowed to fall into a random pile until it is full. Heat-generating wastes could not be accommodated in this concept since there is no mechanism to provide adequate spacing between waste packages. This concept has been demonstrated in Germany, but its nonapplicability to HLW is a serious disadvantage.

**Matrix of drilled holes** — This concept involves emplacing the waste in holes drilled into a geologic formation at a depth of 1000 to 2000 m. Several waste packages would then be lowered down the hole and stacked vertically in a single column. The hole would then be backfilled and plugged. The large number of boreholes that could provide the means for water ingress are a major disadvantage of this concept.

Table 14
TYPICAL PROPERTIES OF SALT, GRANITE, SHALE, AND BASALT[7-10]

| Property | Salt | Granite | Shale[a] | Basalt |
|---|---|---|---|---|
| Density ($g/cm^3$) | 2.1 | 2.6 | 2.6 | 3.0 |
| Natural moisture content (%) | 0.08 | 0 | 2.0 | 0 |
| Strength ($lb/in.^2$) | | | | |
| Compressive | 3,200 | 26,500 | 4,000 (4,000) | 40,000 |
| Tensile | 50 | 1,000 | (0) | 2,300 |
| Coefficient of linear thermal expansion ($\times 10^6/°F$) | 22.2 | 4.5 | 4.5 | 3.0 |
| Thermal conductivity (Btu/hr-ft-°F) | | | | |
| 32°F | 3.53 | 1.65 | 1.12 (0.82) | 0.65 |
| 212°F | 2.43 | 1.48 | 1.02 (0.76) | 0.73 |

[a] Where two numbers are given, they correspond to horizontal and (vertical) values.

Table 15
THERMAL LOAD LIMITS FOR PRECONCEPTUAL REPOSITORY DESIGNS[12]

| | Thermal load limit (controlling factor)[a] | | | |
|---|---|---|---|---|
| | Salt | Granite | Shale | Basalt |
| Canister limits during retrieval period[b] (kW) | | | | |
| Vitrified glass HLW | 3.2 (A) | 1.7(A) | 1.2(A) | 1.3(A) |
| Calcined HLW | 2.6(A) | 1.6(A) | 1.1(A) | 1.1(A) |
| Canister limits after retrieval period[b] (kW) | | | | |
| Vitrified glass HLW | 3.6(B) | 2.3(B) | 1.6(B) | 1.9(B) |
| Calcined HLW | 2.9(B) | 2.0(B) | 1.5(B) | 1.7(B) |
| Local areal thermal loading limits (kW/acre[c]) | | | | |
| 5-year retrieval-HLW | 150(C,D) | 190(C) | 130(C) | 190(C) |
| 5-year retrieval-spent fuel | 150(C)[d] | 190(C)[e] | 130(C)[e] | 190(C)[f] |
| 25-year retrieval-spent fuel | 36(C) | 80(C) | 55(C) | 80(C) |

[a] Controlling factors: A = canister temp limit, B = waste temp limit, C = room closure during operational phase, and D = earth surface movement, long term.
[b] Based on 5-year retrievability and assumes a 6-in. annulus of crushed rock around sleeve.
[c] Acreage includes rooms and adjacent pillars, but not corridors, buttress pillars, receiving areas, etc.
[d] Although 150 kW/acre can be tolerated locally, on the average over the entire repository only 60 kW/acre can be tolerated because of D.
[e] This limit requires that the annulus around the spent fuel sleeve not be backfilled with crushed rock and that the resulting very-near-field rock temperatures not result in excessive fracturing of the rock.
[f] The basalt repository design is based on this limit; however, later analyses indicate that the controlling factor might be the spent fuel cladding temperature and that the local areal thermal loading limit might be as low as 115 kW/acre.

**Superdeep holes** — This concept is basically the same as the matrix of drilled holes above except that here the holes are 10,000 to 20,000 m deep to maximize the distance between potential hydrological or geological changes and the waste. The primary problem with this concept is the lack of demonstrated technology for the drilling of superdeep holes wide enough for waste packages, although the secure plugging of the many boreholes and thermal effects following closure also appear to be potential problems.

**Hydrofracture emplacement** — Hydrofracture emplacement involves pumping a mixture of liquid radioactive waste and a grout or cement into a geologic formation a few hundred meters deep. The waste can be pumped into the formation under high pressure or it can be pumped into a preformed cavity. This process has been used in the U.S. to dispose of nearly 2 million gallons of intermediate-level waste.

**Deep-well injection** — Deep-well injection is basically the same as hydrofracturing except that the depth is increased to about 1000 and a grout may not be used. This concept has been investigated in the U.S. and the Federal Republic of Germany for the disposal of tritium. It has been used in the Soviet Union to dispose of over 300 million gallons of intermediate-level waste over an 8-year period.

**Rock melting concepts** — There are several rock melting concepts involving both liquid and solidified heat-generating strata. After the cavity is filled, the dry waste then generates sufficient heat to melt the surrounding rock. The melt zone increases for about 100 years and then slowly begins to shrink until all of the waste is finally encased in massive pieces of solidified rock. Both liquid and solid wastes have been proposed as the starting materials.

### C. Surface Storage

Surface storage has been proposed as a means for disposing of long-lived wastes and it is a likely disposal method for shorter-lived wastes, particularly the noble gases.

The basic concept is to construct some type of facility that will contain the wastes under any reasonable circumstance. In the past, this has meant tanks with active cooling for the small amount of HLW from commercial spent fuel reprocessing and the large amount from U.S. nuclear weapons production. Some high-level wastes are also stored as solid calcines in cooled bins. The tanks and bins have been proposed by a few for long-term storage, but the threat of leaks from the liquid-containing tanks and the necessity for constant maintenance makes this concept unacceptable over the very long-term.

Nevertheless, surface storage does have a place in the nuclear fuel cycle and in waste management. A large storage facility is a likely place to dispose of cylinders of noble gases containing $^{85}Kr$ since this waste is difficult to immobilize, generates an appreciable amount of heat, and decays to innocuous levels in about 100 years. Thus, a surface facility must be constructed to prevent the release of the noble gases should a cylinder fail; cooling must be maintained for a few decades.

A second possible use of surface storage facilities is to provide a means for holding spent fuel. As was noted in Section IV.B.2, the alternative of retrievable geologic disposal promises to be both troublesome and expensive, so surface storage looks attractive in this application. A surface storage facility for spent fuel could resemble a large swimming pool where spent fuel assemblies are stored in racks and the shielding and cooling is provided by the water in the pool. These pools would be virtually identical to those in use today at reactor plants to provide temporary storage for the spent fuel. An alternative type of spent fuel storage resembles a thick-walled concrete pipe. The pipe would be arranged vertically out in the open either above ground or with its top level with the ground surface. Metal internals would be put inside the pipe and a single fuel assembly is placed in it. The concrete pipe would provide shielding and the internals would facilitate natural convective cooling with air.

Finally, surface storage might be useful for high-level waste to allow the decay heat production rate to decline as much as possible before the waste is sent to disposal. This would maximize the space utilization in the respository and somewhat ease handling problems during emplacement.

### D. Other Disposal Methods

There are three other waste disposal concepts that have been considered as alternatives to geologic disposal in continental formations. These alternatives are

1. Seabed disposal
2. Ice sheet disposal
3. Extraterrestrial disposal

Although seabed and extraterrestrial disposal are being investigated at the present time, none of these alternatives has evolved to the point where it seriously challenges geologic isolation in continental formations as the waste disposal concept of choice.

#### *1. Seabed Disposal*

The technology of emplacing wastes in the seabed is a relatively straightforward extension of existing technology. The emplacement can be accomplished by dropping a suitably shaped waste package and allowing it to penetrate the sediment on the ocean bottom or by drilling holes and emplacing one or more waste packages in each hole.

The areas of technical uncertainty surrounding seabed disposal are concerned with (1) identifying oceanic areas that are stable, tranquil, remote, and of limited use as a source of useful materials, (2) identifying mechanisms for breaching the waste containers, and (3) developing a methodology for predicting the migration of radioactive isotopes in the sediments and the ocean itself. There would also be international, legal, or diplomatic problems to be resolved before waste can be disposed of in international waters.

#### *2. Ice Sheet Disposal*

Ice sheet disposal involves emplacing the waste in Antarctica instead of continental or seabed formations. Although proposals have been put forth that ''suspend'' the waste from cables, the most investigated concept would simply allow the waste to melt through the ice cap to bedrock. The time required for HLW to melt through is about 10 years. The principal difficulties here are the unpredictability of the ice sheet's behavior in the future, the possibility that there is water that moves toward the ocean between the ice sheet and bedrock, and the possibility that the heat from the waste may cause glacial ''surges'' in which the glaciers move forward relatively rapidly. This concept may also suffer from some of the potential international difficulties mentioned above for the seabed disposal concept.

#### *3. Extraterrestrial Disposal*

Extraterrestrial waste disposal would take the most troublesome portions of the radioactive wastes and transport them into a low-earth orbit using the space shuttle being developed by the National Aeronautics and Space Administration (NASA). The waste would then be transferred to an unmanned space tug for the trip to its final destination. Final destinations that have been proposed include high-earth orbit, a solar orbit other than that of the earth, solar system escape, and solar impact. The first two of these would require a velocity change of 7 to 9 km/sec. Solar impact would require a change of 7 to 24 km/sec. Energy requirements are proportional to the square of the velocity change.

The relatively small payload and high cost of space vehicles means that only a portion of the radioactive wastes actually generated can be subjected to extraterrestrial disposal. Thus, it would clearly be desirable to use this disposal method to dispose of the most troublesome portions of the radioactive wastes. These portions are generally

### Table 16
### LIKELY ULTIMATE DISPOSAL METHODS FOR RADIOACTIVE WASTES

| Waste type | Disposal type[a] | Rationale and comments |
|---|---|---|
| High-level | Deep | Very high activity and heat generation; high actinide concentrations |
| Intermediate-level and low-level TRU | Deep | Moderate to low activity and potentially significant amounts of actinides |
| Fuel assembly structural materials | Deep | Moderate to high activity and significant amounts of surface actinide contamination |
| Iodine | Deep | Very long-lived, highly concentrated |
| Carbon | Deep | Long-lived, highly concentrated |
| Spent fuel | Deep | Very high activity and heat generation; very high actinide concentrations |
| Intermediate-level and low-level non-TRU | Surface | Moderate to low activity and insignificant amounts of actinides |
| Tritium | Surface | Must be contained for only 100 years; relatively low toxicity |
| Noble gases ($^{85}Kr$) | Surface | Must be contained for only 100 years; active containment and maintenance will probably be required |
| Uranium mill tailings | Surface | Very low actinide concentrations; no transuranium elements |

[a] "Deep" includes all deep geologic, seabed, ice sheet, and extraterrestrial disposal alternatives; "surface" includes shallow land burial, surface storage facilities, and tailings piles.

considered to be the long-lived actinides, possibly with the inclusion of iodine and carbon wastes. These materials not only represent the bulk of the long-term waste toxicity, but also require relatively little shielding, thus increasing the net waste payload. However, the recovery of the actinides free of fission products necessitates the implementation of waste partitioning — the additional processing of radioactive wastes to recover the actinides. Partitioning will be discussed in more detail in Section V.

### E. What Waste Goes Where?

Up to this point the discussion of the origins and treatment of the various waste types has been separate from the discussion of the waste disposal alternatives. However, the disposal alternative required for each of the waste types is important because there is a considerable difference in cost between near-surface alternatives such as shallow-land burial and deep geologic disposal in a salt mine. The principal criterion for deciding what waste goes where is the concentration of long-lived isotopes in the waste. Among the more significant isotopes that can occur in significant quantities in radioactive wastes and that are considered to be long-lived are $^{14}C$, $^{129}I$, and the actinides. Based on this criterion, the likely disposition of each of the waste types is given in Table 16.

The principal question to be resolved concerning the ultimate disposition of radioactive wastes is at what point does the transuranic actinide element concentration become great enough that the waste is in the "TRU" category, requiring deep disposal, instead of the "non-TRU" category, which may be disposed of by shallow land burial. In the past, the boundary between TRU and non-TRU was taken to be $10^{-8}$ Ci of long-lived alpha emitters per gram of waste material (i.e., 10 nCi/g). However, it was soon discovered that it was very difficult or impossible to measure the long-lived alpha emitter content of some of the wastes at these low levels. This resulted in large amounts of ILW and LLW having to be assumed to be TRU wastes even though it was reasonably certain that they were not. Current indications are that the $10^{-8}$ Ci/g limit may be raised to about $10^{-7}$ to $10^{-6}$ Ci/g. The basis for this change is an improved analysis of the potential mechanisms and consequences of the release of radioactive isotopes from a typical shallow land burial site.

## V. PARTITIONING AND TRANSMUTATION (P-T) OF RADIOACTIVE WASTES

Partitioning is a chemical treatment designed to reduce the amounts of selected undesirable long-lived isotopes in radioactive wastes. Transmutation is a nuclear process whereby long-lived isotopes are converted to shorter-lived or stable isotopes by bombardment with subatomic particles such as neutrons from nuclear power reactors. Partitioning and transmutation (P-T), when taken together, form a waste treatment concept that would significantly reduce the amounts of certain long-lived, toxic species normally present in radioactive wastes and converting them to shorter-lived or less toxic species. P-T is a waste treatment option, since the residual material after partitioning still requires disposal. It is important to note that fuel reprocessing and plutonium recycle must be a part of any P-T fuel cycle since not to do so would defeat the basic P-T premise of reducing the actinide content of the wastes.

### A. Candidates for P-T

The candidate elements for P-T must be capable of being almost totally recovered from radioactive wastes, must be amenable to transmutation, and must contribute significantly to the long-term risk represented by the waste. Application of these criteria results in the principal candidates for P-T being the actinide elements, iodine (i.e., $^{129}I$), and technetium (i.e., $^{99}Tc$). Carbon-14 has too low a transmutation rate to be considered as a P-T candidate, and the other long-lived isotopes do not represent a significant long-term risk by virtue of their low concentration in the waste or their low biological uptake.

### B. Partitioning

As stated in the discussion in the previous section, the principal candidates for partitioning are iodine, technetium, and the actinides. The partitioning of iodine is relatively straightforward since >99% of it is volatilized during the fuel dissolution steps. Additional development work is needed to provide methods to ensure maximum volatilization from the dissolver, to select the best process for recovering the iodine from the off-gas, and to convert it to a compound appropriate for transmutation.

Technetium occurs in the liquid HLW solution and as insoluble solids in the fuel dissolver. Although technetium partitioning processes have not been studied to any significant extent, there is no known reason why such processes could not be developed if needed.

Partitioning the actinides is much more complex than isolating iodine and technetium because the actinide elements have somewhat different chemistries, and they appear in many different fuel cycle wastes, not just one. This means that each waste type must be subjected to more than one partitioning process because of the variety of elements and valence states and the various forms of different wastes.

For liquid TRU wastes the general approach used in partitioning is first to apply additional Purex solvent extraction processing to recover neptunium and more uranium and plutonium. The high-level liquid waste that results from these operations is then treated with a different solvent to recover any residual uranium and plutonium plus virtually all of the trivalent actinides and lanthanides. This product is then routed to an ion exchange step where the actinides are separated from the lanthanides, which are returned to the high-level liquid waste. Solid wastes such as HEPA filters, incinerator ashes, and fuel cladding are leached with a nitric-hydrofluoric acid mixture to recover residual actinides, which are then stripped from the acid using solvent extraction. Based on conceptual flowsheets, it appears that it is possible to reduce the total amount of actinides left in the wastes by a factor of about 30 and the levels of neptunium and the transplutonium isotopes by a factor of up to 1000.

### C. Transmutation

Although transmutation of the actinides and iodine can be accomplished in any device producing subatomic particles (e.g., cyclotrons, fusion reactors, etc.), most studies in the past have concentrated on commercial nuclear power reactors as transmutation devices. The reason for this is that the power reactors are demonstrated technology and are available in large enough numbers to handle the actinides and iodine being recycled.

Conceptually, transmutation is relatively simple, involving the incorporation of the elements to be transmuted into fuel elements which are inserted into the reactor. Both thermal power reactors (PWRs and BWRs) and liquid metal fast breeder reactors (LMFBRs) have been considered as transmutation devices. Transmutation appears to be feasible in either type of reactor, although the fast reactor is the superior transmutation device for actinides. This is because the actinides being recycled are comprised mainly of neptunium, americium, and curium which are relatively nonfissile in a thermal reactor. The transmutation rate of actinides in the reactors is about 6% per full-power year in thermal reactors and 9% per full-power year in fast reactors.

Iodine and technetium transmute more readily in a thermal reactor than in a fast reactor since only a single neutron capture is required for destruction. Calculated transmutation rates are about 3% per full-power year for iodine and 11% per full-power year for technetium.

### D. Description of a Generic P-T Fuel Cycle

A generic fuel cycle including provisions for actinide P-T is shown schematically in Figure 9. Spent fuel from a reactor is reprocessed and partitioned, resulting in the production of actinide-depleted wastes for disposal and a stream of actinides to be recycled.

The actinides, along with enriched uranium or plutonium are manufactured into fuel elements in a fabrication plant with provisions for partitioning to minimize the losses of these materials to the fabrication plant wastes. This fuel is then sent to the reactor where it is irradiated, transmuting the actinides. Spent fuel from the reactor is sent to the reprocessing plant, thus completing the cycle. The amounts of the heavier actinides circulating in the fuel cycle can become quite large after several recycles.

### E. Benefits and Penalties of P-T

The following discussion qualitatively describes the principal benefits and penalties of actinide P-T based on a comprehensive study[13] of the concept.

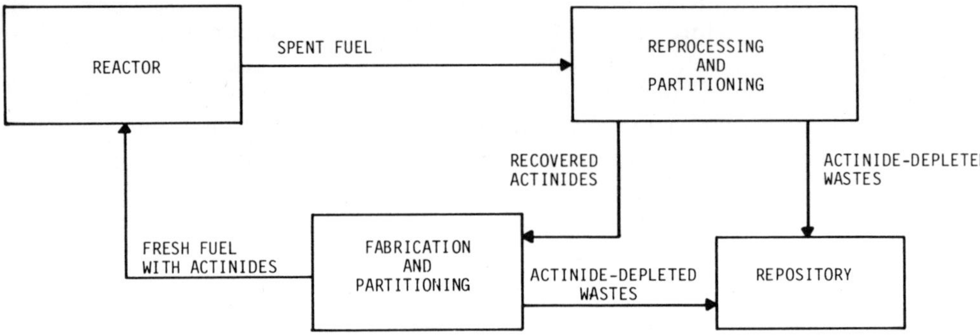


FIGURE 9. Generic P-T fuel cycle.

**Benefits** — By far the greatest benefit of P-T is that it would reduce the average amount of uranium and plutonium going to the repository by a factor of roughly 30. Furthermore, the amounts of some materials such as neptunium and the transplutonium elements that are contained in the wastes can be reduced by a factor ranging up to 1000. However, the overall reduction in long-term risk from the repository would be very small because the actinides migrate through the geosphere so slowly.

**Penalties** — Some of the major penalties that would be expected to accrue from the implementation of P-T are

1. Large and costly chemical processing facilities would be required to partition the actinides from the radioactive wastes. The partitioning facilities for the reprocessing plant would probably be about the same size as the reprocessing plant itself.
2. The additional processing that partitioning entails would increase the volume of radioactive wastes produced by the nuclear fuel cycle.
3. Recycling the actinides would increase the neutron activity of the fuel, thus necessitating both additional shielding in some fuel cycle facilities and the shipping of fresh fuel in shielded shipping casks.
4. The additional processing steps could mean greater routine releases of radioactivity and chemicals from the reprocessing and fabrication plants and greater risks of accidental releases from these facilities.
5. The fabrication of reactor fuels containing substantial amounts of neptunium and the transplutonium isotopes has not been demonstrated, although such a process can probably be developed. The fabrication of reactor fuels or targets containing iodine has also not been developed and may pose a formidable problem due to the low melting point and volatility of iodine and its compounds, the corrosive nature of iodine, and the large volumes of xenon gas resulting from the decay of the iodine transmutation products. The fabrication of technetium appears feasible, although considerable study and development would be needed.

In summary, it has been found that the contemporary costs and risks of implementing actinide P-T outweigh the long-term risk reduction from the waste repository to the point that the net integrated risk is actually increased by this option. The case is less clear for iodine and technetium P-T since they have not been studied as much as the actinides. However, if it is assumed that (1) low-cost, low-risk partitioning processes can be developed and (2) that very conservative long-term risk analyses continue to be used, then there may be incentives for technetium and iodine P-T.

### Table 17
### FUEL CYCLE WASTES PROJECTED FOR THE YEAR 2000[14,15]

| | Annual generation | | Total accumulated at repository | | |
|---|---|---|---|---|---|
| Category of wastes | ($10^3$ $m^3$) | Activity (MCi) | Volume ($10^3$ $m^3$) | Activity (MCi) | Metric tons of actinides |
| High-level solidified | 0.88 | 24,000 | 2.2 | 5,900[a] | 150 |
| Spent fuel | 418 | 24,000 | 10.5 | 12,100[a] | 32,000 |
| Cladding hulls | 0.78 | 86 | 4.5 | 260 | 31 |
| Intermediate-level solid | 3.9 | 1.3 | 20 | 5.8 | 0.16 |
| Low-level solid | 270 | 11 | 2,800 | 40 | 4.5 |
| Noble gases | 0.029[b] | 73 | 0.29[b] | 550 | |
| Iodine | 0.013 | 0.00034 | 0.13 | 0.0033 | |
| Carbon-14 | [c] | 0.005 | [c] | 0.058 | |
| Tritium (solidified) | 65 | 5.2 | 730 | 38 | |
| Uranium ore tailings | 20,000 | 0.24 | 300,000 | 3.4 | 48,000[d] |

[a] Shipped to repository 10 years after generation.
[b] Pressurized at 2200 psi.
[c] Waste form not yet determined, therefore, volume not given.
[d] Comprised of ~25% natural uranium and ~75% stable lead with traces of radioactive daughters.

## VI. WASTE PROJECTIONS

The final subject of this chapter is an estimate of the volumes and activities of the various wastes discussed in previous sections. The results of a typical projection of the radioactive wastes to be produced in the future are summarized in Table 17. The waste volumes and activities are given for two conditions: the annual production in the year 2000 and the total amount accumulated in a repository (or surface disposal for short-lived wastes) by the year 2000. Both the annual production rate and the accumulations depend heavily on the amount of nuclear power assumed to be produced, which was taken to be 510,000 MW(e) in the year 2000.

The most significant aspect of these projections is that nearly all of the radioactivity is concentrated in a relatively small volume of either high-level waste or spent fuel, which are mutually exclusive in this table. For example, if the thermal effects could be ignored, all of the HLW produced through the year 2000 could be placed in a cube about 13 m on a side. The lower activity wastes such as intermediate-level, low-level, and tritium tend to have much larger volumes and very much lower activities than the very active HLW and structural material wastes.

The spent fuel is somewhat more active than the HLW in the repository because all of the plutonium is contained in the fuel; this results in a much slower decline in activity. The spent fuel also contains a large mass of actinides, principally consisting of the unburned uranium originally in the fresh fuel.

The ore tailings contain roughly the same amount of actinides as the spent fuel. The difference between the two is, of course, that virtually all of the activity is from the decay daughters of uranium, which is diluted in a tremendous volume of rock, whereas the spent fuel contains significant amounts of concentrated, transuranium isotopes.

## REFERENCES


1. Alternatives for Managing Wastes from Reactors and Post-Fission Operations in the LWR Fuel Cycle (also known as the Technical Alternatives Document), Rep. ERDA-76-43, 5 Vols., U.S. Energy Research and Development Administration, Washington, D.C., May 1976.
2. Annual Report on the Characterization of High-Level Waste Glasses, Rep. PNL-2625, Battelle Pacific Northwest Laboratory, Seattle, June 1978.
3. **McCarthy, G.J. and Grutzeck, M.W.,** Preliminary Evaluation of the Characteristics of Nuclear Wastes Relevant to Geologic Isolation in Basalt, Rep. RHO-C-12, National Technical Information Service, U.S. Department of Commerce, Springfield, Va., May 1978.
4. **Kibbey, A.H., Godbee, H.W., and Blomeke, J.O.,** Management of waste from the nuclear fuel cycle, lecture presented at the Georgia Institute of Technology, Atlanta, as a part of a short course on Basic Radiological Health, April 2, 1976.
5. Proc. Int. Symp. Management of Wastes from the LWR Fuel Cycle, Rep. CONF-76-0701, U.S. Energy Research and Development Administration, Washington, D.C., July 1976.
6. **Jump, M.J.,** Current practices for disposal of solid low-level radioactive waste, radioactive wastes from the nuclear fuel cycle, *AIChE Symp. Ser.*, 72, 154, 1976.
7. Office of Waste Isolation, Technical Support for GEIS: Radioactive Waste Isolation in Geologic Formations, Volume 4, Baseline Rock Properties — Salt, Rep. Y/OWI/TM-36/4, National Technical Information Service, U.S. Department of Commerce, Springfield, Va., April 1978.
8. Office of Waste Isolation, Technical Support for GEIS: Radioactive Waste Isolation in Geologic Formations, Volume 5, Baseline Rock Properties — Granite, Rep. Y/OWI/TM-36/5, National Technical Information Service, U.S. Department of Commerce, Springfield, Va., April 1978.
9. Office of Waste Isolation, Technical Support for GEIS: Radioactive Waste Isolation in Geologic Formations, Volume 6, Baseline Rock Properties — Shale, Rep. Y/OWI/TM-36/6, National Technical Information Service, U.S. Department of Commerce, Springfield, Va., April 1978.
10. Office of Waste Isolation, Technical Support for GEIS: Radioactive Waste Isolation in Geologic Formations, Volume 7, Baseline Rock Properties — Basalt, Rep. Y/OWI/TM-36/7, National Technical Information Service, U.S. Department of Commerce, Springfield, Va., April 1978.
11. **Bredehoeft, J.D., England, A.W., Stewart, D.B., Trask, N.J., and Winograd, I.J.,** Geologic Disposal of High-Level Radioactive Wastes — Earth Science Perspectives, Geological Survey Circular 779, 1978.
12. Office of Waste Isolation, Technical Support for GEIS: Radioactive Waste Isolation in Geologic Formations, Volume 1, Executive Summary, Rep. Y/OWI/TM-36/1, National Technical Information Service, U.S. Department of Commerce, Springfield, Va., April 1978.
13. **Croff, A. G., Blomeke, J. O., and Finney, B. C.,** Actinide Partitioning-Transmutation Program Final Report. I. Overall Assessment, Rep. ORNL-5566, Oak Ridge National Laboratory, Oak Ridge, Tenn., June 1980.
14. **Kee, C.W., Croff, A.G., and Blomeke, J.O.,** Updated Projections of Radioactive Wastes to be Generated by the U.S. Nuclear Power Industry, Rep. ORNL/TM-5427, Oak Ridge National Laboratory, Oak Ridge, Tenn., December 1976.
15. **Alexander, C.W., Kee, C.W., Croff, A.G., and Blomeke, J.O.,** Projections of Spent Fuel to be Discharged by the U.S. Nuclear Power Industry, Rep. ORNL/TM-6008, Oak Ridge National Laboratory, Oak Ridge, Tenn., October 1977.

# INDEX

## A



## B

C

## D

## E

## F

## G



## H

# I

## K



## L



## M

## N

## O



## P

## R

## S

# T

## U

## V



## W

## X



## Y



## Z